D0512638

Fluid Flow for Chemical Engineers

Fluid Flow for Chemical Engineers

Second edition

Professor F. A. Holland

Overseas Educational Development Office
University of Salford

Dr R. Bragg

Department of Chemical Engineering
University of Manchester Institute of Science and Technology

Edward Arnold
A member of the Hodder Headline Group
LONDON MELBOURNE AUCKLAND

First published in Great Britain 1973
Published in Great Britain 1995 by
Edward Arnold, a division of Hodder Headline PLC,
338 Euston Road, London NW1 3BH

British Library Cataloguing in Publication Data
A catalogue record for this book is available from the British Library

ISBN 0 340 61058 1

1 2 3 4 5 95 96 97 98 99

Typeset in 10/13pt Plantin by
Wearset, Boldon, Tyne and Wear
Printed and bound in Great Britain by
J. W. Arrowsmith Ltd, Bristol

Contents

List of examples

Chapter 5

Chapter 6

Chapter 7

Chapter 8

Chapter 9

Chapter 10

Preface to the second edition

In preparing the second edition of this book, the authors have been concerned to maintain or expand those aspects of the subject that are specific to chemical and process engineering. Thus, the chapter on gas–liquid two-phase flow has been greatly extended to cover flow in the bubble regime as well as to provide an introduction to the homogeneous model and separated flow model for the other flow regimes. The chapter on non-Newtonian flow has also been extended to provide a greater emphasis on the Rabinowitsch–Mooney equation and its modification to deal with cases of apparent wall slip often encountered in the flow of suspensions. An elementary discussion of viscoelasticity has also been given.

A second aim has been to make the book more nearly self-contained and to this end a substantial introductory chapter has been written. In addition to the material provided in the first edition, the principles of continuity, momentum of a flowing fluid, and stresses in fluids are discussed. There is also an elementary treatment of turbulence.

Throughout the book there is more explanation than in the first edition. One result of this is a lengthening of the text and it has been necessary to omit the examples of applications of the Navier–Stokes equations that were given in the first edition. However, derivation of the Navier–Stokes equations and related material has been provided in an appendix.

The authors wish to acknowledge the help given by Miss S. A. Petherick in undertaking much of the word processing of the manuscript for this edition.

It is hoped that this book will continue to serve as a useful undergraduate text for students of chemical engineering and related disciplines.

F. A. Holland
R. Bragg

May 1994

Nomenclature

a	blade width, m
a	propagation speed of pressure wave in equation 10.39, m/s
A	area, m^2
b	width, m
c	speed of sound, m/s
C	couple, N m
C	Chezy coefficient $(2g/f)^{1/2}$, $m^{1/2}/s$
C	constant, usually dimensionless
C	solute concentration, kg/m^3 or $kmol/m^3$
C_d	drag coefficient or discharge coefficient, dimensionless
C_p	specific heat capacity at constant pressure, J/(kg K)
C_v	specific heat capacity at constant volume, J/(kg K)
d	diameter, m
d_e	equivalent diameter of annulus, $D_i - d_o$, m
D	diameter, m
De	Deborah number, dimensionless
e	roughness of pipe wall, m
E	efficiency function $\left(\frac{1}{P_A/V}\right)\left(\frac{1}{t}\right)$, m^3/J
E	total energy per unit mass, J/kg or m^2/s^2
$E\ddot{o}$	Eötvos number, dimensionless
f	Fanning friction factor, dimensionless
F	energy per unit mass required to overcome friction, J/kg
F	force, N
Fr	Froude number, dimensionless
g	gravitational acceleration, 9.81 m/s^2
G	mass flux, kg/(s m^2)
h	head, m
H	height, m
H	specific enthalpy, J/kg
He	Hedström number, dimensionless

I_T	tank turnovers per unit time in equation 5.8, s^{-1}
j	volumetric flux, m/s
j_f	basic friction factor $j_f = f/2$, dimensionless
$\mathcal{J}$	molar diffusional flux in equation 1.70, kmol/(m^2 s)
k	index of polytropic change, dimensionless
k	proportionality constant in equation 5.1, dimensionless
K	consistency coefficient, Pa s^n
K	number of velocity heads in equation 2.23
K	proportionality constant in equation 2.64, dimensionless
K_c	parameter in Carman–Kozeny equation, dimensionless
K'	consistency coefficient for pipe flow, Pa s^n
KE	kinetic energy flow rate, W
L	length of pipe or tube, m
ℓ	mixing length, m
ln	$\log_e$, dimensionless
log	$\log_{10}$, dimensionless
m	mass of fluid, kg
M	mass flow rate of fluid, kg/s
Ma	Mach number, dimensionless
n	power law index, dimensionless
n'	flow behaviour index in equation 3.26, dimensionless
N	rotational speed, rev/s or rev/min
NPSH	net positive suction head, m
p	pitch, m
P	pressure, Pa
P_A	agitator power, W
P_B	brake power, W
P_E	power, W
Po	power number, dimensionless
q	heat energy per unit mass, J/kg
q	heat flux in equation 1.69, W/m^2
Q	volumetric flow rate, m^3/s
r	blade length, m
r	radius, m
r_f	recovery factor in equation 6.85
R	universal gas constant, 8314.3 J/(kmol K)
R	radius of viscometer element
R'	specific gas constant, J/(kg K)
Re	Reynolds number, dimensionless
RMM	relative molecular mass conversion factor, kg/kmol

s	distance, m
s	scale reading in equation 8.39, dimensionless
s	slope, $\sin\theta$, dimensionless
S	cross-sectional flow area, m^2
S_0	surface area per unit volume, m^{-1}
t	time, s
T	temperature, K
T_0	stagnation temperature in equation 6.85, K
u	volumetric average velocity, m/s
$u_{G,L}$	characteristic velocity in equation 7.29, m/s
u_t	terminal velocity, m/s
u_T	tip speed, m/s
U	internal energy per unit mass, J/kg or m^2/s^2
v	point velocity, m/s
V	volume, m^3
V	specific volume, m^3/kg
w	weight fraction, dimensionless
W	work per unit mass, J/kg or m^2/s^2
We	Weber number, dimensionless
x	distance, m
X	Martinelli parameter in equation 7.84, dimensionless
y	distance, m
Y	yield number for Bingham plastic, dimensionless
z	distance, m
Z	compressibility factor, dimensionless
α	velocity distribution factor in equation 1.14, dimensionless
α	void fraction, dimensionless
β	coefficient of rigidity of Bingham plastic in equation 1.73, Pa s
γ	ratio of heat capacities C_p/C_v, dimensionless
$\dot{\gamma}$	shear rate, s^{-1}
ε	eddy kinematic viscosity, m^2/s
ε	void fraction of continuous phase, dimensionless
η	efficiency, dimensionless
λ	relaxation time, s
μ	dynamic viscosity, Pa s
ν	kinematic viscosity, m^2/s
ρ	density, kg/m^3
σ	surface tension, N/m
τ	shear stress, Pa
ϕ	power function in equation 5.18, dimensionless

ϕ	square root of two–phase multiplier, dimensionless
ψ	pressure function in equation 6.108, dimensionless
ψ	correction factor in equation 9.12, dimensionless
ω	angular velocity, rad/s
ω	vorticity in equation A26, s^{-1}

Subscripts

a	referring to apparent
a	referring to accelerative component
A	referring to agitator
b	referring to bed or bubble
c	referring to coarse suspension, coil, contraction or critical
d	referring to discharge side
e	referring to eddy, equivalent or expansion
f	referring to friction
G	referring to gas
i	referring to inside of pipe or tube
L	referring to liquid
m	referring to manometer liquid, or mean
mf	referring to minimum fluidization
M	referring to mixing
N	referring to Newtonian fluid
o	referring to outside of pipe or tube
p	referring to pipe or solid particle
r	referring to reduced
s	referring to sonic, suction or system
sh	referring to static head component
t	referring to terminal
t	referring to throat
T	referring to tank, total or tip
v	referring to vapour
V	referring to volume
w	referring to pipe or tube wall
W	referring to water
y	referring to yield point

1 Fluids in motion

1.1 Units and dimensions

Mass, length and time are commonly used primary units, other units being derived from them. Their dimensions are written as M, L and T respectively. Sometimes force is used as a primary unit. In the Système International d'Unités, commonly known as the SI system of units, the primary units are the kilogramme kg, the metre m, and the second s. A number of derived units are listed in Table 1.1.

1.2 Description of fluids and fluid flow

1.2.1 *Continuum hypothesis*

Although gases and liquids consist of molecules, it is possible in most cases to treat them as continuous media for the purposes of fluid flow calculations. On a length scale comparable to the mean free path between collisions, large rapid fluctuations of properties such as the velocity and density occur. However, fluid flow is concerned with the macroscopic scale: the typical length scale of the equipment is many orders of magnitude greater than the mean free path. Even when an instrument is placed in the fluid to measure some property such as the pressure, the measurement is not made at a point—rather, the instrument is sensitive to the properties of a small volume of fluid around its measuring element. Although this measurement volume may be minute compared with the volume of fluid in the equipment, it will generally contain millions of molecules and consequently the instrument measures an average value of the property. In almost all fluid flow problems it is possible to select a measurement volume that is very small compared with the flow field yet contains so many molecules that the properties of individual molecules are averaged out.

Table 1.1

Quantity	*Derived unit*	*Symbol*	*Relationship to primary units*
Force	newton	N	kg m/s^2
Work, energy, quantity of heat	joule	J	N m
Power	watt	W	J/s = N m/s
Area	square metre		m^2
Volume	cubic metre		m^3
Density	kilogramme per cubic metre		kg/m^3
Velocity	metre per second		m/s
Acceleration	metre per second per second		m/s^2
Pressure	pascal, or newton per square metre	Pa	N/m^2
Surface tension	newton per metre		N/m
Dynamic viscosity	pascal second, or newton second per square metre	Pa s	N s/m^2
Kinematic viscosity	square metre per second		m^2/s

It follows from the above facts that fluids can be treated as continuous media with continuous distributions of properties such as the pressure, density, temperature and velocity. Not only does this imply that it is unnecessary to consider the molecular nature of the fluid but also that meaning can be attached to spatial derivatives, such as the pressure gradient dP/dx, allowing the standard tools of mathematical analysis to be used in solving fluid flow problems.

Two examples where the continuum hypothesis may be invalid are low pressure gas flow in which the mean free path may be comparable to a linear dimension of the equipment, and high speed gas flow when large changes of properties occur across a (very thin) shock wave.

1.2.2 *Homogeneity and isotropy*

Two other simplifications that should be noted are that in most fluid flow problems the fluid is assumed to be homogeneous and isotropic. A homogeneous fluid is one whose properties are the same at all locations and this is usually true for single–phase flow. The flow of gas–liquid mixtures and of solid–fluid mixtures exemplifies heterogeneous flow problems.

A material is isotropic if its properties are the same in all directions. Gases and simple liquids are isotropic but liquids having complex, chain–like molecules, such as polymers, may exhibit different properties in different directions. For example, polymer molecules tend to become partially aligned in a shearing flow.

1.2.3 *Steady flow and fully developed flow*

Steady processes are ones that do not change with the passage of time. If ϕ denotes a property of the flowing fluid, for example the pressure or velocity, then for steady conditions

$$\frac{\partial \phi}{\partial t} \equiv 0 \tag{1.1}$$

for all properties. This does not imply that the properties are constant: they may vary from location to location but may not change at any fixed position.

Fully developed flow is flow that does not change along the direction of flow. An example of developing and fully developed flow is that which occurs when a fluid flows into and through a pipe or tube. Along most of the length of the pipe, there is a constant velocity profile: there is a maximum at the centre–line and the velocity falls to zero at the pipe wall. In the case of laminar flow of a Newtonian liquid, the fully developed velocity profile has a parabolic shape. Once established, this fully developed profile remains unchanged until the fluid reaches the region of the pipe exit. However, a considerable distance is required for the velocity profile to develop from the fairly uniform velocity distribution at the pipe entrance. This region where the velocity profile is developing is known as the entrance length. Owing to the changes taking place in the developing flow in the entrance length, it exhibits a higher pressure gradient. Developing flow is more difficult to analyse than fully developed flow owing to the variation along the flow direction.

1.2.4 *Paths, streaklines and streamlines*

The pictorial representation of fluid flow is very helpful, whether this be done by experimental flow visualization or by calculating the velocity field. The terms 'path', 'streakline' and 'streamline' have different meanings.

Consider a flow visualization study in which a small patch of dye is injected instantaneously into the flowing fluid. This will 'tag' an element of the fluid and, by following the course of the dye, the path of the tagged element of fluid is observed. If, however, the dye is introduced continuously, a streakline will be observed. A streakline is the locus of all particles that have passed through a specified fixed point, namely the point at which the dye is injected.

A streamline is defined as the continuous line in the fluid having the property that the tangent to the line is the direction of the fluid's velocity at that point. As the fluid's velocity at a point can have only one direction, it follows that streamlines cannot intersect, except where the velocity is zero. If the velocity components in the x, y and z coordinate directions are v_x, v_y, v_z, the streamline can be calculated from the equation

$$\frac{dx}{v_x} = \frac{dy}{v_y} = \frac{dz}{v_z} \tag{1.2}$$

This equation can be derived very easily. Consider a two-dimensional flow in the x–y plane, then the gradient of the streamline is equal to dy/dx. However, the gradient must also be equal to the ratio of the velocity components at that point v_y/v_x. Equating these two expressions for the gradient of the streamline gives the first and second terms of equation 1.2. This relationship is not restricted to two-dimensional flow. In three-dimensional flow the terms just considered are the gradient of the projection of the streamline on to the x–y plane. Similar terms apply for each of the three coordinate planes, thus giving equation 1.2.

Although in general, particle paths, streaklines and streamlines are different, they are all the same for steady flow. As flow visualization experiments provide either the particle path or the streakline through the point of dye injection, interpretation is easy for steady flow but requires caution with unsteady flow.

1.3 Types of flow

1.3.1 *Laminar and turbulent flow*

If water is caused to flow steadily through a transparent tube and a dye is continuously injected into the water, two distinct types of flow may be observed. In the first type, shown schematically in Figure 1.1(a), the streaklines are straight and the dye remains intact. The dye is observed to spread very slightly as it is carried through the tube; this is due to molecular diffusion. The flow causes no mixing of the dye with the surrounding water. In this type of flow, known as laminar or streamline flow, elements of the fluid flow in an orderly fashion without any macroscopic intermixing with neighbouring fluid. In this experiment, laminar flow is observed only at low flow rates. On increasing the flow rate, a markedly different type of flow is established in which the dye streaks show a chaotic, fluctuating type of motion, known as turbulent flow, Figure 1.1(b). A characteristic of turbulent flow is that it promotes rapid mixing over a length scale comparable to the diameter of the tube. Consequently, the dye trace is rapidly broken up and spread throughout the flowing water.

In turbulent flow, properties such as the pressure and velocity fluctuate rapidly at each location, as do the temperature and solute concentration in flows with heat and mass transfer. By tracking patches of dye distributed across the diameter of the tube, it is possible to demonstrate that the liquid's velocity (the time–averaged value in the case of turbulent flow) varies across the diameter of the tube. In both laminar and turbulent flow the velocity is zero at the wall and has a maximum value at the centre–line. For laminar flow the velocity profile is a parabola but for turbulent flow the profile is much flatter over most of the diameter.

If the pressure drop across the length of the tube were measured in these experiments it would be found that the pressure drop is proportional to the flow rate when the flow is laminar. However, as shown in Figure 1.2, when the flow is turbulent the pressure drop increases more rapidly, almost as the square of the flow rate. Turbulent flow has the advantage of

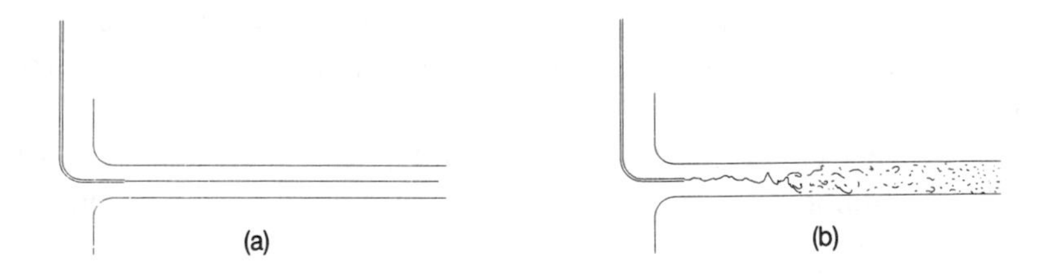

Figure 1.1
Flow regimes in a pipe shown by dye injection
(a) *Laminar flow* (b) *Turbulent flow*

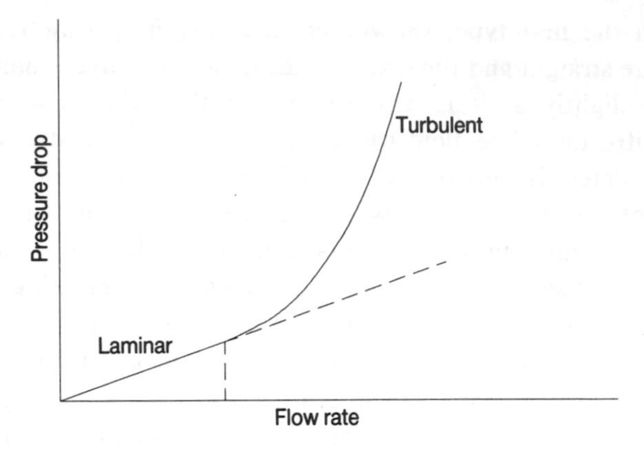

Figure 1.2
The relationship between pressure drop and flow rate in a pipe

promoting rapid mixing and enhances convective heat and mass transfer. The penalty that has to be paid for this is the greater power required to pump the fluid.

Measurements with different fluids, in pipes of various diameters, have shown that for Newtonian fluids the transition from laminar to turbulent flow takes place at a critical value of the quantity $\rho u d_i/\mu$ in which u is the volumetric average velocity of the fluid, d_i is the internal diameter of the pipe, and ρ and μ are the fluid's density and viscosity respectively. This quantity is known as the Reynolds number *Re* after Osborne Reynolds who made his celebrated flow visualization experiments in 1883:

$$Re = \frac{\rho u d_i}{\mu} \tag{1.3}$$

It will be noted that the units of the quantities in the Reynolds number cancel and consequently the Reynolds number is an example of a dimensionless group: its value is independent of the system of units used.

The volumetric average velocity is calculated by dividing the volumetric flow rate by the flow area ($\pi d_i^2/4$).

Under normal circumstances, the laminar–turbulent transition occurs at a Reynolds number of about 2100 for Newtonian fluids flowing in pipes.

1.3.2 *Compressible and incompressible flow*

All fluids are compressible to some extent but the compressibility of liquids is so low that they can be treated as being incompressible. Gases are much more compressible than liquids but if the pressure of a flowing gas changes little, and the temperature is sensibly constant, then the density will be nearly constant. When the fluid density remains constant, the flow is described as incompressible. Thus gas flow in which pressure changes are small compared with the average pressure may be treated in the same way as the flow of liquids.

When the density of the gas changes significantly, the flow is described as compressible and it is necessary to take the density variation into account in making flow calculations. When the pressure difference in a flowing gas is made sufficiently large, the gas speed approaches, and may exceed, the speed of sound in the gas. Flow in which the gas speed is greater than the local speed of sound is known as supersonic flow and that in which the gas speed is lower than the sonic speed is called subsonic flow. Most flow of interest to chemical engineers is subsonic and this is also the type of flow of everyday experience. Sonic and supersonic gas flow are encountered most commonly in nozzles and pressure relief systems. Some rather startling effects occur in supersonic flow: the relationships of fluid velocity and pressure to flow area are the opposite of those for subsonic flow. This topic is discussed in Chapter 6. Unless specified to the contrary, it will be assumed that the flow is subsonic.

1.4 Conservation of mass

Consider flow through the pipe–work shown in Figure 1.3, in which the fluid occupies the whole cross section of the pipe. A mass balance can be written for the fixed section between planes 1 and 2, which are normal to the axis of the pipe. The mass flow rate across plane 1 into the section is equal to $\rho_1 Q_1$ and the mass flow rate across plane 2 out of the section is equal to $\rho_2 Q_2$, where ρ denotes the density of the fluid and Q the volumetric flow rate.

Thus, a mass balance can be written as

mass flow rate in = mass flow rate out
+ rate of accumulation within section

that is

$$\rho_1 Q_1 = \rho_2 Q_2 + \frac{\partial}{\partial t}(\rho_{av} V)$$

or

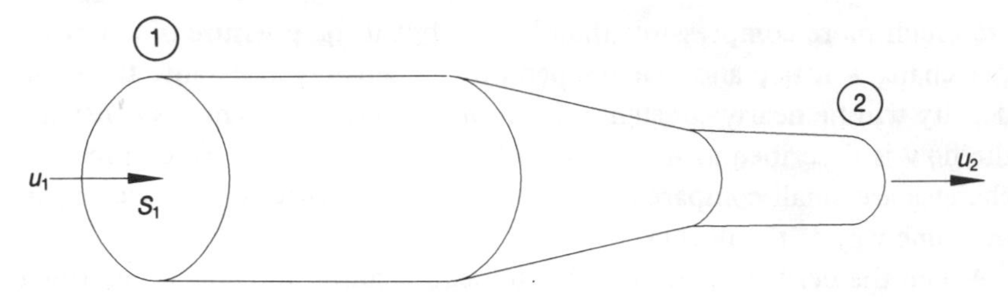

Figure 1.3
Flow through a pipe of changing diameter

$$\rho_1 Q_1 = \rho_2 Q_2 + V \frac{\partial \rho_{av}}{\partial t} \tag{1.4}$$

where V is the constant volume of the section between planes 1 and 2, and ρ_{av} is the density of the fluid averaged over the volume V. This equation represents the conservation of mass of the flowing fluid: it is frequently called the 'continuity equation' and the concept of 'continuity' is synonymous with the principle of conservation of mass.

In the case of unsteady compressible flow, the density of the fluid in the section will change and consequently the accumulation term will be non–zero. However, for steady compressible flow the time derivative must be zero by definition. In the case of incompressible flow, the density is constant so the time derivative is zero even if the flow is unsteady.

Thus, for incompressible flow or steady compressible flow, there is no accumulation within the section and consequently equation 1.4 reduces to

$$\rho_1 Q_1 = \rho_2 Q_2 \tag{1.5}$$

This simply states that the mass flow rate into the section is equal to the mass flow rate out of the section.

In general, the velocity of the fluid varies across the diameter of the pipe but an average velocity can be defined. If the cross–sectional area of the pipe at a particular location is S, then the volumetric flow rate Q is given by

$$Q = uS \tag{1.6}$$

Equation 1.6 defines the volumetric average velocity u: it is the uniform velocity required to give the volumetric flow rate Q through the flow area S. Substituting for Q in equation 1.5, the zero accumulation mass balance becomes

$$\rho_1 u_1 S_1 = \rho_2 u_2 S_2 \tag{1.7}$$

This is the form of the Continuity Equation that will be used most frequently but it is valid only when there is no accumulation. Although Figure 1.3 shows a pipe of circular cross section, equations 1.4 to 1.7 are valid for a cross section of any shape.

1.5 **Energy relationships and the Bernoulli equation**

The total energy of a fluid in motion consists of the following components: internal, potential, pressure and kinetic energies. Each of these energies may be considered with reference to an arbitrary base level. It is also convenient to make calculations on unit mass of fluid.

Internal energy This is the energy associated with the physical state of the fluid, ie, the energy of the atoms and molecules resulting from their motion and configuration [Smith and Van Ness (1987)]. Internal energy is a function of temperature. The internal energy per unit mass of fluid is denoted by U.

Potential energy This is the energy that a fluid has by virtue of its position in the Earth's field of gravity. The work required to raise a unit mass of fluid to a height z above an arbitrarily chosen datum is zg, where g is the acceleration due to gravity. This work is equal to the potential energy of unit mass of fluid above the datum.

Pressure energy This is the energy or work required to introduce the fluid into the system without a change of volume. If P is the pressure and V is the volume of mass m of fluid, then PV/m is the pressure energy per unit mass of fluid. The ratio m/V is the fluid density ρ. Thus the pressure energy per unit mass of fluid is equal to P/ρ.

Kinetic energy This is the energy of fluid motion. The kinetic energy of unit mass of the fluid is $v^2/2$, where v is the velocity of the fluid relative to some fixed body.

Total energy Summing these components, the total energy E per unit mass of fluid is given by the equation

$$E = U + zg + \frac{P}{\rho} + \frac{v^2}{2} \tag{1.8}$$

where each term has the dimensions of force times distance per unit mass, ie $(ML/T^2)L/M$ or L^2/T^2.

Consider fluid flowing from point 1 to point 2 as shown in Figure 1.4. Between these two points, let the following amounts of heat transfer and work be done per unit mass of fluid: heat transfer q to the fluid, work W_i done on the fluid and work W_o done by the fluid on its surroundings. W_i and W_o may be thought of as work input and output. Assuming the conditions to be steady, so that there is no accumulation of energy within the fluid between points 1 and 2, an energy balance can be written per unit mass of fluid as

$$E_1 + W_i + q = E_2 + W_o$$

or, after rearranging

$$E_2 = E_1 + q + W_i - W_o \tag{1.9}$$

A flowing fluid is required to do work to overcome viscous frictional forces so that in practice the quantity W_o is always positive. It is zero only for the theoretical case of an inviscid fluid or ideal fluid having zero viscosity. The work W_i may be done on the fluid by a pump situated between points 1 and 2.

If the fluid has a constant density or behaves as an ideal gas, then the internal energy remains constant if the temperature is constant. If no heat transfer to the fluid takes place, q=0. For these conditions, equations 1.8 and 1.9 may be combined and written as

$$\left(z_2 g + \frac{P_2}{\rho_2} + \frac{v_2^2}{2}\right) = \left(z_1 g + \frac{P_1}{\rho_1} + \frac{v_1^2}{2}\right) + W_i - W_o \tag{1.10}$$

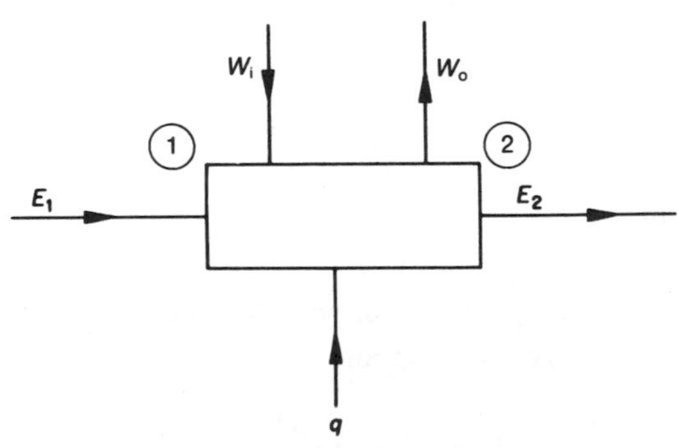

Figure 1.4

Energy balance for fluid flowing from location 1 to location 2

For an inviscid fluid, ie frictionless flow, and no pump, equation (1.10) becomes

$$\left(z_2 g + \frac{P_2}{\rho_2} + \frac{v_2^2}{2}\right) = \left(z_1 g + \frac{P_1}{\rho_1} + \frac{v_1^2}{2}\right) \tag{1.11}$$

Equation 1.11 is known as Bernoulli's equation.

Dividing throughout by g, these equations can be written in a slightly different form. For example, equation 1.10 can be written as

$$\left(z_2 + \frac{P_2}{\rho_2 g} + \frac{v_2^2}{2g}\right) = \left(z_1 + \frac{P_1}{\rho_1 g} + \frac{v_1^2}{2g}\right) + \frac{W_i}{g} - \frac{W_o}{g} \tag{1.12}$$

In this form, each term has the dimensions of length. The terms z, $P/(\rho g)$ and $v^2/(2g)$ are known as the potential, pressure and velocity heads, respectively. Denoting the work terms as heads, equation 1.12 can also be written as

$$\left(z_2 + \frac{P_2}{\rho_2 g} + \frac{v_2^2}{2g}\right) = \left(z_1 + \frac{P_1}{\rho_1 g} + \frac{v_1^2}{2g}\right) + \Delta h - h_f \tag{1.13}$$

where Δh is the head imparted to the fluid by the pump and h_f is the head loss due to friction. The term Δh is known as the total head of the pump.

Equation 1.13 is simply an energy balance written for convenience in terms of length, ie heads. The various forms of the energy balance, equations 1.10 to 1.13, are often called Bernoulli's equation but some people reserve this name for the case where the right hand side is zero, ie when there is no friction and no pump, and call the forms of the equation including the work terms the 'extended' or 'engineering' Bernoulli equation.

The various forms of energy are interchangeable and the equation enables these changes to be calculated in a given system. In deriving the form of Bernoulli's equation without the work terms, it was assumed that the internal energy of the fluid remains constant. This is not the case when frictional dissipation occurs, ie there is a head loss h_f. In this case h_f represents the conversion of mechanical energy into internal energy and, while internal energy can be recovered by heat transfer to a cooler medium, it cannot be converted into mechanical energy.

The equations derived are valid for a particular element of fluid or, the conditions being steady, for any succession of elements flowing along the same streamline. Consequently, Bernoulli's equation allows changes along a streamline to be calculated: it does not determine how conditions, such as the pressure, vary in other directions.

Bernoulli's equation is based on the principle of conservation of energy and, in the form in which the work terms are zero, it states that the total mechanical energy remains constant along a streamline. Fluids flowing along different streamlines have different total energies. For example, for laminar flow in a horizontal pipe, the pressure energy and potential energy for an element of fluid flowing in the centre of the pipe will be virtually identical to those for an element flowing near the wall, however, their kinetic energies are significantly different because the velocity near the wall is much lower than that at the centre. To allow for this and to enable Bernoulli's equation to be used for the fluid flowing through the whole cross section of a pipe or duct, equation 1.13 can be modified as follows:

$$\left(z_2+\frac{P_2}{\rho_2 g}+\frac{u_2^2}{2g\alpha}\right)=\left(z_1+\frac{P_1}{\rho_1 g}+\frac{u_1^2}{2g\alpha}\right)+\Delta h-h_f \qquad (1.14)$$

where u is the volumetric average velocity and α is a dimensionless correction factor, which accounts for the velocity distribution across the pipe or duct. For the relatively flat velocity profile that is found in turbulent flow, α has a value of approximately unity. In Chapter 2 it is shown that α has a value of $\frac{1}{2}$ for laminar flow of a Newtonian fluid in a pipe of circular section.

As an example of a simple application of Bernoulli's equation, consider the case of steady, fully developed flow of a liquid (incompressible) through an inclined pipe of constant diameter with no pump in the section considered. Bernoulli's equation for the section between planes 1 and 2 shown in Figure 1.5 can be written as

$$\left(z_2+\frac{P_2}{\rho g}+\frac{u_2^2}{2g\alpha}\right)=\left(z_1+\frac{P_1}{\rho g}+\frac{u_1^2}{2g\alpha}\right)-h_f \qquad (1.15)$$

For the conditions specified, $u_1=u_2$, and α has the same value because the flow is fully developed. The terms in equation 1.15 are shown schematically in Figure 1.5. The total energy E_2 is less than E_1 by the frictional losses h_f. The velocity head remains constant as indicated and the potential head increases owing to the increase in elevation. As a result the pressure energy, and therefore the pressure, must decrease. It is important to note that this upward flow occurs because the upstream pressure P_1 is sufficiently high (compare the two pressure heads in Figure 1.5). This high pressure would normally be provided by a pump upstream of the section considered; however, as the pump is not in the section there must be no pump head term Δh in the equation. The effect of the pump is already manifest in the high pressure P_1 that it has generated.

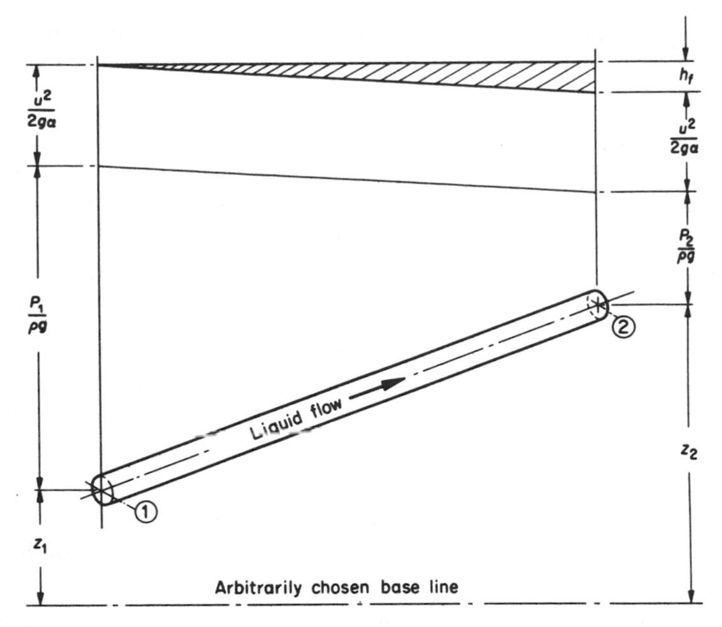

Figure 1.5
Diagrammatic representation of heads in a liquid flowing through a pipe

The method of calculating frictional losses is described in Chapter 2. It may be noted here that losses occur as the fluid flows through the plain pipe, pipe fittings (bends, valves), and at expansions and contractions such as into and out of vessels.

A slightly more general case is incompressible flow through an inclined pipe having a change of diameter. In this case the fluid's velocity and velocity head will change. Rearranging equation 1.15, the pressure drop $P_1 - P_2$ experienced by the fluid in flowing from location 1 to location 2 is given by

$$P_1 - P_2 = \rho g(z_2 - z_1) + \frac{\rho(u_2^2 - u_1^2)}{2\alpha} + \rho g h_f \tag{1.16}$$

Equation 1.16 shows that, in general, the upstream pressure P_1 must be greater than the downstream pressure P_2 in order to raise the fluid, to increase its velocity and to overcome frictional losses.

In some cases, one or more of the terms on the right hand side of equation 1.16 will be zero, or may be negative. For downward flow the hydrostatic pressure *increases* in the direction of flow and for decelerating flow the loss of kinetic energy produces an increase in pressure (pressure recovery).

Denoting the total pressure drop $(P_1 - P_2)$ by ΔP, it can be written as

$$\Delta P = \Delta P_{sh} + \Delta P_a + \Delta P_f \tag{1.17}$$

where ΔP_{sh}, ΔP_a, ΔP_f are respectively the static head, accelerative and frictional components of the total pressure drop given in equation 1.16. Equation 1.16 shows that each component of the pressure drop is equal to the corresponding change of head multiplied by ρg.

An important application of Bernoulli's equation is in flow measurement, discussed in Chapter 8. When an incompressible fluid flows through a constriction such as the throat of the Venturi meter shown in Figure 8.5, by continuity the fluid velocity must increase and by Bernoulli's equation the pressure must fall. By measuring this change in pressure, the change in velocity can be determined and the volumetric flow rate calculated.

Applications of Bernoulli's equation are usually straightforward. Often there is a choice of the locations 1 and 2 between which the calculation is made: it is important to choose these locations carefully. All conditions must be known at each location. The appropriate choice can sometimes make the calculation very simple. A rather extreme case is discussed in Example 1.1.

Example 1.1

The contents of the tank shown in Figure 1.6 are heated by circulating the liquid through an external heat exchanger. Bernoulli's equation can be used to calculate the head Δh that the pump must generate. It is assumed here that the total losses h_f have been calculated. Locations A and B might be considered but these are unsuitable because the flow changes in the region of the inlet and outlet and the conditions are therefore unknown.

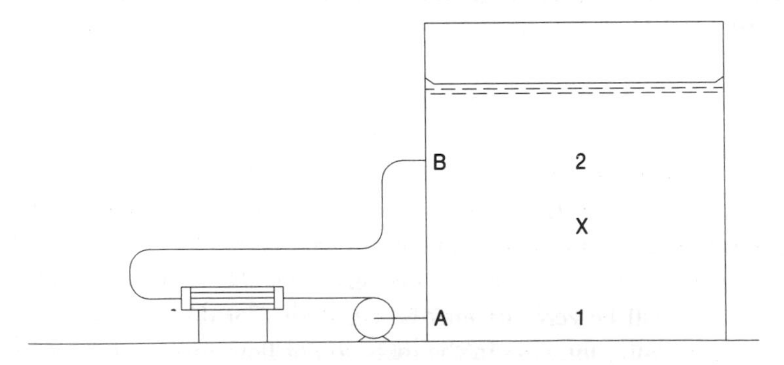

Figure 1.6

Recirculating liquid: application of Bernoulli's equation

For a recirculating flow like this, the fluid's destination is the same as its origin so the two locations can be chosen to be the same, for example the point marked X. In this case equation 1.14 reduces to

$$\Delta h = h_f$$

showing that the pump is required simply to overcome the losses. There is no change in the potential, pressure and kinetic energies of the liquid because it ends with a height, pressure and speed identical to those with which it started.

An alternative is to choose locations 1 and 2 as shown. These points are in the bulk of the liquid where the liquid's speed is negligibly small. Applying Bernoulli's equation between points 1 and 2 gives the pump head as

$$\Delta h = \left(z_2 + \frac{P_2}{\rho g} \right) - \left(z_1 + \frac{P_1}{\rho g} \right) + h_f \tag{1.18}$$

As the liquid in the main part of the tank is virtually stationary, the pressure difference between point 1 and point 2 is just the hydrostatic pressure difference:

$$P_1 - P_2 = (z_2 - z_1)\rho g$$

Substituting this pressure difference in equation 1.18 gives the result $\Delta h = h_f$ as found before.

Example 1.2

Water issues from the nozzle of a horizontal hose–pipe. The hose has an internal diameter of 60 mm and the nozzle tapers to an exit diameter of 20 mm. If the gauge pressure at the connection between the nozzle and the pipe is 200 kPa, what is the flow rate? The density of water is 1000 kg/m^3.

Calculations

The pressure is given at the connection of the nozzle to the pipe so this will be taken as location 1. The flow is caused by the fact that this pressure is greater than the pressure of the atmosphere into which the jet discharges. The pressure in the jet at the exit from the nozzle will be very nearly the same as the atmospheric pressure so the exit plane can be taken as location 2. (Note that when a liquid discharges into another liquid the flow is much more complicated and there are large frictional losses.) Friction is negligible in a short tapering nozzle. The nozzle is horizontal so $z_1 = z_2$ and for turbulent flow $\alpha = 1.0$. With these simplifications and the fact that there is no pump in the section, Bernoulli's equation reduces to

$$\frac{P_2}{\rho g}+\frac{u_2^2}{2g}=\frac{P_1}{\rho g}+\frac{u_1^2}{2g}$$

Thus

$$u_2^2-u_1^2 = 2(P_1-P_2)/\rho$$

The fluid pressure P_2 at the exit plane is the atmospheric pressure, ie zero gauge pressure. Therefore

$$u_2^2-u_1^2 = 2(2\times 10^5\ \text{Pa})/(1000\ \text{kg/m}^3) = 400\ \text{m}^2/\text{s}^2$$

By continuity

$$u_1 r_1^2 = u_2 r_2^2$$

therefore

$$u_2 = u_1 r_1^2/r_2^2 = 9u_1$$

Thus

$$80u_1^2 = 400\ \text{m}^2/\text{s}^2$$

and hence

$$u_1 = 2.236\ \text{m/s and } u_2 = 20.12\ \text{m/s}$$

The volumetric discharge rate can be calculated from either velocity and the corresponding diameter. Using the values for the pipe

$$Q = u_1\pi d_1^2/4 = (2.236\ \text{m/s})(3.142)(6\times 10^{-2}\ \text{m})^2/4 = \underline{6.32\times 10^{-3}\ \text{m}^3/\text{s}}$$

Note that in this example the pressure head falls by $(P_1-P_2)/(\rho g)$ which is equal to 20.4 m, and the velocity head increases by the same amount. It is clear that if the nozzle were not horizontal, the difference in elevation between points 1 and 2 would be negligible compared with these changes.

1.5.1 *Pressure terminology*

It is appropriate here to define some pressure terms. Consider Bernoulli's equation for frictionless flow with no pump in the section:

$$\left(z_2\,g+\frac{P_2}{\rho_2}+\frac{v_2^2}{2}\right)=\left(z_1\,g+\frac{P_1}{\rho_1}+\frac{v_1^2}{2}\right) \tag{1.11}$$

This is for flow along a streamline, not through the whole cross-section.

Consider the case of incompressible, horizontal flow. Equation 1.11 shows that if a flowing element of fluid is brought to rest ($v_2 = 0$), the pressure P_2 is given by

$$P_2 = P_1 + \frac{\rho v_1^2}{2} \tag{1.19}$$

In coming to rest without losses, the fluid's kinetic energy is converted into pressure energy so that the pressure P_2 of the stopped fluid is greater than the pressure P_1 of the flowing fluid by an amount $\frac{1}{2}\rho v_1^2$.

For a fluid having a pressure P and flowing at speed v, the quantity $\frac{1}{2}\rho v^2$ is known as the dynamic pressure and $P+\frac{1}{2}\rho v^2$ is called the total pressure or the stagnation pressure. The pressure P of the flowing fluid is often called the static pressure, a potentially misleading name because it is not the same as the hydrostatic pressure.

Clearly, if the dynamic pressure can be measured by stopping the fluid, the upstream velocity can be calculated. Figure 8.7 shows a device known as a Pitot tube, which may be used to determine the velocity of a fluid at a point. The tube is aligned pointing into the flow, consequently the fluid approaching it is brought to rest at the nose of the Pitot tube. By placing a pressure tapping at the nose of the Pitot tube, the pressure at the stagnation point can be measured. If the pressure in the undisturbed fluid upstream of the Pitot tube and that at the stagnation point at the nose are denoted by P_1 and P_2 respectively, then they are related by equation 1.19. It will be seen from this example why the total pressure is also called the stagnation pressure. The so-called static pressure of the flowing fluid can be measured by placing a pressure tapping either in the wall of the pipe as shown or in the wall of the Pitot tube just downstream of the nose; in the latter case the device is known as a Pitot-static tube. By placing the opening parallel to the direction of flow, the fluid flows by undisturbed and its undisturbed pressure is measured. This undisturbed pressure is the static pressure. As the gradient of the static pressure will usually be very low, placing the static pressure tapping as described will give a good measure of the static pressure upstream of the Pitot tube. Thus the pressure difference $P_2 - P_1$ can be measured and the fluid's velocity v_1 calculated from equation 1.19. If the Pitot tube is tracked across the pipe or duct, the velocity profile may be determined.

1.6 Momentum of a flowing fluid

Although Newton's second law of motion

$$\text{net force} = \text{rate of change of momentum}$$

applies to an element of fluid, it is difficult to follow the motion of such an element as it flows. It is more convenient to formulate a version of Newton's law that can be applied to a succession of fluid elements flowing through a particular region, for example flowing through the section between planes 1 and 2 in Figure 1.3.

To understand how an appropriate momentum equation can be derived, consider first a stationary tank into which solid masses are thrown, Figure 1.7a. Momentum is a vector and each component can be considered separately; here only the x-component will be considered. Each mass has a velocity component v_x and mass m so its x-component of momentum as it enters the tank is equal to mv_x. As a result of colliding with various parts of the tank and its contents, the added mass is brought to rest and loses the x-component of momentum equal to mv_x. As a result there is an impulse on the tank, acting in the x-direction. Consider now a stream of masses, each of mass m and with a velocity component v_x. If a steady state is achieved, the rate of destruction of momentum of the added masses must be equal to the rate at which momentum is added to the tank by their entering it. If n masses are added in time t, the rate of addition of mass is nm/t and the rate of addition of x-component momentum is $(nm/t)v_x$. It is convenient to denote the rate of addition of mass by M, so the rate of addition of x-momentum is Mv_x.

Figure 1.7b shows the corresponding process in which a jet of liquid flows into the tank. In this case, the rate of addition of mass M is simply the mass flow rate. If the x-component of the jet's velocity is v_x then the rate of 'flow' of x-momentum into the tank is Mv_x. Note that the mass flow rate M is a scalar quantity and is therefore always positive. The momentum is a vector quantity by virtue of the fact that the velocity is a vector.

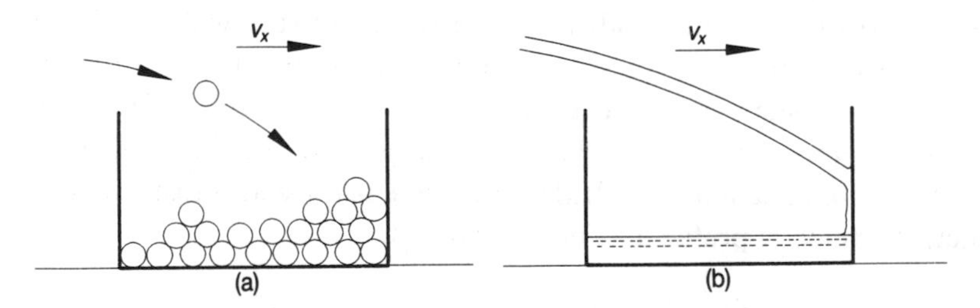

Figure 1.7
Momentum flow into a tank
(a) *Discrete mass* (b) *Flowing liquid*

When each mass is brought to rest its momentum is destroyed and a corresponding impulse is thereby imposed on the tank. As the input of a succession of masses increases towards a steady stream, the impulses merge into a steady force. This is also the case with the stream of liquid: the fluid's momentum is destroyed at a constant rate and by Newton's second law there must be a force acting on the fluid equal to the rate of change of its momentum. If there is no accumulation of momentum within the tank, the jet's momentum must be destroyed at the same rate as it flows into the tank. The rate of change of momentum of the jet can be expressed as

$$\begin{aligned}\text{rate of change of momentum} &= \text{final momentum flow rate} - \text{initial momentum flow rate} \\ &= 0 - Mv_x \qquad (1.20)\end{aligned}$$

Consequently, a force equal to $-Mv_x$ is required to retard the jet, ie a force of magnitude Mv_x acting in the negative x-direction. By Newton's third law of motion, there must be a reaction of equal magnitude acting on the tank in the positive x-direction.

Similarly, if a jet of liquid were to issue from the tank with a velocity component v_x and mass flow rate M, there would be a reaction $-Mv_x$ acting on the tank.

Consider the momentum change that occurs when a fluid flows steadily through the pipe-work shown in Figure 1.3. It will be assumed that the axial velocity component is uniform over the cross section and equal to u. This is a good approximation for turbulent flow. The x-momentum flow rate into the section across plane 1 is equal to M_1u_1 and that out of the section across plane 2 is equal to M_2u_2. By continuity, $M_1 = M_2 = M$. From equation 1.20, the rate of change of momentum is given by

$$\begin{aligned}\text{rate of change of momentum} &= \text{change of flow of momentum} \\ &= Mu_2 - Mu_1 \qquad (1.21)\end{aligned}$$

Although the fluid flows continuously through the section, the change of momentum is the same as if the fluid were brought to rest in the section then ejected from it. Consequently, Newton's second law of motion can be written as

$$\begin{aligned}\text{net force acting on the fluid} &= \text{rate of change of momentum} \\ &= \text{momentum flow rate out of section} \\ &\quad - \text{momentum flow rate into secton} \\ &= Mu_2 - Mu_1 \qquad (1.22)\end{aligned}$$

Thus a force equal to $M(u_2 - u_1)$ must be applied to the fluid. This force is measured as positive in the positive x-direction. These equations are valid when there is no accumulation of momentum within the section.

When accumulation of momentum occurs within the section, the momentum equation must be written as

$$\begin{aligned}
\text{net force acting on the fluid} &= \text{rate of change of momentum} \\
&= \text{momentum flow rate out of section} \\
&\quad - \text{momentum flow rate into secton} \\
&\quad + \text{rate of accumulation within section} \\
&= M_2u_2 - M_1u_1 + V\frac{\partial}{\partial t}(\rho_{av}u_{av}) \qquad (1.23)
\end{aligned}$$

In the last term of equation 1.23, the averages are taken over the fixed volume V of the section. This term is simply the rate of change of the momentum of the fluid instantaneously contained in the section. It is clear that accumulation of momentum may occur with unsteady flow even if the flow is incompressible. In general, the mass flow rates M_1 and M_2 into and out of the section need not be equal but, by continuity, they must be equal for incompressible or steady compressible flow.

When there is no accumulation of momentum, equation 1.23 reduces to equation 1.22.

It is instructive to substitute for the mass flow rate in the momentum equation. For the case of no accumulation of momentum

$$\begin{aligned}
\text{rate of change of momentum} &= Mu_2 - Mu_1 \\
&= (\rho_2u_2S_2)u_2 - (\rho_1u_1S_1)u_1 \\
&= \rho_2u_2^2S_2 - \rho_1u_1^2S_1 \qquad (1.24)
\end{aligned}$$

Note that the momentum flow rate is proportional to the square of the fluid's velocity.

Example 1.3

In which directions do the forces arising from the change of fluid momentum act for steady incompressible flow in the pipe-work shown in Figure 1.3?

Calculations

The rate of change of momentum is given by:

$$M_2u_2 - M_1u_1 = M(u_2 - u_1)$$

By continuity

$$\rho_1 u_1 S_1 = \rho_2 u_2 S_2$$

But

$$\rho_1 = \rho_2 \text{ and } S_2 < S_1$$

Therefore

$$u_2 > u_1$$

Thus, the rate of change of momentum is positive and by Newton's law a positive force must act on the fluid in the section, ie a force in the positive x-direction. If the flow were reversed, the force would be reversed.

The above example shows the effect of a change in pipe diameter, and therefore flow area, on the momentum flow rate. It is clear that for steady, fully developed, incompressible flow in a pipe of constant diameter, the fluid's momentum must remain constant. However, it is possible for the fluid's momentum to change even in a straight pipe of constant diameter. If the (incompressible) flow were accelerating, as during the starting of flow, the momentum flow rates into and out of the section would be equal but there would be an accumulation of momentum within the section. (The mass of fluid in the section would remain constant but its velocity would be increasing.) Consequently, a force must act on the fluid in the direction of flow.

Now consider the case of steady, compressible flow in a straight pipe. As the gas flows from high pressure to lower pressure it expands and, by continuity, it must accelerate. Consequently, the momentum flow rate increases along the length of the pipe, although the mass flow rate remains constant.

In these examples, a pressure gradient is required to provide the increase in the fluid's momentum.

Example 1.4

Determine the magnitude and direction of the reaction on the bend shown in Figure 1.8 arising from changes in the fluid's momentum. The pipe is horizontal and the flow may be assumed to be steady and incompressible.

Calculations

It is necessary to consider both x and y components of the fluid's momentum.

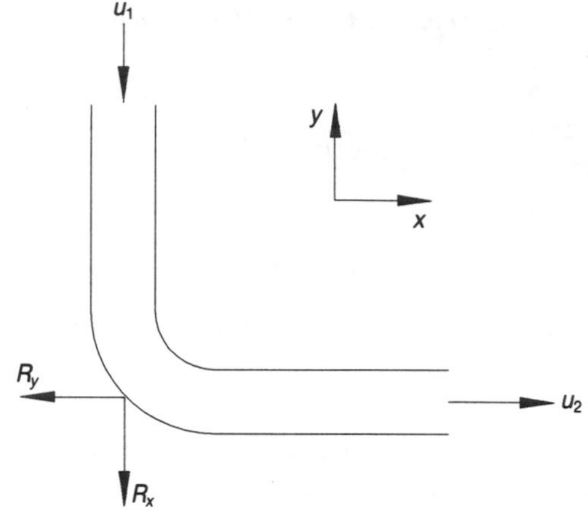

Figure 1.8
Reaction components acting on a pipe bend due to the change in fluid momentum

y-component:

$$y\text{-momentum flow rate out} = 0$$
$$y\text{-momentum flow rate in} = M(-u_1) = -Mu_1$$

Note: the minus sign arises from the fact that the fluid flows in the negative y-direction.

Thus the rate of change of y-momentum is Mu_1 and the force acting on the fluid in the y-direction is equal to Mu_1. There is therefore a reaction R_y of magnitude Mu_1 acting on the pipe in the negative y-direction.

x-component:

$$x\text{-momentum flow rate out} = Mu_2$$
$$x\text{-momentum flow rate in} = 0$$

Thus, the rate of change of x-momentum is Mu_2 and the force acting on the fluid in the x-direction is equal to Mu_2. A reaction R_x of magnitude Mu_2 acts on the bend in the negative x-direction.

If the pipe is of constant diameter, then $S_1 = S_2$ and by continuity $u_2 = u_1 = u$. Thus, the magnitude of each component of the reaction is equal to Mu, so the total reaction R acts at 45° and has magnitude $\sqrt{2}Mu$. This reaction is that due to the change in the fluid's momentum; in general other forces will also act, for example that due to the pressure of the fluid.

1.6.1 *Laminar flow*

The cases considered so far are ones in which the flow is turbulent and the velocity is nearly uniform over the cross section of the pipe. In laminar flow the curvature of the velocity profile is very pronounced and this must be taken into account in determining the momentum of the fluid.

The momentum flow rate over the cross sectional area of the pipe is easily determined by writing an equation for the momentum flow through an infinitesimal element of area and integrating the equation over the whole cross section. The element of area is an annular strip having inner and outer radii r and $r+\delta r$, the area of which is $2\pi r\delta r$ to the first order in δr. The momentum flow rate through this area is $2\pi r\delta r.\rho v^2$ so the momentum flow rate through the whole cross section of the pipe is equal to

$$2\pi\rho\int_0^{r_i} rv^2\,\mathrm{d}r \qquad (1.25)$$

where r_i is the internal radius of the pipe.

It is shown in Example 1.9 that the velocity profile for laminar flow of a Newtonian fluid in a pipe of circular section is parabolic and can be expressed in terms of the volumetric average velocity u as:

$$v = 2u\left(1-\frac{r^2}{r_i^2}\right) \qquad (1.67)$$

Therefore the momentum flow rate is equal to

$$8\pi\rho u^2\int_0^{r_i} r\left(1-\frac{r^2}{r_i^2}\right)^2 \mathrm{d}r = \frac{4}{3}\pi r_i^2\,\rho u^2 \qquad (1.26)$$

If the velocity had the uniform value u, the momentum flow rate would be $\pi r_i^2\rho u^2$. Thus for laminar flow of a Newtonian fluid in a pipe the momentum flow rate is greater by a factor of 4/3 than it would be if the same fluid with the same mass flow rate had a uniform velocity. This difference is analogous to the different values of α in Bernoulli's equation (equation 1.14).

Example 1.5

A Newtonian liquid in laminar flow in a horizontal tube emerges into the air as a jet from the end of the tube. What is the relationship between the diameters of the jet and the tube?

Calculations
It is assumed that the Reynolds number is sufficiently high for the fluid's momentum to be dominant and consequently the momentum flow rate in the jet will be the same as that in the tube. On emerging from the tube, there is no wall to maintain the liquid's parabolic velocity profile and consequently the jet develops a uniform velocity profile.

Equating the momentum of the liquid in the tube to that in the jet gives

$$\frac{4}{3}\pi r_i^2 \rho u_1^2 = \pi r_j^2 \rho u_2^2$$

where u_1, u_2 are the volumetric average velocities in the tube and jet respectively and r_i, r_j the radii of the tube and the jet. By continuity:

$$r_i^2 u_1 = r_j^2 u_2$$

Therefore

$$4u_1 = 3u_2$$

or

$$\frac{r_j}{r_i} = \left(\frac{3}{4}\right)^{1/2} = 0.866$$

Thus, the jet must have a smaller diameter than the tube in order for momentum to be conserved. This result is valid when the liquid's momentum is dominant. At very low Reynolds numbers, viscous stresses are dominant and the velocity profile starts to change even before the exit plane: in this case the jet diameter is slightly larger than the tube diameter.

1.6.2 *Total force due to flow*

In the preceding examples, cases in which there is a change in the momentum of a flowing fluid have been considered and the reactions on the pipe-work due solely to changes of fluid momentum have been determined. Sometimes it is required to make calculations of all forces acting on a piece of equipment as a result of the presence of the fluid and its flow through the equipment; this is illustrated in Example 1.6.

Example 1.6
Figure 1.9 illustrates a nozzle at the end of a hose-pipe. It is convenient to align the x-coordinate axis along the axis of the nozzle. The y-axis is perpendicular to the x-axis as shown and the x–y plane is vertical.

It is necessary first to define the region or 'control volume' for which the momentum equation is to be written. In this example, it is convenient to select the fluid within the nozzle as that control volume. The control volume is defined by drawing a 'control surface' over the inner surface of the nozzle and across the flow section at the nozzle inlet and the outlet. In this way, the nozzle itself is excluded from the control volume and external forces acting on the body of the nozzle, such as atmospheric pressure, are not involved in the momentum equation. This interior control surface is shown in Figure 1.9(a).

If the volume of the nozzle is V, a force ρVg due to gravity acts vertically downwards on the fluid. This force can be resolved into components $-\rho Vg \sin\theta$ acting in the positive x-direction and $-\rho Vg \cos\theta$ in the positive y-direction. The pressure of the liquid in the nozzle exerts a force in the x-direction but, owing to symmetry, the force components due to this pressure are zero in the y and z directions (excluding the hydrostatic pressure variation, which has already been accounted for by the weight of the fluid). The pressure P_1 of the fluid outside the control volume at plane 1 exerts a force P_1S_1 in the positive x-direction *on* the control volume. Similarly, at plane 2 a force of magnitude P_2S_2 is exerted on the control volume but this force acts in the negative x-direction.

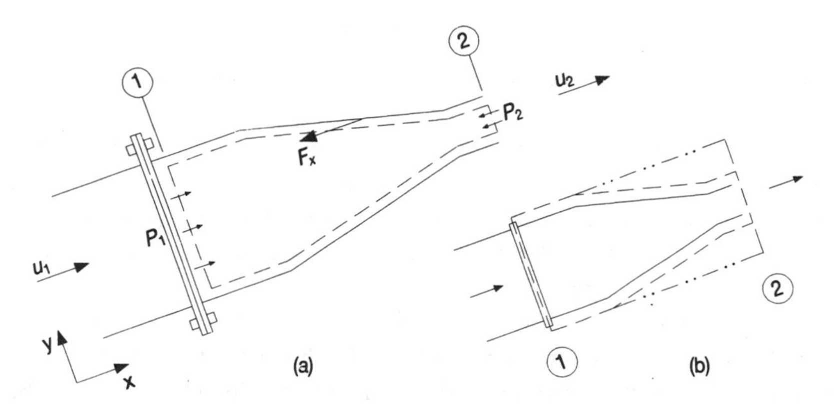

Figure 1.9
Forces acting on a nozzle inclined at angle θ to the horizontal
(a) *Internal control volume.* (b) *Two possible external control volumes*

As before, the rate of change of the fluid's x-component of momentum is $M(u_2-u_1)$, so the net force acting on the fluid in the x-direction is equal to $M(u_2-u_1)$. There is no change of momentum in the y or z directions.

The momentum equation can now be written but it must include the unknown reaction between the fluid and the nozzle. The unknown reaction of the nozzle on the fluid is denoted by F_x and for convenience (to show it acting on the fluid across the control surface) is taken as positive in the negative x-direction. Adding all the forces acting on the fluid in the positive x-direction, the momentum equation is

$$-F_x-\rho Vg\sin\theta+P_1S_1-P_2S_2 = M(u_2-u_1) \qquad (1.27)$$

This is the basic momentum equation for this type of problem in which all forces acting on the interior control volume are considered. Further observations on this particular example are given below.

For a given pressure difference P_1-P_2, the relationship between the velocities can be determined using Bernoulli's equation. Neglecting friction and the small change in elevation

$$P_1-P_2 = \rho(u_2^2-u_1^2)/2$$

Substituting

$$M = \rho u_1S_1 = \rho u_2S_2$$

allows the force acting on the fluid due to its reaction with the nozzle to be determined as

$$F_x = P_2(S_1-S_2)+\rho S_1(u_2-u_1)^2/2-\rho Vg\sin\theta$$

The pressure P_2 at the exit plane of the nozzle is very close to the pressure of the surrounding atmosphere.

In practice, the gravitational term will be negligible, and it is zero when the nozzle is horizontal. Thus, the force F_x is usually positive and therefore acts on the fluid in the negative x-direction. There is an equal and opposite reaction on the nozzle, which in turn exerts a tensile load on the coupling to the pipe.

An alternative approach is to draw the control surface over the outside of the nozzle as shown in Figure 1.9(b). In this case, the weight of the nozzle and the atmospheric pressure acting on its surface must be included. The reaction between the fluid and the nozzle forms equal and opposite *internal* forces and these are therefore excluded from the balance. However, the tension in the coupling generated by this reaction must be included as an external force acting on the control volume. It can be seen that this force is required by the fact that the exterior control surface cuts through the bolts of the coupling. Similarly, if there were a restraining bracket the force exerted by it on the control volume would be incorporated in the force–momentum balance.

In a case such as this, the force of the atmosphere on the surface of the nozzle can be simplified by using a cylindrical control volume shown by the dotted line in Figure 1.9(b). Assuming the thickness of the nozzle wall to be negligible, the pressure forces acting in the x-direction are P_1S_1 at plane 1 and $P_2S_2+P_{atm}(S_1-S_2)$ in the negative x direction at plane 2. By using the cylindrical control volume, these are the only surfaces on which pressure forces act in the x-direction. The area S_1-S_2 is just the projection of the tapered surface area on to the y–z plane.

In all cases the weight of all material within the control volume must be included in the force–momentum balance, although in many cases it will be a small force. Gravity is an external agency and it may be considered to act across the control surface. The momentum flows and all forces crossing the control surface must be included in the balance in the same way that material flows are included in a material balance.

An application of an internal momentum balance to determine the pressure drop in a sudden expansion is given in Section 2.4.

1.7 Stress in fluids

1.7.1 *Stress and strain*

It is necessary to know how the motion of a fluid is related to the forces acting on the fluid. Two types of force may be distinguished: long range forces, such as that due to gravity, and short range forces that arise from the relative motion of an element of fluid with respect to the surrounding fluid. The long range forces are called body forces because they act throughout the body of the fluid. Gravity is the only commonly encountered body force.

In order to appreciate the effect of forces acting on a fluid it is helpful first to consider the behaviour of a solid subjected to forces. Although the deformation behaviour of a fluid is different from that of a solid, the method of describing forces is the same for both.

Figure 1.10 shows two parallel, flat plates of area A. Sandwiched between the plates and bonded to them is a sample of a relatively flexible solid material. If the lower plate is fixed and a force F applied to the upper

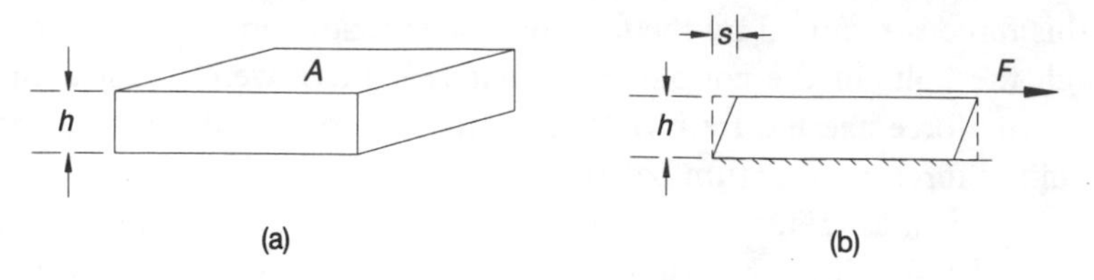

Figure 1.10
Shearing of a solid material
(a) *Sample of area A* (b) *Cross-section showing displacement*

plate as shown in Figure 1.10, the solid material is deformed until the resulting internal forces balance the applied force. In order to keep the lower plate stationary it is necessary that a restraining force of magnitude F acting in the opposite direction be provided by the fixture. The direction of the force F being in the plane of the plate, the sample is subject to a shearing deformation and the force is known as a shear force. The shear force divided by the area over which it acts defines the shear stress τ:

$$\tau = \frac{F}{A} \tag{1.28}$$

As shown in Figure 1.10(b), the horizontal displacement of the solid is proportional to the distance from the fixed plate. If the upper plate is displaced a distance s and the solid has a thickness h then the shear strain γ is defined as

$$\gamma = \frac{s}{h} \tag{1.29}$$

and is uniform throughout the sample. For small strains, γ is the angle in radians through which the sample is deformed.

It has been found experimentally that most solid materials exhibit a particularly simple relationship between the shear stress and the shear strain, at least over part of their range of behaviour:

$$\tau = G\gamma \tag{1.30}$$

The shear stress is proportional to the shear strain and the constant of proportionality G is known as the shear modulus. It does not matter how rapidly the solid is sheared, the shear stress depends only on the amount by which the solid is sheared.

A thin slice of the sample, parallel to the plates, is shown in Figure 1.11. The material above the slice is displaced further than the slice so the

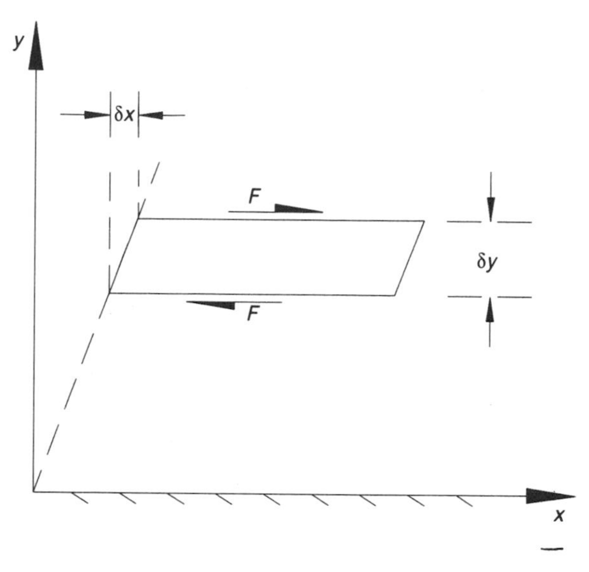

Figure 1.11
Shearing of an element of material

internal force acting on the upper surface of the slice acts in the direction of the applied force. Below the slice, the material is displaced less far and this lower material therefore exerts a force in the opposite direction. If the slice at distance y from the fixed plate is displaced a distance x from its unstressed position, the shear strain γ is equal to x/y. It will be seen that the shear strain can also be written as the displacement gradient:

$$\gamma = \frac{dx}{dy} \tag{1.31}$$

This expression is valid even when the displacement x is not proportional to y, in which case the strain is not uniform throughout the sample.

The behaviour of a fluid is different. If an ordinary liquid is placed between the plates and a constant shearing force F applied to the upper plate, the lower plate being fixed, the upper plate does not come to an equilibrium position but continues to move at a steady speed. The liquid adheres to each plate, ie there is no slip between the liquid and the solid surfaces, and at any instant the deformation of the sample is as shown in Figures 1.10 and 1.11. Thus the liquid sample is continuously sheared when subjected to a constant shear stress. The distinction between a fluid and a solid is that a fluid cannot sustain a shear stress without continuously deforming (ie flowing).

1.7.2 *Newton's law of viscosity*

In contrast to the behaviour of a solid, for a normal fluid the shear stress is independent of the magnitude of the deformation but depends on the *rate of change* of the deformation. Gases and many liquids exhibit a simple linear relationship between the shear stress τ and the rate of shearing:

$$\tau = \mu \frac{\mathrm{d}\gamma}{\mathrm{d}t} = \mu\dot{\gamma} \tag{1.32}$$

This is a statement of Newton's law of viscosity and the constant of proportionality μ is known as the coefficient of dynamic viscosity or, simply, the viscosity, of the fluid. The rate of change of the shear strain is known as the rate of (shear) strain or the shear rate. The coefficient of viscosity is a function of temperature and pressure but is independent of the shear rate $\dot{\gamma}$.

Referring to Figure 1.11, it has been noted that the strain at distance y from the fixed plate can be written as

$$\gamma = \frac{\mathrm{d}x}{\mathrm{d}y} \tag{1.31}$$

Thus, the shear rate $\dot{\gamma}$ is given by

$$\dot{\gamma} = \frac{\mathrm{d}}{\mathrm{d}t}\left(\frac{\mathrm{d}x}{\mathrm{d}y}\right) = \frac{\mathrm{d}}{\mathrm{d}y}\left(\frac{\mathrm{d}x}{\mathrm{d}t}\right)$$

Therefore

$$\dot{\gamma} = \frac{\mathrm{d}v_x}{\mathrm{d}y} \tag{1.33}$$

Equation 1.33 shows that the shear rate at a point is equal to the velocity gradient at that location. Figure 1.12 shows the flow in terms of the velocity component v_x, the magnitude of which is indicated by the length of the arrows.

In order to maintain steady flow, the net force acting on the element in the direction of flow must be zero. It follows that for this type of flow the shear stress acting on the lower face of the element must have the same magnitude but opposite direction to the shear stress acting on the upper face (see Figure 1.11). Consequently, the magnitude of the shear stress τ is the same at all values of y and from equation 1.32 the shear rate $\dot{\gamma}$ must be constant. In this type of flow, generated by moving the solid boundaries but with no pressure gradient imposed, the velocity profile is linear.

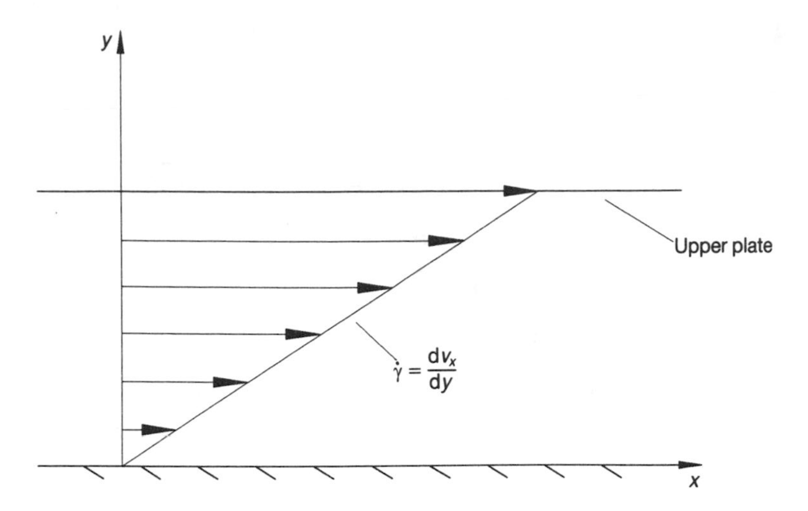

Figure 1.12
Shearing of a fluid showing the velocity gradient

In general, the velocity profile will be curved but as equation 1.33 contains only the *local* velocity gradient it can be applied in these cases also. An example is shown in Figure 1.13. Clearly, as the velocity profile is curved, the velocity gradient is different at different values of y and by equation 1.32 the shear stress τ must vary with y. Flows generated by the application of a pressure difference, for example over the length of a pipe, have curved velocity profiles. In the case of flow in a pipe or tube it is natural to use a cylindrical coordinate system as shown in Figure 1.14.

Newton's law of viscosity applies only to laminar flow. In laminar flow, layers of fluid flow past neighbouring layers without any macroscopic intermixing of fluid between layers. It may help in trying to visualize the two flows in Figures 1.12 and 1.14 if two analogies are considered. The analogy of the flow in Figure 1.12 is the shearing of a pile of writing paper or a pack of playing cards. The layers of liquid are equivalent to the sheets of paper or the individual cards. The analogy of the flow in Figure 1.14 is a telescope or telescopic aerial: if the innermost element is pulled out it drags with it the adjacent element which, in turn, pulls the next element and so on. The resistance to motion caused by the sliding friction between the moving surfaces is equivalent to the flow resistance between layers of fluid caused by viscosity. It should be noted, however, that the origin of viscosity is different: it is caused by molecules diffusing between layers. When a molecule diffuses from one layer to another with a lower velocity it

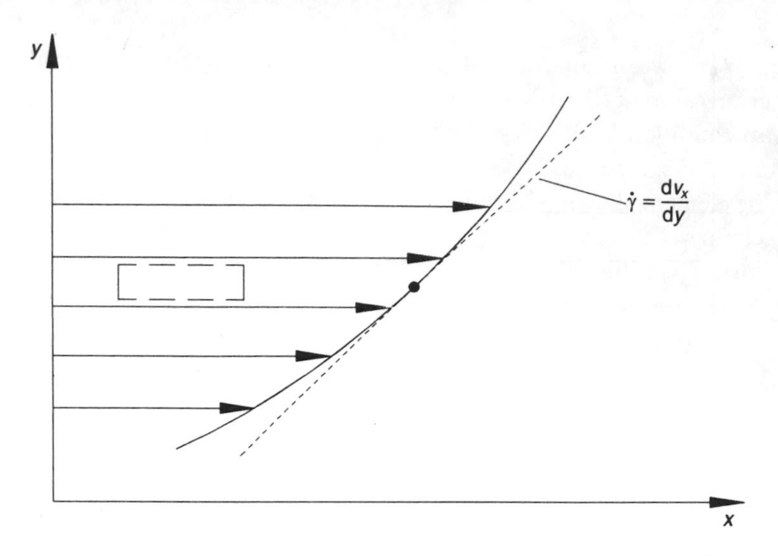

Figure 1.13
The varying shear rate for a curved velocity profile

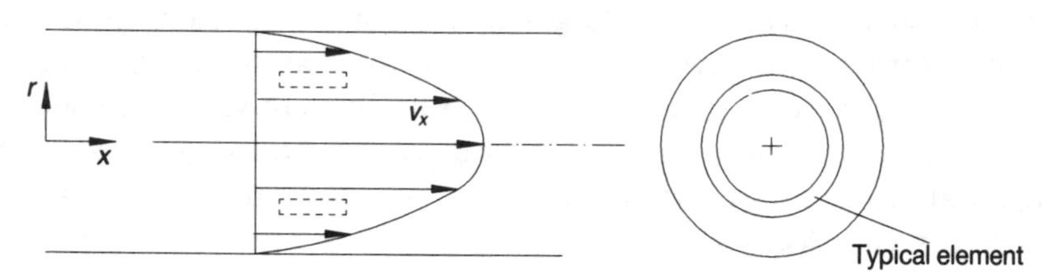

Figure 1.14
Typical velocity profile for flow in a pipe—the annular element shown is used in the analysis of such flows

will carry its momentum to that layer and tend on average to accelerate the slower layer. Similarly, when a molecule diffuses from a slow layer to a faster one it will retard the faster layer. This molecular diffusion occurs at the microscopic scale.

The direction in which the shear stress acts on a specified portion of the fluid depends on the relative motion of the neighbouring fluid. Consider the element of fluid shown as a broken line rectangle in Figure 1.13. The fluid above the element has a higher velocity and consequently drags the element in the direction of flow, while the fluid below the element has a lower velocity and has a retarding action on the element. In the case of the pipe flow shown in Figure 1.14, the flow is caused by the imposition of a higher pressure upstream (ie at the left) than downstream. By virtue of the no-slip condition at the pipe wall, the fluid there must be stationary but layers of fluid closer to the centre line have successively higher velocities. For the element shown, the fluid nearer the wall retards the element while that closer to the centre drags the element in the direction of flow. A steady state is achieved when the difference in the shear forces acting on the element balances the force due to the pressure difference across the element. (Note that Figure 1.14 shows a section on the diameter through the pipe. The element is an annular shell and the two rectangles shown are the two surfaces formed by the section cutting the element.)

Example 1.7
Determine the relationship between the shear stress at the wall and the pressure gradient for steady, fully developed, incompressible flow in a horizontal pipe.

Derivations
Figure 1.15 shows the flow with a suitable element of fluid, extending over the whole cross section of the pipe. For the conditions specified, the fluid's momentum remains constant so the net force acting on the fluid is zero.

Three forces act on the element in the (positive or negative) x-direction: the pressure P_1 pushes the fluid in the direction of flow, the pressure P_2 pushes against the flow, and the frictional drag between the fluid and the pipe wall acts against the flow.

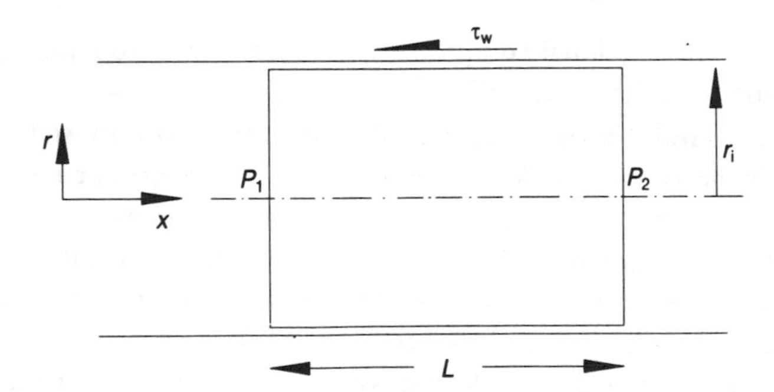

Figure 1.15
An element of fluid extending over the whole of the pipe cross section

The upstream pressure P_1 acts over the cross-sectional area of the element, so that the force acting on the element in the direction of flow is given by

$$\text{force acting in flow direction} = \pi r_i^2 P_1 \tag{1.34}$$

where the radius of the element is the same as the inside radius r_i of the pipe.

The downstream pressure P_2 acts on the element against the flow, as does the drag of the pipe wall on the fluid. The shear stress at the wall is called the wall shear stress and is denoted by τ_w. This shear stress acts over the area of the element in contact with the wall. The force acting against the flow is therefore given by

$$\text{force acting against flow} = \pi r_i^2 P_2 + 2\pi r_i L \tau_w \tag{1.35}$$

where L is the length of the element.

The net force being zero requires that

$$\pi r_i^2 (P_1 - P_2) - 2\pi r_i L \tau_w = 0 \tag{1.36}$$

The wall shear stress τ_w is therefore related to the pressure drop ΔP by

$$\tau_w = \frac{r_i(P_1 - P_2)}{2L} = \frac{r_i}{2}\left(\frac{\Delta P}{L}\right) \tag{1.37}$$

For the conditions specified in Example 1.7, the pressure drop is caused entirely by fluid friction. In general, there will also be static head and accelerative components and in these cases equation 1.37 should be written in the more general form

$$\tau_w = \frac{r_i}{2}\left(\frac{\Delta P_f}{L}\right) = \frac{d_i}{4}\left(\frac{\Delta P_f}{L}\right) \tag{1.38}$$

where ΔP_f is the frictional component of the pressure drop and d_i is the inside diameter of the pipe.

This simple force balance has provided an extremely important result: the wall shear stress for flow in a pipe can be determined from the frictional component of the pressure drop. In practice it is desirable to use the conditions in Example 1.7 so that the frictional component is the only component of the total pressure drop, which can be measured directly.

In Section 1.9 it is explained that the state of stress can be described by nine terms. In the above example, the wall shear stress is a particular value of one stress component, that denoted by τ_{rx}. In this notation, the second subscript denotes the direction in which the stress component acts, here, in the x-direction. The first subscript, here r, indicates that the stress component acts on the surface normal to the r-coordinate direction. (Specifying the normal to a surface is the easiest way of defining the orientation of the surface.) The shear stress component τ_{rx} is caused by the shearing of the liquid such that the velocity component v_x varies in the r-direction.

The wall shear stress τ_w is just the value of τ_{rx} at the wall of the pipe.

Example 1.8

How does the shear stress vary with radial location for the flow in Example 1.7?

Derivation

The variation of the shear stress τ_{rx} with radial coordinate r can be determined by making a force balance similar to that in Example 1.7 but using an element extending from the centre-line to a general radial distance r.

In this case, the force balance, equivalent to equation 1.36, can be written as

$$\pi r^2(P_1 - P_2) - 2\pi r L \tau_{rx} = 0 \qquad (1.39)$$

The shear stress τ_{rx} at distance r from the centre-line is therefore given by

$$\tau_{rx} = \frac{r(P_1 - P_2)}{2L} = \frac{r}{2}\left(\frac{\Delta P}{L}\right) \qquad (1.40)$$

Equation 1.40 shows that the shear stress varies linearly with radial location, from zero on the centre line to a maximum at the wall. The value at the wall is τ_w and putting $r = r_i$ in equation 1.40 therefore gives equation 1.37. Again, in general the pressure drop ΔP should be replaced by the frictional component of the pressure drop ΔP_f.

Combining equations 1.37 and 1.40 gives

$$\frac{\tau_{rx}}{\tau_w} = \frac{r}{r_i} \qquad (1.41)$$

which again demonstrates the linear variation of the shear stress with radial position.

The reason for this variation of the shear stress is easily understood. For steady flow there must be a balance between the force due to the pressure difference and the shear force. As shown in equation 1.39, the pressure force is proportional to r^2 but the shear force to r, so to maintain the balance it is necessary for the shear stress τ_{rx} to be proportional to r. (Note that ΔP is uniform over the whole cross section.)

It is important to note that in deriving the shear stress distribution no assumption was made as to whether the fluid was Newtonian or whether the flow was laminar. In the case of turbulent flow, it is the time-averaged values of τ_{rx} and τ_w that are given by equations 1.40 and 1.41. In Section 1.13 these time-averaged stresses will be denoted by $\bar{\tau}_{rx}$ and $\bar{\tau}_w$.

1.8 Sign conventions for stress

In analysing flow problems it is usual to select one coordinate axis (the x-coordinate axis in the above examples) to be parallel to the flow with the coordinate value increasing in the direction of flow so that the fluid's velocity is positive. If the velocity component v_x varies in the y-direction, it is normal to define the shear rate in terms of the change of v_x in the *positive* y-direction, so

$$\dot{\gamma} = \frac{dv_x}{dy} \tag{1.42}$$

and, similarly, for cylindrical coordinates

$$\dot{\gamma} = \frac{dv_x}{dr} \tag{1.43}$$

If v_x increases with y as in Figure 1.12 or Figure 1.13, the velocity gradient and therefore the shear rate are positive, but if v_x decreases with increasing y or r, as in Figure 1.14, the velocity gradient and shear rate are negative.

When analysing simple flow problems such as laminar flow in a pipe, where the form of the velocity profile and the directions in which the shear stresses act are already known, no formal sign convention for the stress components is required. In these cases, force balances can be written with the shear forces incorporated according to the directions in which the shear stresses physically act, as was done in Examples 1.7 and 1.8. However, in order to derive general equations for an arbitrary flow field it is necessary to adopt a formal sign convention for the stress components.

Before describing the two sign conventions that may be used, it may be helpful to consider a loose analogy with elementary mechanics. It is required to calculate the acceleration of a car from its mass and the forces acting on the car, all of which are known. It is necessary to evaluate the net force acting in the direction of motion and this can be done in two equivalent ways:

(i) net force = sum of forces acting in direction of motion
(ii) net force = propulsive forces − retarding forces

In (i) all forces are taken as acting in the direction of motion so the propulsive forces are positive but the retarding forces must be entered as negative quantities. In (ii), the fact that the retarding forces act opposite to the direction of motion has been incorporated into the equation so positive numbers are to be entered. Of course the result will be the same. Two different conventions have been adopted with regard to how the retarding forces are specified and the sign of the numbers entered must agree with the convention being used.

There are two sign conventions for stress components as illustrated in Figure 1.16. These diagrams are drawn for the shearing that occurs when there is a gradient in the y-direction of the x-component of the velocity, and show the directions in which the shear stress components τ_{yx} are taken as positive. Figure 1.16(a) shows the positive sign convention and Figure 1.16(b) the negative sign convention.

The way to remember the conventions is as follows. For the positive sign convention, the stress component acting on the element's face at the higher y-value (the upper face) is taken as positive in the positive x-direction; for the negative sign convention the same component is taken as positive in the negative x-direction. In each convention, the stress component acting on the opposite face is taken as positive in the opposite direction.

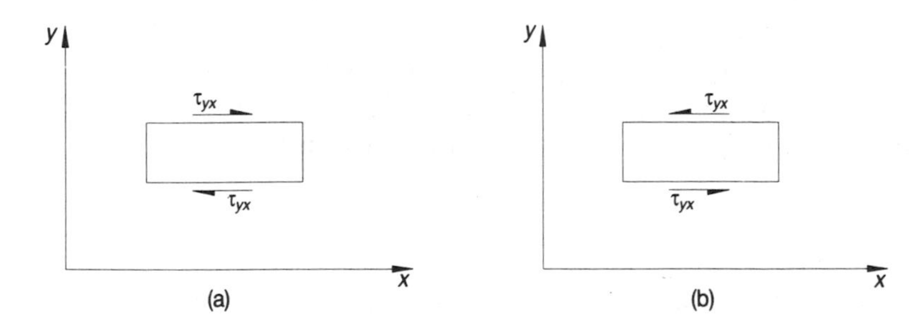

Figure 1.16
Sign conventions for stress
(a) *Positive sign convention* (b) *Negative sign convention*

With the positive sign convention, Newton's law of viscosity is expressed as

Positive convention: $$\tau_{yx} = \mu \frac{dv_x}{dy} \tag{1.44a}$$

but with the negative sign convention it must be expressed as

Negative convention: $$\tau_{yx} = -\mu \frac{dv_x}{dy} \tag{1.44b}$$

τ_{yx} is the shear stress component associated with the velocity gradient dv_x/dy and corresponding equations hold for other components.

In cylindrical coordinates, the velocity gradient dv_x/dr generates the shear stress component τ_{rx} and Newton's law must be expressed in the two sign conventions as:

Positive convention: $$\tau_{rx} = \mu \frac{dv_x}{dr} \tag{1.45a}$$

Negative convention: $$\tau_{rx} = -\mu \frac{dv_x}{dr} \tag{1.45b}$$

It is important to appreciate that the sign conventions do not dictate the direction in which a stress physically acts, they simply specify the directions chosen for measuring the stresses. If a stress physically acts in the opposite direction to that specified in the convention being used, then the stress will be found to have a negative value, just as in elementary mechanics where a negative force reflects the fact that it acts in the opposite direction to that taken as positive.

Both sign conventions are used in the fluid flow literature and consequently the reader should be able to work in either, as appropriate. The negative sign convention is convenient for flow in pipes because the velocity gradient dv_x/dr is negative and therefore the shear stress components turn out to be positive indicating that they physically act in the directions assumed in the sign convention. This is illustrated in Example 1.9.

Example 1.9

Determine the shear stress distribution and velocity profile for steady, fully developed, laminar flow of an incompressible Newtonian fluid in a horizontal pipe. Use a cylindrical shell element and consider both sign conventions. How should the analysis be modified for flow in an annulus?

Derivations

The velocity profile must have a form like that shown in Figure 1.17. The velocity is zero at the pipe wall and increases to a maximum at the centre. From Example 1.8, it is known that the shear stress vanishes on the centre-line $r = 0$, so from Newton's law of viscosity (equation 1.45) the velocity gradient must be zero at the centre.

The general arrangement of a representative element of fluid is shown in Figure 1.17. A cylindrical shell of fluid of length L has its inner and outer cylindrical surfaces at radial distances r and $r+\delta r$ respectively, where δr represents an infinitesimally small increment in r.

Using the negative sign convention for stress components, the shear stress acting on the outer surface of the element (the higher value of r) must be measured in the negative x-direction and that on the inner surface in the positive x-direction, as indicated in Figure 1.17.

In accordance with the sign convention, the forces acting on the fluid in the positive x-direction are the force due to the upstream pressure P_1 and the shear force on the inner surface of the element. Those acting in the negative x-direction are the force due to the pressure P_2 and the shear force on the outer surface of the element. The cross-sectional area of the element, on which the pressure acts, is equal to $2\pi r\delta r$. As the fluid's momentum remains constant, the net force acting on the fluid is zero:

$$2\pi r\delta r\,.\,P_1 + 2\pi rL\,.\,\tau_{rx}\big|_r - 2\pi r\delta r\,.\,P_2 - 2\pi(r+\delta r)L\,.\,\tau_{rx}\big|_{r+\delta r} = 0 \quad (1.46)$$

The notation $\tau_{rx}\big|_r$ denotes the value of τ_{rx} at radial distance r, ie on the inner surface, and $\tau_{rx}\big|_{r+\delta r}$ denotes the value at radial distance $r+\delta r$, ie on the outer surface of the element.

Rearranging and dividing by the volume of the element $2\pi rL\delta r$,

$$\frac{r\tau_{rx}\big|_r - (r+\delta r)\tau_{rx}\big|_{r+\delta r}}{r\delta r} + \frac{P_1 - P_2}{L} = 0 \quad (1.47)$$

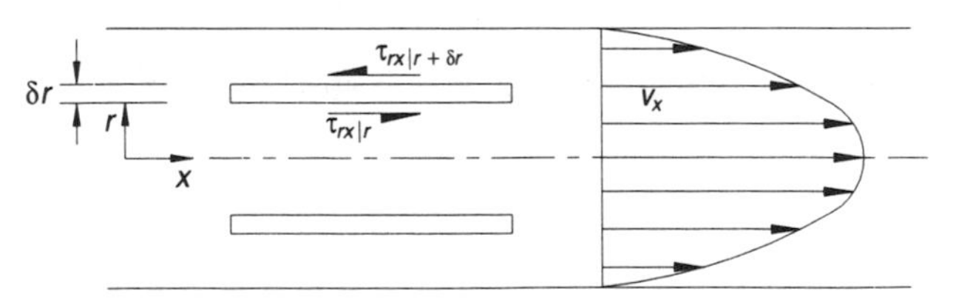

Figure 1.17

Laminar flow in a pipe showing a typical fluid element and the velocity profile. The negative sign convention for stress components is shown

On letting δr tend to zero,

$$\lim_{\delta r \to 0}\left(\frac{(r+\delta r)\tau_{rx}|_{r+\delta r} - r\tau_{rx}|_r}{r\delta r}\right) \equiv \frac{1}{r}\frac{\mathrm{d}}{\mathrm{d}r}(r\tau_{rx}) \tag{1.48}$$

therefore

$$\frac{1}{r}\frac{\mathrm{d}}{\mathrm{d}r}(r\tau_{rx}) = \frac{P_1 - P_2}{L} \tag{1.49}$$

Integrating

$$\tau_{rx} = \left(\frac{P_1 - P_2}{L}\right)\frac{r}{2} + \frac{A_1}{r} \tag{1.50}$$

The shear stress must remain finite at $r = 0$ so $A_1 = 0$. Thus the shear stress distribution is given by

$$\tau_{rx} = \left(\frac{P_1 - P_2}{L}\right)\frac{r}{2} = \left(\frac{\Delta P}{L}\right)\frac{r}{2} \tag{1.51}$$

as found in Example 1.8. The upstream pressure P_1 is greater than P_2 so τ_{rx} is positive, indicating that the shear stress components physically act in the directions employed in the sign convention and shown in Figure 1.17.

In order to determine the velocity profile, it is necessary to substitute for τ_{rx} using Newton's law of viscosity, thereby introducing the velocity gradient. For the negative sign convention, Newton's law is given by equation 1.45b and substituting for τ_{rx} in equation 1.51 gives

$$-\mu\frac{\mathrm{d}v_x}{\mathrm{d}r} = \left(\frac{\Delta P}{L}\right)\frac{r}{2} \tag{1.52}$$

Integrating

$$-\mu v_x = \left(\frac{\Delta P}{L}\right)\frac{r^2}{4} + B \tag{1.53}$$

where B is a constant of integration. The no-slip boundary condition at the wall is

$$v_x = 0 \quad \text{at} \quad r = r_i$$

therefore

$$B = -\left(\frac{\Delta P}{L}\right)\frac{r_i^2}{4}$$

and the velocity profile is

$$v_x = \frac{1}{4\mu}\left(\frac{\Delta P}{L}\right)(r_i^2 - r^2) \tag{1.54}$$

This is the equation of a parabola.

A slightly different procedure is to substitute for τ_{rx} in equation 1.49 using Newton's law of viscosity. If this is done and the resulting equation integrated twice, equations 1.55 and 1.56 are obtained:

$$-\mu\frac{\mathrm{d}v_x}{\mathrm{d}r} = \left(\frac{P_1 - P_2}{L}\right)\frac{r}{2} + \frac{A_2}{r} \tag{1.55}$$

$$-\mu v_x = \left(\frac{P_1 - P_2}{L}\right)\frac{r^2}{4} + A_2 \ln r + B \tag{1.56}$$

There are two constants of integration in equation 1.56 so two boundary conditions are required. The first is the no-slip condition at $r = r_i$ and the second is that the velocity gradient is zero at $r = 0$. Using the latter condition in equation 1.55 shows that $A_2 = 0$ so that equation 1.56 becomes identical to equation 1.53. The no-slip boundary condition gives the value of B as before.

In determining the flow in a whole pipe, as above, it is unnecessary to use the infinitesimal shell element: the method used in Example 1.8, with an element extending from the centre to a general position r, is preferable because it is simpler. Where the infinitesimal cylindrical shell element is required is for flow in an annulus, for example between $r = r_1$ and $r = r_2$. This is necessary because the flow region does not extend to the centre-line so a whole cylindrical element cannot be fitted in. In the case of flow in an annulus, equation 1.56 is valid but the constants of integration must be determined using the boundary conditions that the velocity is zero at *both* walls. (Note that this specifies a value of v_x at two different values of r and therefore provides two boundary conditions as required.)

As an illustration of the fact that the two sign conventions give the same results, the equivalents of equations 1.46 to 1.53 can be written for the positive sign convention. In the positive sign convention, the shear stress acting on the outer surface of the element is measured in the positive x-direction and that on the inner surface is measured in the negative x-direction. This is the opposite of the directions shown in Figure 1.17. The force balance, equivalent to equation 1.46 is now

$$2\pi r\delta r . P_1 + 2\pi(r + \delta r)L . \tau_{rx}|_{r+\delta r} - 2\pi r\delta r . P_2 - 2\pi rL . \tau_{rx}|_r = 0 \tag{1.57}$$

This leads to

$$-\frac{1}{r}\frac{\mathrm{d}}{\mathrm{d}r}(r\tau_{rx}) = \frac{P_1 - P_2}{L} \tag{1.58}$$

and

$$-\tau_{rx} = \left(\frac{P_1 - P_2}{L}\right)\frac{r}{2} = \left(\frac{\Delta P}{L}\right)\frac{r}{2} \tag{1.59}$$

Equations 1.58 and 1.59 are equivalent to equations 1.49 and 1.51. It will be noted that the only difference is the sign of each term containing the shear stress τ_{rx}. P_1 being greater than P_2, equation 1.59 shows that τ_{rx} is negative, which indicates that the shear stress components act physically in the opposite directions to those employed in the sign convention. This is in agreement with the findings when using the negative sign convention.

As before, in order to determine the velocity profile it is necessary to introduce Newton's law of viscosity but as the positive sign convention is now being used it is necessary to express Newton's law by equation 1.45a:

$$\tau_{rx} = \mu\frac{\mathrm{d}v_x}{\mathrm{d}r} \tag{1.45a}$$

When this is used to substitute for τ_{rx} in equation 1.59, equation 1.60 is obtained:

$$-\mu\frac{\mathrm{d}v_x}{\mathrm{d}r} = \left(\frac{\Delta P}{L}\right)\frac{r}{2} \tag{1.60}$$

Equation 1.60 is identical to equation 1.52. If the derivation were continued, the same velocity profile would be obtained as when using the negative sign convention.

In general, with the different sign conventions, equations involving stress components have opposite signs in the two conventions. On substituting the *appropriate* form of Newton's law of viscosity, the sign difference cancels giving identical equations for the velocity profile.

Although the sign conventions have been illustrated in Example 1.9 using a cylindrical shell element, it should not be thought that they apply only to infinitesimal elements like this and not to 'whole' cylindrical elements as used in Examples 1.7 and 1.8. The cylindrical element is a special case of the cylindrical shell element in which the inner surface has been allowed to shrink to, and disappear on, the centre-line. Figure 1.17 shows the directions in which τ_{rx} is taken as positive when using the negative sign convention. If the inner surface is allowed to shrink to the centre-line, only the outer surface, with τ_{rx} taken as positive in the negative x-direction remains. With the positive sign convention, τ_{rx} on that surface is taken as positive in the positive x-direction. This may be the easiest way to remember the sign conventions.

1.9 Stress components

In the preceding section, only one stress component was considered and that component was the only one of direct importance in the simple flow considered. The force acting at a point in a fluid is a vector and can be resolved into three components, one in each of the coordinate directions. Consequently the stress acting on each face of an element of fluid can be represented by three stress components, as shown in Figure 1.18 for the negative sign convention.

All the stress components shown are taken as positive in the directions indicated. Each stress component is written with two subscripts, the first denoting the face on which it acts and the second the direction in which the stress acts. Thus the stress component τ_{yx} acts on a face normal to the y-axis and in the x-direction. There are two τ_{yx} terms, one acting on the left face (at y) and the other on the right face (at $y + \delta y$). With the negative sign convention for stress components, the components τ_{yx}, τ_{yy}, τ_{yz} at the higher y-value (the right face) are taken as positive in the negative x, y, z directions respectively. At the lower y-value (the left face) the stress components are taken as positive in the opposite directions. The same rule applies for the faces normal to the y and z coordinate directions. For

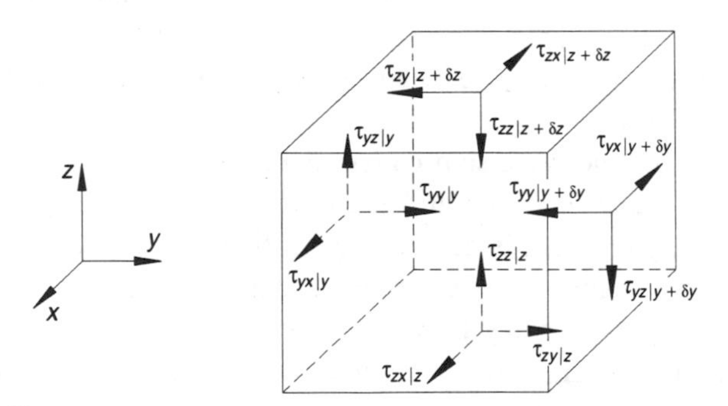

Figure 1.18

Negative sign convention for stress components. The diagram shows the directions in which components are taken as positive. The components acting on the faces normal to the x-axis have been omitted for clarity

clarity, the stress components acting on the face normal to the x-axis are omitted from Figure 1.18.

When using the positive sign convention, the direction of each stress component that is taken as positive is the opposite of that shown in Figure 1.18.

Independent of the sign convention used, the stress components can be classified into two types: those that act tangentially to the face of the element and those that act normal to the face. Tangential components such as τ_{xy}, τ_{yx}, τ_{yz} tend to cause shearing and are called shear stress components (or simply shear stresses). In contrast, the stress components τ_{xx}, τ_{yy}, τ_{zz} act normal to the face of the element and are therefore called normal stress components (or normal stresses). Although there are six shear stress components, it is easily shown that $\tau_{ij} = \tau_{ji}$ for $i \neq j$; for example, $\tau_{yx} = \tau_{xy}$. Thus there are three independent shear stress components and three independent normal stress components.

The pressure acting on a surface in a static fluid is the normal force per unit area, ie the normal stress. The pressure of the surrounding fluid acts inwards on each face of a fluid element. Consequently, with the negative sign convention the normal stress components may be identified with the pressure. With the positive sign convention, the normal stress components may be identified with the negative of the pressure: positive normal stresses correspond to tension with this convention.

In the case of a *flowing* fluid the mechanical pressure is not necessarily the same as the thermodynamic pressure as is the case in a static fluid. The pressure in a flowing fluid is defined as the average of the normal stress components. In the case of inelastic fluids, the normal stress components are equal and therefore, with the negative sign convention, equal to the pressure. It is for this reason that the pressure can be used in place of the normal stress when writing force balances for inelastic liquids, as was done in Examples 1.7–1.9.

Both positive and negative sign conventions for stress components are used in the fluid flow literature and each convention has its advantages. When the velocity gradient is negative, as in flow in a pipe, the negative sign convention is slightly more convenient than the positive sign convention. In addition, the analogy between momentum transfer and heat or mass transfer has the same sign when the negative convention is used. The negative sign convention for stress components will be used throughout the remainder of this book.

1.10 Volumetric flow rate and average velocity in a pipe

The volumetric flow rate is determined by writing the equation for the volumetric flow rate across an infinitesimal element of the flow area then integrating the equation over the whole flow area, ie the cross-sectional area of the pipe. It is necessary to use an infinitesimal element of the flow area because the velocity varies over the cross section. Over the infinitesimal area, the velocity may be taken as uniform, and the variation with r is accommodated in the integration.

A typical element is a strip perpendicular to the axis (and therefore perpendicular to v_x) with an inner radius r and outer radius $r+\delta r$ where δr is infinitesimally small. To the first order in δr, the area of this element is equal to $2\pi r\delta r$. The volumetric flow rate δQ across this area is therefore

$$\delta Q = 2\pi r\delta r . v_x \tag{1.61}$$

The total volumetric flow rate through the pipe is obtained by integrating the element equation over the whole cross section of the pipe, that is from $r = 0$ to $r = r_i$:

$$Q = 2\pi \int_0^{r_i} r v_x \, \mathrm{d}r \tag{1.62}$$

This expression applies to any symmetric flow in a pipe.

The volumetric average velocity u is that velocity which, if uniform over the flow area S, would give the volumetric flow rate and is therefore defined by

$$Q = uS \tag{1.6}$$

Thus, for flow in a pipe,

$$u = \frac{Q}{\pi r_i^2} = \frac{2}{r_i^2} \int_0^{r_i} r v_x \, \mathrm{d}r \tag{1.63}$$

This is a general expression and is valid for any symmetric velocity profile.

1.10.1 *Laminar Newtonian flow in a pipe*

In the case of laminar flow of a Newtonian fluid in a pipe, the velocity profile is given by equation 1.54 so the volumetric flow rate is

$$Q = \frac{\pi}{2\mu}\left(\frac{\Delta P}{L}\right)\int_0^{r_i} r(r_i^2 - r^2)\,\mathrm{d}r$$

$$= \frac{\pi r_i^4}{8\mu}\left(\frac{\Delta P}{L}\right) \tag{1.64}$$

This last result is known as the Hagen–Poiseuille equation.

The volumetric average velocity u could be determined from equation 1.63 but as the expression for Q has already been found it is more convenient to determine u by dividing equation 1.64 by the flow area πr_i^2:

$$u = \frac{r_i^2}{8\mu}\left(\frac{\Delta P}{L}\right) = \frac{d_i^2}{32\mu}\left(\frac{\Delta P}{L}\right) \tag{1.65}$$

On putting $r = 0$ in equation 1.54, the maximum velocity is

$$v_{\max} = \frac{r_i^2}{4\mu}\left(\frac{\Delta P}{L}\right) \tag{1.66}$$

Thus, for a parabolic velocity profile in a pipe, the volumetric average velocity is half the centre-line velocity and the equation for the velocity profile can be written as:

$$v = v_{\max}\left(1 - \frac{r^2}{r_i^2}\right) = 2u\left(1 - \frac{r^2}{r_i^2}\right) \tag{1.67}$$

1.11 Momentum transfer in laminar flow

Using the negative sign convention for stress components, Newton's law of viscosity can be written as

$$\tau_{yx} = -\mu\frac{\mathrm{d}v_x}{\mathrm{d}y} \tag{1.44b}$$

for the stress component τ_{yx} that results from a velocity gradient $\mathrm{d}v_x/\mathrm{d}y$. For constant density, this can be written in the form

$$\tau_{yx} = -\frac{\mu}{\rho}\frac{\mathrm{d}}{\mathrm{d}y}(\rho v_x) = -\nu\frac{\mathrm{d}}{\mathrm{d}y}(\rho v_x) \tag{1.68}$$

In equation 1.68, ρv_x is the fluid's momentum per unit volume. The quantity μ/ρ, denoted by ν, is known as the kinematic viscosity. In SI units ν has the units m^2/s.

Heat conduction is described by Fourier's law and diffusion by Fick's first law:

$$\text{heat flux} \qquad q_y = -k\frac{\mathrm{d}T}{\mathrm{d}y} = -\frac{k}{\rho C_p}\frac{\mathrm{d}}{\mathrm{d}y}(\rho C_p T) \qquad (1.69)$$

$$\text{molar diffusional flux} \qquad \mathcal{J}_{Ay} = -\mathscr{D}\frac{\mathrm{d}C_A}{\mathrm{d}y} \qquad (1.70)$$

Equation 1.70 shows that the molar diffusional flux of component A in the y-direction is proportional to the concentration gradient of that component. The constant of proportionality is the molecular diffusivity $\mathscr{D}$. Similarly, equation 1.69 shows that the heat flux is proportional to the gradient of the quantity $\rho C_p T$, which represents the concentration of thermal energy. The constant of proportionality $k/\rho C_p$, which is often denoted by α, is the thermal diffusivity and this, like $\mathscr{D}$, has the units m^2/s.

By analogy, equation 1.68 shows that a shear stress component can be interpreted as a flux of momentum because it is proportional to the gradient of the 'concentration' of momentum. In particular, τ_{yx} is the flux in the y-direction of the fluid's x-component of momentum. Furthermore, the kinematic viscosity ν can be interpreted as the diffusivity of momentum of the fluid.

Consider the flow shown in Figure 1.12. Owing to the motion of the upper plate, the fluid in contact with it has a higher velocity than that below it and consequently it has higher momentum per unit volume (ρv_x). This layer of fluid in contact with the plate exerts a shear force on the lower fluid causing its motion and therefore providing its momentum. This may be compared with a thermal analogy: thermal conduction from the upper plate to the lower one through a material between them. If the upper plate is hotter than the lower one, there will be a temperature gradient analogous to the velocity gradient in the flow example. Heat will be conducted from the top to the bottom, analogous to the transfer of momentum in fluid flow. If the two plates are maintained at constant but different temperatures, the steady heat conduction corresponds to the steady momentum transfer and the constant heat flux corresponds to the constant shear stress (momentum flux).

The kinematic viscosity ν is of more fundamental importance than the dynamic viscosity μ and it is appropriate to consider typical values of both these quantities, as shown in Table 1.2.

Table 1.2 Viscosities of fluids at 25 °C

Fluid	μ (Pa s)	ν (m^2/s)
Air	1.8×10^{-5}	1.5×10^{-5}
Water	1.0×10^{-3}	1.0×10^{-6}
Mercury	1.5×10^{-3}	1.1×10^{-7}
Castor oil	0.99	1.0×10^{-3}

1.12 Non-Newtonian behaviour

For a Newtonian fluid, the shear stress is proportional to the shear rate, the constant of proportionality being the coefficient of viscosity. The viscosity is a property of the material and, at a given temperature and pressure, is constant. Non-Newtonian fluids exhibit departures from this type of behaviour. The relationship between the shear stress and the shear rate can be determined using a viscometer as described in Chapter 3. There are three main categories of departure from Newtonian behaviour: behaviour that is independent of time but the fluid exhibits an apparent viscosity that varies as the shear rate is changed; behaviour in which the apparent viscosity changes with time even if the shear rate is kept constant; and a type of behaviour that is intermediate between purely liquid-like and purely solid-like. These are known as time-independent, time-dependent, and viscoelastic behaviour respectively. Many materials display a combination of these types of behaviour.

The term viscosity has no meaning for a non-Newtonian fluid unless it is related to a particular shear rate $\dot{\gamma}$. An apparent viscosity μ_a can be defined as follows (using the negative sign convention for stress):

$$\mu_a \equiv -\frac{\tau}{\dot{\gamma}} \tag{1.71}$$

In the simplest case, that of time-independent behaviour, the shear stress depends only on the shear rate but not in the proportional manner of a Newtonian fluid. Various types of time-independent behaviour are shown in Figure 1.19(a), in which the shear stress is plotted against the shear rate on linear axes. The absolute values of shear stress and shear rate are plotted so that irrespective of the sign convention used the curves always lie in the first quadrant.

From such a flow curve, the apparent viscosity can be calculated at any

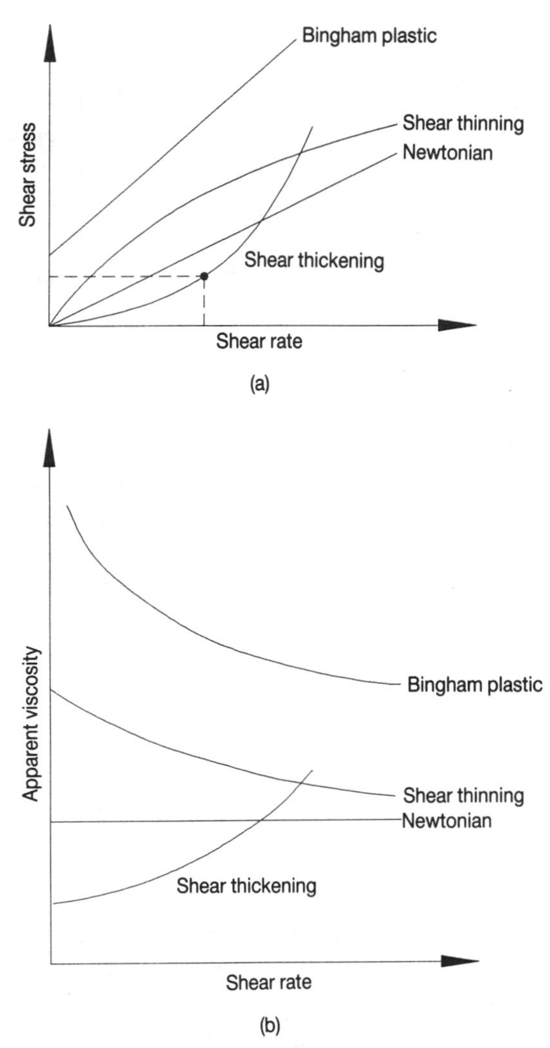

Figure 1.19
Flow curves for time–independent fluids
(a) *Shear stress against shear rate* (b) *Apparent viscosity against shear rate*

point such as that indicated, and Figure 1.19(b) shows the apparent viscosity for each type of behaviour corresponding to the curves in Figure 1.19(a). Fluids for which the apparent viscosity decreases with increasing shear rate are called shear thinning or pseudoplastic fluids, while those with the opposite behaviour are known as shear thickening fluids. As a guide, dilute and moderately concentrated suspensions and solutions of macromolecules exhibit shear thinning behaviour, the suspended matter or the molecules tending to become aligned with the flow. Shear thickening behaviour occurs most commonly with highly concentrated suspensions, in which progressively stronger interactions occur between the suspended particles as the shear rate increases. Examples of shear thinning fluids are polymer solutions and melts, dilute suspensions and paper pulp; and shear thickening materials include starch, and concentrated pigment suspensions such as paint and ink.

Some very concentrated suspensions are dilatant. If, in such a suspension, the particles are closely packed, then when the suspension is sheared the particles have to adopt a greater spacing in order to move past neighbouring particles and as a result the suspension expands, ie it dilates. Dilatant materials tend to be shear thickening but it does not follow that shear thickening behaviour is necessarily due to dilatancy. Consequently, dilatancy should not be used as a synonym for shear thickening behaviour.

It should be noted that for shear thinning and shear thickening behaviour the shear stress–shear rate curve passes through the origin. This type of behaviour is often approximated by the 'power law' and such materials are called 'power law fluids'. Using the negative sign convention for stress components, the power law is usually written as

$$\tau = -K\dot{\gamma}^n \qquad (1.72a)$$

but, rigorously, it should be written as

$$\tau = -K\dot{\gamma}\,|\,\dot{\gamma}\,|^{\,n-1} \qquad (1.72b)$$

The latter form is required to reflect the fact that the direction of the shear stress must reverse when the shear rate is reversed, and to overcome objections such as $\dot{\gamma}^n$, and therefore τ, having imaginary values when $\dot{\gamma}$ is negative. The power n is known as the power law index or flow behaviour index, and K as the consistency coefficient.

Clearly, shear thinning behaviour corresponds to $n<1$ and shear thickening behaviour to $n>1$. The special case, $n = 1$, is that of Newtonian behaviour and in this case the consistency coefficient K is identical to the viscosity μ. Values of n for shear thinning fluids often extend to 0.5 but less commonly can be as low as 0.3 or even 0.2, while values of n for shear thickening behaviour usually extend to 1.2 or 1.3.

Several objections can be raised against the power law, for example the consistency coefficient is not a genuine physical property, which is clear from its units, Pa s^n. In addition, there is a discontinuity in the gradient of the $\tau-\dot{\gamma}$ curve at the origin, ie a kink in the curve, for values of n other than unity. More importantly, with many fluids it is impossible to fit a single value of n to the flow curve over a wide range of shear rates, partly because most fluids tend to Newtonian behaviour at very low deformation rates. Nevertheless, the flow behaviour of some materials can be represented quite well by the power law and, provided it is appreciated that it is merely a curve-fitting tool, its use can be very helpful in making practical engineering flow calculations.

A different kind of time-independent behaviour is that characterized by materials known as Bingham plastics, which exhibit a yield stress τ_y. If subject to a shear stress smaller than the yield stress, they retain a rigid structure and do not flow. It is only at stresses in excess of the yield value that flow occurs. In the case of a Bingham plastic, the shear rate is proportional to shear stress in excess of the yield stress:

$$\begin{aligned} \tau-\tau_y &= -\beta\dot{\gamma} \qquad && \tau \geq \tau_y \\ \dot{\gamma} &= 0 && \tau < \tau_y \end{aligned} \tag{1.73}$$

The apparent viscosity becomes infinite as the shear stress is reduced to the yield value because below the non-zero yield stress there is no flow. As the shear rate is increased, the apparent viscosity tends to the value β, which is equal to the gradient of the flow curve.

Some materials can be modelled well by modifying the power law to include a yield stress; this is known as the Herschel–Bulkley model:

$$\begin{aligned} \tau-\tau_y &= -K\dot{\gamma}\,|\,\dot{\gamma}\,|^{\,n-1} \qquad && \tau \geq \tau_y \\ \dot{\gamma} &= 0 && \tau < \tau_y \end{aligned} \tag{1.74}$$

Under conditions of steady fully developed flow, molten polymers are shear thinning over many orders of magnitude of the shear rate. Like many other materials, they exhibit Newtonian behaviour at very low shear rates; however, they also have Newtonian behaviour at very high shear rates as shown in Figure 1.20. The term pseudoplastic is used to describe this type of behaviour. Unfortunately, the same term is frequently used for shear thinning behaviour, that is the falling viscosity part of the full curve for a pseudoplastic material. The whole flow curve can be represented by the Cross model [Cross (1965)]:

$$\frac{\mu_a-\mu_\infty}{\mu_0-\mu_\infty} = \frac{1}{1+(\dot{\gamma}/\dot{\gamma}_m)^n} \tag{1.75}$$

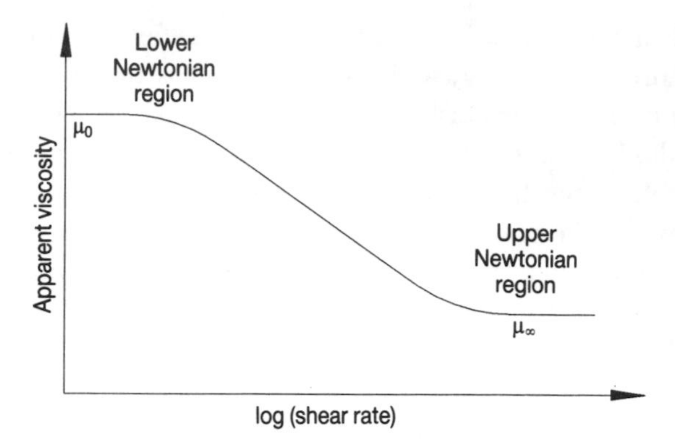

Figure 1.20
Variation of apparent viscosity with shear rate for a polymer

In equation 1.75, μ_0 and μ_∞ are the values of the apparent viscosity for the lower and upper Newtonian regions respectively. The constant $\dot{\gamma}_m$ is the shear rate evaluated at the mean apparent viscosity $(\mu_0 + \mu_\infty)/2$.

The second category, time-dependent behaviour, is common but difficult to deal with. The best known type is the thixotropic fluid, the characteristic of which is that when sheared at a constant rate (or at a constant shear stress) the apparent viscosity decreases with the duration of shearing. Figure 1.21 shows the type of flow curve that is found. The apparent viscosity continues to fall during shearing so that if measurements are made for a series of increasing shear rates and then the series is reversed, a hysteresis loop is observed. On repeating the measurements, similar behaviour is seen but at lower values of shear stress because the apparent viscosity continues to fall.

This decreasing of the apparent viscosity during constant rate shearing reflects a breaking down of the structure of the material as a result of the shearing. Eventually, a dynamic equilibrium is reached where the rate of breakdown is balanced by the rate of the simultaneous reformation of the structure. Consequently, a minimum value of the apparent viscosity is reached for a given constant shear rate. Another aspect of the reversibility of the structural changes is that if, after shearing, a thixotropic material is allowed to stand for several hours the original viscosity will be recovered. Sometimes when thixotropic fluids, which are often dispersions of solids in liquids, are prepared the apparent viscosity is very high and, on standing after shearing, only partial recovery of the original viscosity is observed. This reflects a permanent change in the material brought about by shearing; it might be, for example, the result of incomplete dispersion.

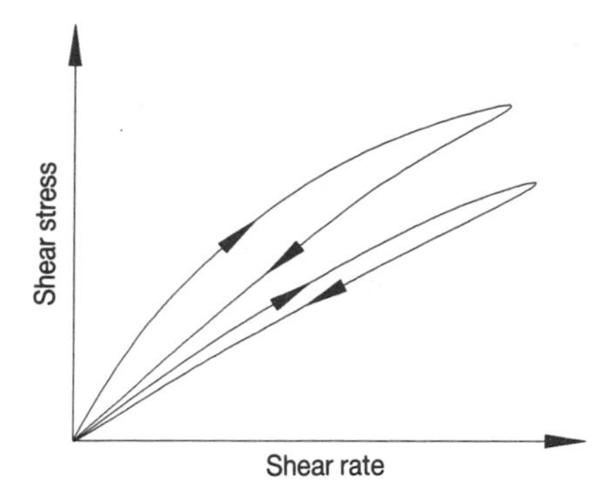

Figure 1.21
Flow curves for a thixotropic fluid

That recovery which does occur reflects thixotropy: true thixotropy is reversible.

Many food preparations and some paints are deliberately designed to be thixotropic so that the solid matter remains in suspension when the product is standing but, on being shaken, the apparent viscosity falls and the product can be poured.

The opposite behaviour, increasing apparent viscosity during shear is called rheopexy and is far less common than thixotropy. With a rheopectic fluid small shearing motions facilitate the formation of a structure but above a critical value breakdown occurs. If shearing is rapid, the structure does not form. In general, the apparent viscosity of a rheopectic fluid increases with time to a maximum value for a given constant shear rate. On being left to stand, most rheopectic fluids revert very quickly to their original viscosity. Examples of rheopectic fluids are aqueous gypsum suspensions, and sols of bentonite and vanadium pentoxide.

The final main category of non-Newtonian behaviour is viscoelasticity. As the name implies, viscoelastic fluids exhibit a combination of ordinary liquid-like (viscous) and solid-like (elastic) behaviour. The most important viscoelastic fluids are molten polymers but other materials containing macromolecules or long flexible particles, such as fibre suspensions, are viscoelastic. An everyday example of purely viscous and viscoelastic behaviour can be seen with different types of soup. When a 'thin', watery soup is stirred in a bowl and the stirring then stopped, the soup continues to flow round the bowl and gradually comes to rest. This is an example of purely viscous behaviour. In contrast, with certain 'thick' soups, on cessation of stirring the soup rapidly slows down and then recoils slightly.

The rapid stopping may be due merely to the higher apparent viscosity but the recoil is a manifestation of solid-like behaviour: a purely viscous material cannot recoil.

The simplest model that can show the most important aspects of viscoelastic behaviour is the Maxwell fluid. A mechanical model of the Maxwell fluid is a viscous element (a piston sliding in a cylinder of oil) in series with an elastic element (a spring). The total extension of this mechanical model is the sum of the extensions of the two elements and the rate of extension is the sum of the two rates of extension. It is assumed that the same form of combination can be applied to the shearing of the Maxwell fluid.

Using the negative sign convention, the equation for this model can be written by simply combining the rheological equation for a Hookean linear elastic solid

$$\tau = -G\gamma_e \tag{1.76}$$

with that for a Newtonian liquid

$$\tau = -\mu\dot{\gamma}_v \tag{1.77}$$

where subscripts e and v denote the elastic and viscous elements respectively. The rate of strain of the elastic element is therefore given by

$$\dot{\tau} = -G\dot{\gamma}_e \tag{1.78}$$

Taking the rate of strain of the fluid as the sum of the rates of strain $(\dot{\gamma}_v + \dot{\gamma}_e)$ of the two elements gives

$$\frac{\tau}{\mu} + \frac{\dot{\tau}}{G} = -\dot{\gamma} \tag{1.79}$$

or

$$\tau + \lambda\dot{\tau} = -\mu\dot{\gamma} \tag{1.80}$$

where $\lambda = \mu/G$. The first term in equation 1.80 represents the viscous contribution and the second the elastic contribution. The constant λ has the dimensions of time. If a Maxwell fluid is sheared at a constant rate, producing a shear stress τ_0, and the shearing is then stopped at time t_0, the shear stress decays exponentially:

$$\tau = \tau_0 \mathrm{e}^{-(t-t_0)/\lambda} \tag{1.81}$$

Consequently, λ is called the relaxation time: it is the time taken for the shear stress to fall to $1/e$ times the initial value.

By inspecting equations 1.79 and 1.80, it is clear that purely viscous behaviour corresponds to $\lambda = 0$ and purely elastic behaviour is approached as $\lambda \rightarrow \infty$. The relaxation time has a physical origin, being related to the time taken for molecules or particles to change orientation in response to the applied stress. In Chapter 3 it is shown that the response of a viscoelastic fluid depends on how rapidly it is deformed in relation to the relaxation time: when the deformation is very rapid and the fluid does not have time to relax, it exhibits largely elastic behaviour. If the same fluid is deformed slowly, it has time to relax and may exhibit mainly viscous behaviour. This can be observed with the material known as 'Crazy putty', a high polymer that is malleable at room temperature, rather like Plasticine. If a piece of this material is formed into a cylinder and slowly pulled out it flows forming a long strand. This is liquid-like behaviour. On repeating the experiment but pulling the sample rapidly, the material snaps like a weak solid. Also, if a ball of the material is dropped on to a hard surface, it bounces, again exhibiting elastic behaviour.

Although some real materials fall into just one of the three categories described above, most exhibit a combination of more than one type of behaviour. In practice, thixotropic materials are also shear thinning. Suspensions may be shear thinning or shear thickening depending on both the concentration and the shear rate. Concentrated suspensions may also have a significant relaxation time and therefore exhibit viscoelastic behaviour. This has a significance in relation to the definition and measurement of the yield stress. The material has to relax in order to yield, so if measurements are made rapidly a curve such as that shown in Figure 1.19 for the Bingham plastic may be determined. If, however, the material is subject to a constant low shear stress maintained for a long time, the material will relax to some extent and a lower value of the yield stress will be determined.

1.13 Turbulence and boundary layers

Most examples of flow in nature and many in industry are turbulent. Turbulence is an instability phenomenon caused, in most cases, by the shearing of the fluid. Turbulent flow is characterized by rapid, chaotic fluctuations of all properties including the velocity and pressure. This chaotic motion is often described as being made up of 'eddies' but it is important to appreciate that eddies do not have a purely circular motion.

The word 'eddy' is simply a convenient term to denote an identifiable group of fluid elements having a common motion, whether that motion be shearing, stretching or rotation.

The eddies have a wide range of sizes: in pipe flow the largest eddies are comparable in size to the diameter of the pipe, while the size of the smallest eddies will be typically 1 per cent of the pipe diameter. The various sizes of eddy also have different characteristic speeds and lifetimes. The large eddies are generated by the shearing of the mean (time averaged) flow and they produce smaller eddies which in turn generate yet smaller ones. Energy extracted from the mean flow in the generation of the large eddies is passed on to the successively smaller eddies. The smallest eddies are so small, and their velocity gradients therefore so large, that viscous stresses are dominant and viscosity destroys the smallest eddies, dissipating their kinetic energy by converting it into internal energy of the fluid. This process of passing energy from large to small eddies is known as the 'energy cascade'.

The generation of successively smaller eddies can be explained by vortex stretching. Consider turbulent flow over a flat, solid surface. The mean, that is time-averaged, flow has only one non-zero velocity component v_x parallel to the surface and only one non-zero velocity gradient dv_x/dy, where the y-coordinate direction is normal to the surface. The velocity profile will be like that shown in Figure 1.13 and, owing to the shearing, an element of fluid will rotate in a clockwise direction. The element of fluid can be described as part of a vortex whose axis lies parallel to the z-coordinate direction (out of the page). If this vortex is stretched in the z-direction, ie along its axis, the vortex will contract in the x and y directions. By the principle of conservation of angular momentum, the rotation must speed up. Thus stretching the vortex in the z-direction reduces the length scales in the x and y directions and increases the velocity components in these two directions at the expense of the velocity, and therefore kinetic energy, in the direction of stretching. The contraction of the vortex in the x and y directions will draw in surrounding fluid producing stretching in the x and y directions. Stretching in these two directions will produce stretching in the y and z, and x and z directions respectively. It can be seen that every time vortex stretching occurs it generates vortex stretching in the two orthogonal directions. In this way energy is passed to smaller and smaller eddies. Equally importantly, with successively smaller eddies the turbulence becomes less oriented, that is it becomes more nearly isotropic. Turbulence is *always* three-dimensional, even if the mean flow is not.

The properties of the turbulence are different at the two extremes of the scale of turbulence. The largest eddies, known as the macroscale turbulence, contain most of the turbulent kinetic energy. Their motion is dominated by inertia and viscosity has little direct effect on them. In contrast, at the microscale of turbulence, the smallest eddies are dominated by viscous stresses, indeed viscosity completely smooths out the microscale turbulence.

1.13.1 *Velocity fluctuations and Reynolds stresses*

A record of the axial velocity component v_x for steady turbulent flow in a pipe would look like the trace shown in Figure 1.22. The trace exhibits rapid fluctuations about the mean value, which is determined by averaging the instantaneous velocity over a sufficiently long period of time. Figure 1.22 shows the case in which the mean velocity remains constant; this is therefore known as steady turbulent flow. In unsteady turbulent flow, the mean value changes with time but it is still possible to define a mean value because, in practice, the mean will drift slowly compared with the frequency of the fluctuations.

Owing to the complexity of turbulent flow, it is usually treated as if it were a random process. In addition, it is usually adequate to calculate mean values of flow quantities, but as will be seen these are not always as simple as might be expected. The instantaneous value of the velocity

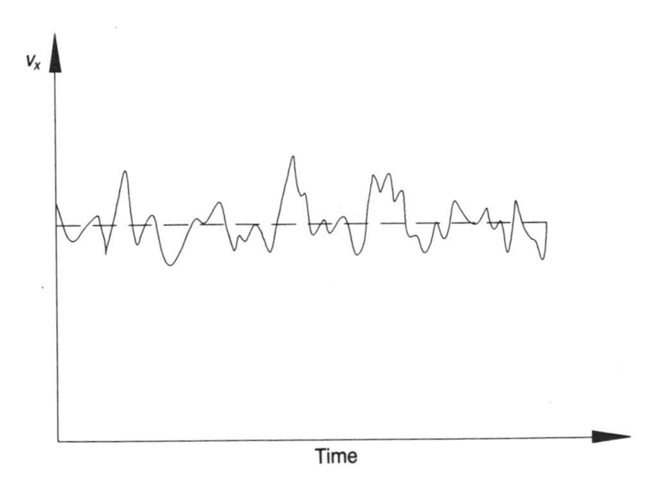

Figure 1.22
Variation of the velocity component v_x in turbulent flow

component v_x at a point can be represented as the sum of the mean velocity $\bar{v}_x$ at that point and the instantaneous fluctuation v'_x from the mean:

$$v_x = \bar{v}_x + v'_x \tag{1.82}$$

For turbulent flow near the axis of a pipe, the fluctuation v'_x will not exceed about ± 10 per cent of the mean value.

The time-averaged value $\bar{v}_x$ can be calculated from

$$\bar{v}_x = \frac{1}{\Delta t} \int_{T-\Delta t/2}^{T+\Delta t/2} v_x \, dt \tag{1.83}$$

This equation is an ordinary definition of an average but it is for the average at time T at the centre of the sampling period Δt. In principle, the average should be defined for $\Delta t \rightarrow \infty$ but in practice the averaging period Δt need only be long compared with the period of the slowest fluctuations. In order to follow the changing mean during unsteady turbulent flow, it is clearly essential to average over short periods.

The square of the instantaneous velocity is given by

$$v_x^2 = (\bar{v}_x + v'_x)^2 = (\bar{v}_x)^2 + 2\bar{v}_x v'_x + (v'_x)^2 \tag{1.84}$$

The time averaged value of v_x^2 is therefore

$$\overline{v_x^2} = \overline{(\bar{v}_x)^2} + \overline{2\bar{v}_x v'_x} + \overline{(v'_x)^2} \tag{1.85}$$

where the overbars denote time averaged quantities. It is obvious that

$$\overline{(\bar{v}_x)^2} = (\bar{v}_x)^2 \tag{1.86}$$

and

$$\overline{v'_x} = 0 \tag{1.87}$$

The latter result follows from the definitions embodied in equations 1.82 and 1.83. However, the mean of the square of the fluctuations is not zero. This is easily seen from the fact that, while the fluctuation takes both positive and negative values, the square of the fluctuation is always positive. Thus

$$\overline{v_x^2} = (\bar{v}_x)^2 + \overline{(v'_x)^2} \tag{1.88}$$

The mean value of the square of the velocity component is equal to the sum of the square of the mean velocity component and the mean of the square of the fluctuation. Corresponding relationships hold for the other two velocity components. It follows from equation 1.88 that turbulent flow possesses more kinetic energy than it would with the same mean velocity but no fluctuations.

Just as the velocity fluctuations give turbulent flow extra kinetic energy, so they generate extra momentum transfer. Consider the transfer of x-component momentum across a plane of area $\delta y \delta z$ perpendicular to the x-coordinate direction. The momentum flow rate is the product of the mass flow rate $(\rho v_x \delta y \delta z)$ across the plane and the velocity component v_x:

$$x\text{-momentum flow rate} = (\rho v_x \delta y \delta z) v_x = \rho v_x^2 \delta y \delta z \tag{1.89}$$

Writing the instantaneous velocity component v_x as the sum of the mean value and the fluctuation

$$x\text{-momentum flow rate} =$$

$$\rho(\bar{v}_x + v'_x)^2 \, \delta y \delta z = \rho[(\bar{v}_x)^2 + 2\bar{v}_x v'_x + (v'_x)^2] \, \delta y \delta z \tag{1.90}$$

Taking the time average and dividing by the area, the mean momentum flux is given by

$$\text{mean } x\text{-momentum flux} = \rho(\bar{\bar{v}}_x)^2 + \rho\overline{(v'_x)^2} \tag{1.91}$$

It has been assumed that the flow is incompressible so that there are no fluctuations of the density. Equation 1.91 shows that the momentum flux consists of a part due to the mean flow and a part due to the velocity fluctuation. The extra momentum flux is proportional to the square of the fluctuation because the momentum is the product of the mass flow rate and the velocity, and the velocity fluctuation contributes to both. The extra momentum flux is equivalent to an extra apparent stress perpendicular to the face, ie a normal stress component. As $\overline{(v'_x)^2}$ is always positive it produces a compressive stress, which is positive in the negative sign convention for stress.

Velocity fluctuations can also cause extra apparent shear stress components. An element of fluid with a non-zero velocity component in the x-direction possesses an x-component of momentum. If this element of fluid also has a non-zero velocity component in the y-direction then as it moves in the y-direction it carries with it the x-component of momentum. The mass flow rate across a plane of area $\delta x \delta z$ normal to the y-coordinate direction is $\rho v_y \delta x \delta z$ and the x-component of momentum per unit mass is v_x, so the rate of transfer of x-momentum in the y-direction is given by the expression

$$\begin{array}{c}\text{rate of } x\text{-momentum transfer} \\ \text{in } y\text{-direction}\end{array} = (\rho v_y \delta x \delta z) v_x \tag{1.92}$$

Writing the instantaneous velocity components v_x, v_y as the sums of the mean values and fluctuations, and taking the time average gives the mean momentum flux as:

$$\begin{aligned} \text{mean } x\text{-momentum flux in } y\text{-direction} &= \overline{\rho(\bar{v}_y + v'_y)(\bar{v}_x + v'_x)} \\ &= \rho\bar{v}_y\bar{v}_x + \overline{\rho v'_y v'_x} \end{aligned} \tag{1.93}$$

In general, the time-averaged value of the product of the fluctuations is non-zero so there is an additional flux of x-momentum in the y-direction due to the velocity fluctuations v'_x and v'_y. This momentum flux is equivalent to an extra apparent shear stress acting in the x-direction on the plane normal to the y-coordinate direction. Consequently, the mean total shear stress for turbulent flow can be written as

$$\overline{\tau_{yx}} = (\tau_{yx})_v + \overline{\rho v'_y v'_x} \tag{1.94}$$

In equation 1.94, $(\tau_{yx})_v$ is the viscous shear stress due to the mean velocity gradient $d\bar{v}_x/dy$ and $\overline{\rho v'_y v'_x}$ is the extra shear stress due to the velocity fluctuations v'_x and v'_y. These extra stress components arising from the velocity fluctuations are known as Reynolds stresses. (Note that if the positive sign convention for stresses were used, the sign of the Reynolds stress would be negative in equation 1.94.)

The reason for the time-averaged product of fluctuations being non-zero in general is illustrated in Figure 1.23. It is easiest to consider the case in which the *mean* velocity is in the x-direction only but, as noted before, the turbulence will be three-dimensional. In Figure 1.23, $\bar{v}_x$ is shown increasing with y. In this case, a negative fluctuation v'_y will cause an element of fluid to move to a region with lower mean velocity component in the x-direction; in so doing it carries its higher momentum with it and consequently tends, on average, to produce a positive velocity fluctuation v'_x in the surrounding fluid. Similarly, for a positive fluctuation v'_y, the element of fluid will move to a region of higher $\bar{v}_x$ and produce a negative velocity fluctuation v'_x on average. Thus the velocity fluctuations v'_x are linked to the fluctuations v'_y and are said to be 'correlated'. The time-averaged product of correlated quantities is non-zero. Not only are the fluctuations linked, it can be seen that in the above illustration with a positive mean velocity gradient, a fluctuation in the y-component of velocity gives rise to a fluctuation of the opposite sign in the x-component of velocity and consequently the Reynolds stress is negative on average.

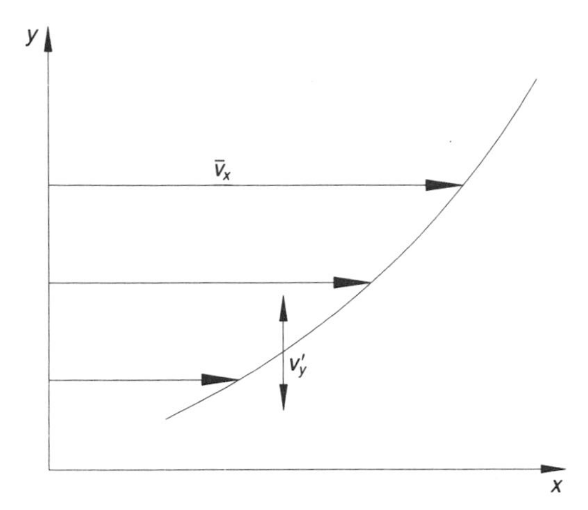

Figure 1.23
Velocity fluctuation in the y-direction across the gradient of the mean velocity

When the mean velocity gradient $d\bar{v}_x/dy$ is negative, a velocity fluctuation v'_y will produce a fluctuation v'_x of the same sign and the Reynolds stress will be positive.

It follows that if an element of fluid moves in the y-direction in a region where the mean velocity gradient $d\bar{v}_x/dy$ is zero, a fluctuation v'_y gives rise, on average, to a zero fluctuation v'_x. The time-average product of the fluctuations (the Reynolds stress) is zero and the fluctuations are said to be uncorrelated.

As noted above, the process of vortex stretching leads to successively smaller eddies being more nearly isotropic. The microscale eddies are statistically isotropic and, as a result

$$\rho\overline{v'_y v'_x} = 0$$

at the microscale. Consequently, the largest eddies contribute overwhelmingly to the Reynolds stresses and the small eddies make an insignificant contribution.

1.13.2 *Transport properties*

It is the large scale eddies that are responsible for the very rapid transport of momentum, energy and mass across the whole flow field in turbulent flow, while the smallest eddies and their destruction by viscosity are responsible for the uniformity of properties on the fine scale. Although it is the fluctuations in the flow that promote these high transfer rates, it is desirable to attempt to relate the mean values of the turbulent fluxes to the corresponding gradients of the mean profiles. In order to do this, it is necessary to introduce the concept of eddy diffusivities. By analogy with the flux equations for purely molecular transfer, equations 1.68 to 1.70, mean turbulent flux equations can be written as

$$\text{momentum flux } \overline{\tau_{yx}} = -\rho(\nu + \varepsilon)\frac{d\bar{v}_x}{dy} \tag{1.95}$$

$$\text{heat flux } \overline{q_y} = -\rho C_p(\alpha + \varepsilon_H)\frac{d\bar{T}}{dy} \tag{1.96}$$

$$\text{molar flux } \overline{J_{Ay}} = -(\mathcal{D} + \varepsilon_D)\frac{d\bar{C}_A}{dy} \tag{1.97}$$

It has been assumed that the density is constant in writing these equations, which are therefore strictly valid only for incompressible flow. ε_D is called the eddy diffusivity and ε_H the eddy thermal diffusivity. Although ε can be interpreted as the eddy diffusivity of momentum, it is usually called the eddy viscosity and sometimes by the better name eddy kinematic viscosity.

The mean profiles of velocity, temperature and solute concentration are relatively flat over most of a turbulent flow field. As an example, in Figure 1.24 the velocity profile for turbulent flow in a pipe is compared with the profile for laminar flow with the same volumetric flow rate. As the turbulent fluxes are very high but the velocity, temperature and concentration gradients are relatively small, it follows that the effective diffusivities $(\nu+\varepsilon)$, $(\alpha+\varepsilon_H)$ and $(\mathcal{D}+\varepsilon_D)$ must be extremely large. In the main part of the turbulent flow, ie away from the walls, the eddy diffusivities are much larger than the corresponding molecular diffusivities:

$$\varepsilon \gg \nu, \quad \varepsilon_H \gg \alpha, \quad \varepsilon_D \gg \mathcal{D}$$

Prandtl's mixing length theory, the basis of which is outlined in Section 2.9, predicts that the three eddy diffusivities are equal. It is important to appreciate that these eddy diffusivities are not genuine physical properties of the fluid; their values vary with position in the flow, as illustrated in Example 1.10.

Example 1.10

Estimate the value of the eddy kinematic viscosity ε as a function of position for turbulent flow of water in a smooth pipe of internal diameter 100 mm. The centre-line velocity is 6.1 m/s and the pressure drop over a

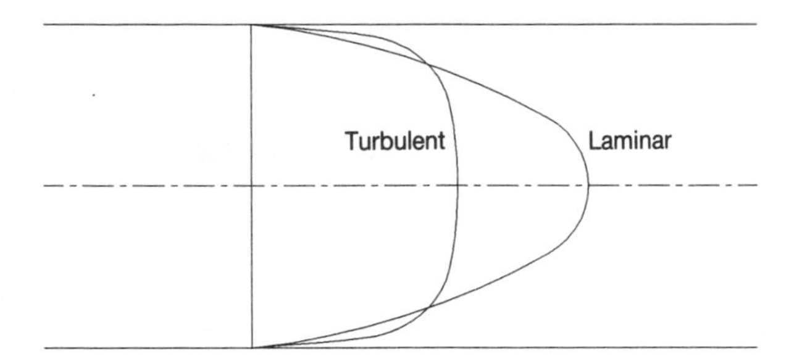

Figure 1.24
Comparison of the time-averaged velocity profile for turbulent flow and the profile for laminar flow at the same volumetric flow rate

length of 8 m is 14 000 Pa. For water, the value of the kinematic viscosity ν is 1×10^{-6} m^2/s. It may be assumed that the mean velocity profile is described by Prandtl's 1/7th power law over most of the cross section.

Calculations
The eddy kinematic viscosity ε is defined by

$$\overline{\tau_{rx}} = -\rho(\nu + \varepsilon)\frac{d\bar{v}_x}{dr} \qquad \text{from (1.95)}$$

Prandtl's 1/7th power law is

$$\frac{\bar{v}_x}{\bar{v}_{\max}} = \left(\frac{y}{r_i}\right)^{1/7}$$

where y is the distance measured from the wall. Thus

$$\bar{v}_x = \bar{v}_{\max}\left(1 - \frac{r}{r_i}\right)^{1/7}$$

and

$$\frac{d\bar{v}_x}{dr} = -\frac{\bar{v}_{\max}}{7r_i}\left(1 - \frac{r}{r_i}\right)^{-6/7} = -17.43\left(1 - \frac{r}{r_i}\right)^{-6/7} \text{ s}^{-1}$$

The wall shear stress is given by

$$\tau_w = \frac{d_i}{4}\left(\frac{\Delta P_f}{L}\right) = \frac{(0.1 \text{ m})(14\,000 \text{ Pa})}{4 \times 8 \text{ m}} = 43.75 \text{ Pa} \qquad \text{from (1.38)}$$

At radial distance r, the shear stress is given by

$$\frac{\overline{\tau_{rx}}}{\tau_w} = \frac{r}{r_i} \qquad \text{from (1.41)}$$

Using these equations, Table 1.3 can be constructed.

Table 1.3

r/r_i	τ (Pa)	$d\bar{v}_x/dr$ (s^{-1})	ε (m^2/s)	ε/ν
0.1	4.38	−19.1	0.23×10^{-3}	230
0.2	8.75	−21.1	0.41×10^{-3}	410
0.4	17.50	−27.0	0.65×10^{-3}	650
0.6	26.25	−38.2	0.68×10^{-3}	680
0.8	35.00	−69.3	0.51×10^{-3}	510
0.9	39.38	−125	0.31×10^{-3}	310

These values of ε may be compared with the value of the molecular kinematic viscosity ν (1×10^{-6} m^2/s). ε is nearly three orders of magnitude larger than ν.

At the wall, $\varepsilon \rightarrow 0$, but this behaviour cannot be calculated from the 1/7th power law, which is not valid near the wall (ie in the viscous sublayer and buffer zone). The equation is also slightly in error at the centre-line where it does not predict the required zero velocity gradient. ε tends to a non-zero value at the centre-line. Although the shear stress and velocity gradient both tend to zero at the centre-line and ε is therefore indeterminate from equation 1.95, it can be determined by applying L'Hôpital's rule [Longwell (1966)].

1.13.3 *Boundary layers*

When a fluid flows past a solid surface, the velocity of the fluid in contact with the wall is zero, as must be the case if the fluid is to be treated as a continuum. If the velocity at the solid boundary were not zero, the velocity gradient there would be infinite and by Newton's law of viscosity, equation 1.44, the shear stress would have to be infinite. If a turbulent stream of fluid flows past an isolated surface, such as an aircraft wing in a large wind tunnel, the velocity of the fluid is zero at the surface but rises with increasing distance from the surface and eventually approaches the velocity of the bulk of the stream. It is found that almost all the change in velocity occurs in a very thin layer of fluid adjacent to the solid surface:

this is known as a boundary layer. As a result, it is possible to treat the turbulent flow as two regions: the boundary layer where viscosity has a significant effect, and the region outside the boundary layer, known as the free stream, where viscosity has no direct influence on the flow. This artificial segregation allows considerable simplification in the analysis of turbulent flow.

Figure 1.25 shows the boundary layer that develops over a flat plate placed in, and aligned parallel to, the fluid having a uniform velocity v_∞ upstream of the plate. Flow over the wall of a pipe or tube is similar but eventually the boundary layer reaches the centre-line. Although most of the change in the velocity component $\bar{v}_x$ parallel to the wall takes place over a short distance from the wall, it does continue to rise and tends gradually to the value v_∞ in the fluid distant from the wall (the free stream). Consequently, if a boundary layer *thickness* is to be defined it has to be done in some arbitrary but useful way. The normal definition of the boundary layer thickness is that it is the distance from the solid boundary to the location where $\bar{v}_x$ has risen to 99 per cent of the free stream velocity v_∞. The locus of such points is shown in Figure 1.25. It should be appreciated that this is a time averaged distance; the thickness of the boundary layer fluctuates owing to the velocity fluctuations.

The boundary layer thickness gradually increases until a critical point is reached at which there is a *sudden* thickening of the boundary layer: this reflects the transition from a laminar boundary layer to a turbulent boundary layer. For both types, the flow outside the boundary layer is completely turbulent. In that part of the boundary layer near the leading edge of the plate the flow is laminar and consequently this is known as a

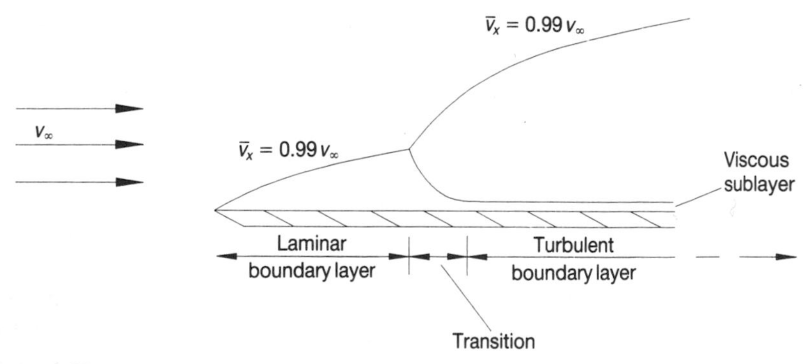

Figure 1.25

Laminar and turbulent boundary layers over a flat plate

laminar boundary layer. After the transition, the structure of the boundary layer is more complex: the flow in most of that part of the boundary layer is turbulent and hence it is called a turbulent boundary layer. However, in the turbulent boundary layer there is a very thin layer of fluid adjacent to the solid surface where turbulent stresses are negligible and the flow is dominated by viscous stresses: this is known as the 'viscous sublayer'. The viscous sublayer used to be called the 'laminar sublayer' but this is an inappropriate name because the flow there is not genuinely laminar; it is subject to disruptions. Just outside the viscous sublayer is a layer known as the 'buffer zone' or 'generation zone'. It is here that most of the turbulent fluctuations are generated.

The transition from a laminar boundary layer to a turbulent boundary layer occurs at the value $Re_x \approx 3.2 \times 10^5$. In the Reynolds number Re_x the free stream velocity v_∞ and the distance x from the leading edge of the plate are used as the characteristic velocity and linear dimension.

As the fluid's velocity must be zero at the solid surface, the velocity fluctuations must be zero there. In the region very close to the solid boundary, ie the viscous sublayer, the velocity fluctuations are very small and the shear stress is almost entirely the viscous stress. Similarly, transport of heat and mass is due to molecular processes, the turbulent contribution being negligible. In contrast, in the outer part of the turbulent boundary layer turbulent fluctuations are dominant, as they are in the free stream outside the boundary layer. In the buffer or generation zone, turbulent and molecular processes are of comparable importance.

In flow in pipes and, indeed, in most types of flow, turbulence is generated by shearing. The shear stress in pipe flow is greatest at the wall, as shown by equation 1.41, and consequently the potential to generate turbulence is greatest at the wall. However, the restraining effect of the solid surface (the necessity for the velocity to be zero) is greatest at the surface. The calming effect of the wall, conveyed to the fluid *via* viscosity, diminishes rapidly with distance, so at a small distance from the wall, where the shear stress remains close to the wall value, the damping effect has fallen significantly and is insufficient to prevent the formation of turbulent eddies: this is the generation zone.

This part of the turbulent boundary layer is rich in coherent structures, ie the flow exhibits features that are not random. Flow visualization studies [Kline *et al* (1967), Praturi and Brodkey (1978), Rashidi and Banerjee (1990)] have revealed a fascinating picture. The first observations indicated the occurrence of fluid motions called 'inrushes' and 'eruptions' or 'bursts' in the fluid very close to the wall. During eruptions, fluid moving downstream slowly, compared with the surrounding fluid, erupts away from the wall into the main part of the turbulent boundary layer, ie the viscous sublayer appears to burst. An inrush is the opposite of a burst: relatively rapidly moving fluid rushes in towards the wall.

More recently, 'horseshoe' or 'hairpin' vortices have been observed. The general picture is that elongated patches of fluid (streaks) having a low mean velocity, but with large fluctuations, appear very close to the wall and faster fluid flows over them forming vortices. The vortices may be shaped as hairpins or, more frequently, as half hairpins (horseshoes and hockey sticks in Banerjee's terminology). As each vortex develops, it becomes larger and eventually the head of the vortex breaks up. By this time, a second vortex may have developed over the streak. The length of a hairpin vortex is of the order of the total boundary layer thickness, while the width is of the order of the thickness of the viscous sublayer [Tritton (1988)]. This suggests that boundary layer vortices derive their vorticity from the viscous sublayer: this vorticity is then advected and stretched to form the hairpin vortices. The exact relationship between hairpin vortices, and inrushes and bursts is not yet fully understood.

In the case of laminar flow in a pipe, work is done by the shear stress component τ_{rx} and the rate of doing work is the viscous dissipation rate, that is the conversion of kinetic energy into internal energy. The rate of viscous dissipation per unit volume at a point, is given by

$$\begin{matrix}\text{rate of viscous} \\ \text{dissipation}\end{matrix} = -\tau_{rx}\frac{\mathrm{d}v_x}{\mathrm{d}r} = \mu\left(\frac{\mathrm{d}v_x}{\mathrm{d}r}\right)^2 \tag{1.98}$$

where the negative sign arises from using the negative sign convention for stresses. As shown by equation 1.98, the dissipation is always positive.

In turbulent flow, there is direct viscous dissipation due to the mean flow: this is given by the equivalent of equation 1.98 in terms of the mean values of the shear stress and the velocity gradient. Similarly, the Reynolds stresses do work but this represents the extraction of kinetic energy from the mean flow and its conversion into turbulent kinetic energy. Consequently this is known as the rate of turbulent energy production:

$$\begin{matrix}\text{rate of turbulent} \\ \text{energy production}\end{matrix} = -\rho\overline{v'_x v'_r}\frac{\mathrm{d}\bar{v}_x}{\mathrm{d}r} \tag{1.99}$$

The turbulent energy, extracted from the mean flow, passes through the energy cascade and is ultimately converted into internal energy by viscous dissipation.

It was noted earlier that the Reynolds stress is negative when the mean velocity gradient is positive and vice versa, consequently turbulent energy production is always positive. Very close to the wall, the Reynolds stress

$$\rho \overline{v'_x v'_r}$$

is small, while far from the wall the Reynolds stress and the mean velocity gradient $d\bar{v}_x/dr$ are small, so in both these regions there is very little production of turbulent energy. In the buffer or generation zone, neither quantity is small and the maximum rate of turbulent energy production occurs in this region.

From equation 1.41, the total shear stress varies linearly from a maximum $\bar{\tau}_w$ at the wall to zero at the centre of the pipe. As the wall is approached, the turbulent component of the shear stress tends to zero, that is the whole of the shear stress is due to the viscous component at the wall. The turbulent contribution increases rapidly with distance from the wall and is the dominant component at all locations except in the wall region. Both components of the mean shear stress necessarily decline to zero at the centre-line. (The mean velocity gradient is zero at the centre so the mean viscous shear stress must be zero, but in addition the velocity fluctuations are uncorrelated so the turbulent component must be zero.)

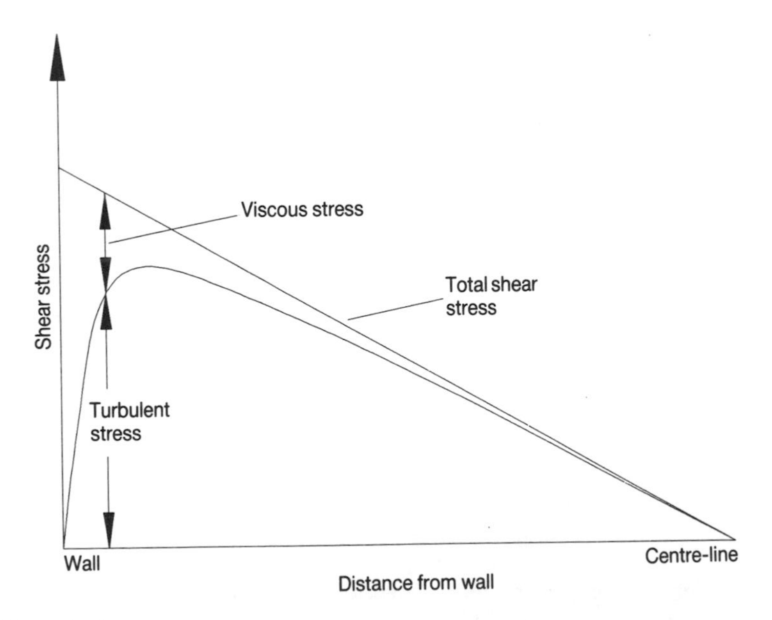

Figure 1.26
Viscous and turbulent contributions to the total shear stress for flow in a pipe

The distribution of the shear stress components is shown schematically in Figure 1.26. For clarity, the magnitude of the viscous stress is exaggerated in Figure 1.26.

References

Cross, M.M., Rheology of non-Newtonian fluids: a new flow equation for pseudoplastic systems, *Journal of Colloid Science*, **20**, pp. 417–37 (1965).

Kline, S.J., Reynolds, W.C., Schraub, F.A. and Runstadler, P.W., The structure of turbulent boundary layers, *Journal of Fluid Mechanics*, **30**, pp. 741–73 (1967).

Longwell, P.A., *Mechanics of Fluid Flow*, p. 337, New York, McGraw-Hill Book Company Inc. (1966).

Praturi, A.K. and Brodkey, R.S., A stereoscopic visual study of coherent structures in turbulent shear flow, *Journal of Fluid Mechanics*, **89**, pp. 251–72 (1978).

Rashidi, M. and Banerjee, S., Streak characteristics and behaviour near wall and interface in open channel flows, *Transactions of ASME, Journal of Fluids Engineering, Series I* **112**, pp. 164–70 (1990).

Smith, J.M. and Van Ness, H.C., *Introduction to Chemical Engineering Thermodynamics*, International edition, Singapore, McGraw-Hill Book Company Inc, pp. 12–17 (1987).

Tritton, D.J., *Physical Fluid Dynamics*, Second edition, Oxford, Oxford University Press, p. 348 (1988).

2 Flow of incompressible Newtonian fluids in pipes and channels

2.1 Reynolds number and flow patterns in pipes and tubes

As mentioned in Chapter 1, the first published work on fluid flow patterns in pipes and tubes was done by Reynolds in 1883. He observed the flow patterns of fluids in cylindrical tubes by injecting dye into the moving stream. Reynolds correlated his data by using a dimensionless group later known as the Reynolds number Re:

$$Re = \frac{\rho u d_i}{\mu} \tag{1.3}$$

In equation 1.3, ρ is the density, μ the dynamic viscosity, and u the mean velocity of the fluid; d_i is the inside diameter of the tube. Any consistent system of units can be used in this equation. The Reynolds number is also frequently written in the form

$$Re = \frac{G d_i}{\mu} \tag{2.1}$$

where $G = \rho u$. Clearly, G is the mass flow rate per unit area. It is usually called the mass flux but sometimes the mass velocity. By definition

$$G = \frac{M}{S} \tag{2.2}$$

where M is the mass flow rate of fluid and S is the cross-sectional flow area in the pipe or tube.

Reynolds found that as he increased the fluid velocity in the tube, the flow pattern changed from laminar to turbulent at a Reynolds number value of about 2100. Later investigators have shown that under certain conditions, eg with very smooth conduits, laminar flow can exist at very much higher Reynolds numbers. These special conditions are not normally encountered in process equipment.

2.2 Shear stress in a pipe

Consider steady, fully developed flow in a straight pipe of length L and internal diameter d_i. As shown in Example 1.8, a force balance on a cylindrical element of the fluid can be written as

$$\pi r^2 \Delta P_f - 2\pi r L \tau_{rx} = 0 \tag{2.3}$$

where ΔP_f is the frictional component of the pressure drop over the pipe length L. In the case of fully developed flow in a horizontal pipe ΔP_f is the only component of the pressure drop, see equations 1.16 and 1.17. Rearranging equation 2.3, the shear stress is given by

$$\tau_{rx} = \frac{r}{2}\left(\frac{\Delta P_f}{L}\right) \tag{2.4}$$

A special case of equation 2.4 is the shear stress τ_w at the wall

$$\tau_w = \frac{r_i}{2}\left(\frac{\Delta P_f}{L}\right) = \frac{d_i}{4}\left(\frac{\Delta P_f}{L}\right) \tag{2.5}$$

Equation 2.5 shows that the value of τ_w can be determined if the pressure gradient is measured; this is how values of the friction factor discussed in Section 2.3 have been found. Alternatively, if τ_w can be predicted, the pressure drop can be calculated.

From equations 2.4 and 2.5, the shear stress distribution can be written as

$$\frac{\tau_{rx}}{\tau_w} = \frac{r}{r_i} \tag{2.6}$$

The shear stress varies linearly from zero at the centre–line to a maximum value τ_w at the pipe wall.

Equations 2.3 to 2.6 are true, irrespective of the nature of the fluid. They are also valid for both laminar and turbulent flow. In the latter case, the shear stress is the total shear stress comprising the viscous stress and the Reynolds stress.

2.3 Friction factor and pressure drop

Rearranging equation 2.5, the frictional pressure drop is given as

$$\Delta P_f = \frac{4\tau_w L}{d_i} \tag{2.7}$$

Owing to its complexity, turbulent flow does not admit of the simple solutions available for laminar flow and the approach to calculating the pressure drop is based on empirical correlations.

It was noted in Section 1.3 that the frictional pressure drop for turbulent flow in a pipe varies as the square of the flow rate at very high values of *Re*. At lower values of *Re* the pressure drop varies with flow rate, and therefore with *Re*, to a slightly lower power which gradually increases to the value 2 as *Re* increases. The pressure drop in turbulent flow is also proportional to the density of the fluid. This suggests writing equation 2.7 in the form

$$\Delta P_f = 4\frac{L}{d_i}\left(\frac{\tau_w}{\rho u^2}\right)\rho u^2 \tag{2.8}$$

In the range where ΔP_f varies exactly as u^2 the quantity $\tau_w/\rho u^2$ must be constant, while at lower values of *Re* the value of $\tau_w/\rho u^2$ will not quite be constant but will decrease slowly with increasing *Re*. Consequently, $\tau_w/\rho u^2$ is a useful quantity with which to correlate pressure drop data. A slightly different form of equation 2.8 is obtained by replacing the two occurrences of ρu^2 with $\frac{1}{2}\rho u^2$:

$$\Delta P_f = 4\frac{L}{d_i}\left(\frac{\tau_w}{\frac{1}{2}\rho u^2}\right)\frac{1}{2}\rho u^2 \tag{2.9}$$

The quantity $\frac{1}{2}\rho u^2$ will be recognized as the kinetic energy per unit volume of the fluid.

The term $\tau_w/(\frac{1}{2}\rho u^2)$ in equation 2.9 defines a quantity known as the Fanning friction factor f, thus

$$f \equiv \frac{\tau_w}{\frac{1}{2}\rho u^2} \tag{2.10}$$

It will be appreciated that the factor of $\frac{1}{2}$ in equation 2.10 is arbitrary and various other friction factors are in use. For example, in the first edition of this book the basic friction factor denoted by j_f was used. This is defined by

$$j_f \equiv \frac{\tau_w}{\rho u^2} \tag{2.11}$$

Thus

$$j_f = f/2 \tag{2.12}$$

When using j_f, the pressure drop is given by equation 2.8. Using the

Fanning friction factor, which is defined by equation 2.10, equation 2.9 may be written as

$$\Delta P_f = 4f\left(\frac{L}{d_i}\right)\frac{\rho u^2}{2} = \frac{2fL\rho u^2}{d_i} \tag{2.13}$$

This is the basic equation from which the frictional pressure drop may be calculated. It is valid for all types of fluid and for both laminar and turbulent flow. However, the value of f to be used does depend on these conditions.

Although it is unnecessary to use the friction factor for laminar flow, exact solutions being available, it follows from equation 1.65 that for laminar flow of a Newtonian fluid in a pipe, the Fanning friction factor is given by

$$f = \frac{16}{Re} \tag{2.14}$$

For turbulent flow of a Newtonian fluid, f decreases gradually with Re, which must be the case in view of the fact that the pressure drop varies with flow rate to a power slightly lower than 2.0. It is also found with turbulent flow that the value of f depends on the relative roughness of the pipe wall. The relative roughness is equal to e/d_i where e is the absolute roughness and d_i the internal diameter of the pipe. Values of absolute roughness for various kinds of pipes and ducts are given in Table 2.1.

Table 2.1

Material	*Absolute roughness e (in m)*
Drawn tubing	0.000 001 5
Commercial steel and wrought iron	0.000 045
Asphalted cast iron	0.000 12
Galvanized iron	0.000 15
Cast iron	0.000 26
Wood stave	0.000 18–0.0009
Concrete	0.000 30–0.0030
Riveted steel	0.0009–0.009

Values of the friction factor are traditionally presented on a friction factor chart such as that shown in Figure 2.1. It will be noted that the

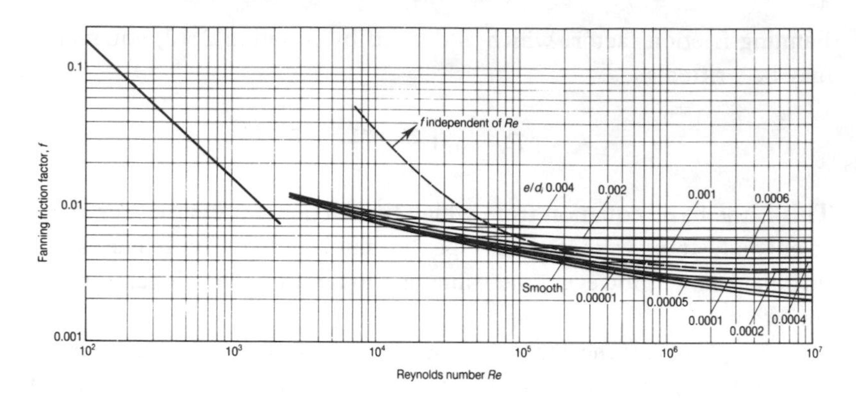

Figure 2.1
Friction factor chart for Newtonian fluids. (See Friction Factor Charts on page 349.)

greater the relative roughness, the higher the value of f for a given value of Re. At high values of Re, the friction factor becomes independent of Re; this is true for the region of the chart above and to the right of the broken line. The reason for this behaviour is discussed at the end of Section 2.9.

In the region of transition between laminar and turbulent flow, the flow is rather unpredictable and caution should be exercised in relying on the value of f used.

Considerable effort has been expended in trying to find algebraic expressions to relate f to Re and e/d_i. For turbulent flow in smooth pipes, the simplest expression is the Blasius equation:

$$f = 0.079 Re^{-0.25} \tag{2.15}$$

This equation is valid for the range of Re from 3000 to 1×10^5.

Similarly, the Drew equation

$$f = 0.00140 + 0.125 Re^{-0.32} \tag{2.16}$$

is good for Re from 3000 to at least 3×10^6.

The most widely accepted relationship for turbulent flow in smooth pipes is the von Kármán equation

$$\frac{1}{f^{1/2}} = 4.0 \log(f^{1/2} Re) - 0.40 \tag{2.17}$$

This equation is very accurate but has the disadvantage of being implicit in f.

For completely rough pipes (above the broken line on the chart) f is given by

$$\frac{1}{f^{1/2}} = 4.06 \log\left(\frac{d_i}{e}\right) + 2.16 \tag{2.18}$$

A very useful correlation has been given by Haaland (1983):

$$\frac{1}{f^{1/2}} = -3.6 \log\left[\left(\frac{e}{3.7d_i}\right)^{1.11} + \frac{6.9}{Re}\right] \tag{2.19}$$

Equation 2.19 has the advantages of giving f explicitly and being adequately accurate over the whole range of turbulent flow.

Use of the friction factor chart or a correlation such as equation 2.19 enables calculation of the frictional pressure drop for a specified flow rate from equation 2.13.

The inverse problem is to determine the flow rate for a given pressure drop. For turbulent flow, this is not so straightforward because the value of f is unknown until the flow rate, and hence Re, are known. The traditional solution to this problem is to use the plot of fRe^2 against Re or $\frac{1}{2}fRe^2$ against Re shown in Figure 2.2.

The reason for using this combination can be seen by rearranging equation 2.13 as follows:

$$f = \frac{d_i \Delta P_f}{2L\rho u^2} \tag{2.20}$$

Thus, the unknown u can be eliminated by multiplying by Re^2 to give

$$fRe^2 = \frac{d_i \Delta P_f}{2L\rho u^2}\left(\frac{\rho u d_i}{\mu}\right)^2 = \frac{d_i^3 \rho \Delta P_f}{2L\mu^2} \tag{2.21}$$

The method of determining the mean velocity, and hence the flow rate, is as follows. Calculate fRe^2 from equation 2.21 from the known values of ΔP_f, ρ, d_i, L and μ. Read the corresponding value of Re from Figure 2.2 for the known value of e/d_i. Hence calculate u from the definition of Re.

Example 2.1

Calculate the frictional pressure drop for a commercial steel pipe with the following characteristics:

length L	= 30.48 m
inside diameter d_i	= 0.0526 m
pipe roughness e	= 0.000045 m
steady liquid flow rate Q	= 9.085 m^3/h
liquid dynamic viscosity μ	= 0.01 Pa s
liquid density ρ	= 1200 kg/m^3

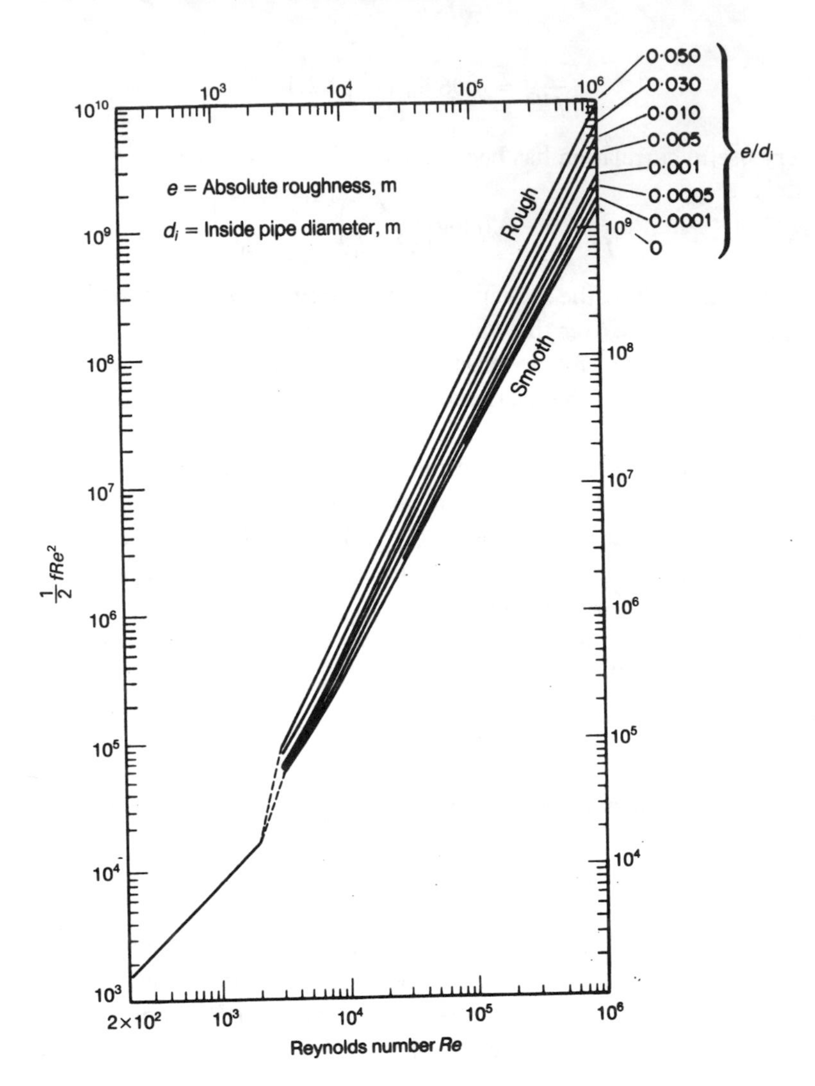

Figure 2.2
Plot of $\frac{1}{2} fRe^2$ against Reynolds number

Calculations

$$\text{mean velocity } u = \frac{Q}{\pi d_i^2/4} \qquad \text{(from 1.6)}$$

From the given values

$$\frac{\pi d_i^2}{4} = \frac{(3.142)(0.0526 \text{ m})^2}{4} = 0.002\,173 \text{ m}^2$$

$$Q = \frac{9.085 \text{ m}^3/\text{h}}{3600 \text{ s/h}} = 0.002\,524 \text{ m}^3/\text{s}$$

Therefore

$$u = \frac{0.002\,524 \text{ m}^3/\text{s}}{0.002\,173 \text{ m}^2} = 1.160 \text{ m/s}$$

The Reynolds number is given by

$$Re = \frac{\rho u d_i}{\mu} \quad (1.3)$$

Substituting the given values

$$Re = \frac{(1200 \text{ kg/m}^3)(1.160 \text{ m/s})(0.0526 \text{ m})}{0.01 \text{ Pa s}} = 7322$$

Relative roughness is given by

$$\frac{e}{d_i} = \frac{0.000\,045 \text{ m}}{0.0526 \text{ m}} = 0.000\,856$$

From the graph of f against Re in Figure 2.1, $f = 0.0084$ for $Re = 7322$ and $e/d_i = 0.000\,856$.

The frictional pressure drop is given by

$$\Delta P_f = 4f\left(\frac{L}{d_i}\right)\frac{\rho u^2}{2} \quad (2.13)$$

From the given values

$$\left(\frac{L}{d_i}\right) = \frac{30.48 \text{ m}}{0.0526 \text{ m}} = 579.5$$

and

$$\frac{\rho u^2}{2} = \frac{(1200 \text{ kg/m}^3)(1.160 \text{ m/s})^2}{2} = 807.4 \text{ N/m}^2$$

Therefore

$$\Delta P_f = 4(0.0084)(579.5)(807.4 \text{ N/m}^2)$$
$$= 15\,720 \text{ N/m}^2 = \underline{15\,720 \text{ Pa}}$$

Example 2.2

Estimate the steady mean velocity for a commercial steel pipe with the following characteristics:

length L	$= 30.48$ m
inside diameter d_i	$= 0.0526$ m
wall roughness e	$= 0.000045$ m
frictional pressure drop ΔP	$= 15720$ N/m^2
liquid dynamic viscosity μ	$= 0.01$ Pa s
liquid density ρ	$= 1200$ kg/m^2

Calculations

$$fRe^2 = \frac{d_i^3 \rho \Delta P_f}{2L\mu^2} \qquad (2.21)$$

Substituting the given values

$$\frac{d_i^3 \rho \Delta P_f}{2L\mu^2} = \frac{(0.0526 \text{ m})^3 (1200 \text{ kg/m}^3)[15720 \text{ kg/(s}^2 \text{ m)}]}{2(30.48 \text{ m})(0.01 \text{ Pa s})^2}$$

$$= 4.503 \times 10^5$$

Relative roughness is given by

$$\frac{e}{d_i} = \frac{0.000045 \text{ m}}{0.0526 \text{ m}} = 0.000856$$

From the graph of fRe^2 against Re in Figure 2.2, $Re = 7200$ for $fRe^2 = 4.503 \times 10^5$ and $e/d_i = 0.000856$.

Rearranging equation 1.3 gives

$$\text{mean velocity } u = \frac{Re\mu}{d_i \rho}$$

Substituting the given values

$$u = \frac{(7200)[0.01 \text{ kg/(s m)}]}{(0.0526 \text{ m})(1200 \text{ kg/m}^3)}$$

$$= \underline{1.141 \text{ m/s}}$$

The slight difference between this and the mean velocity in Example 2.1 is due to error in reading the graphs in Figures 2.1 and 2.2.

Given a suitable algebraic correlation such as equation 2.19, the friction factor chart might be considered obsolete. Both f and fRe^2 can be represented algebraically as functions of Re allowing both types of calculations to be done. In the case of the inverse problem, that is the calculation of the flow rate for a specified pressure drop, an alternative is to use an iterative calculation, a procedure that is particularly attractive with a pocket calculator or a spreadsheet. Using equation 2.19 for f, the procedure is as follows:

Start: Guess Re (hence u)

1 Calculate f from $$\frac{1}{f^{1/2}} = -3.6 \log\left[\left(\frac{e}{3.7d_i}\right)^{1.11} + \frac{6.9}{Re}\right]$$

2 Calculate ΔP_f from $\Delta P_f = 2fL\rho u^2/d_i$

3 Compare calculated ΔP_f with specified ΔP_f

STOP if close enough

4 Estimate new value of Re for next iteration:

$$\text{new } Re = (\text{current } Re)\left(\frac{\text{specified } \Delta P_f}{\text{current } \Delta P_f}\right)^{1/2}$$

5 Return to 1.

Applying this procedure to Example 2.2, using an initial guess of $Re = 15\,000$ gives the series of values shown in Table 2.2.

Table 2.2

Re	f	ΔP_f (Pa)	*Relative error*
15 000	0.007 27	57 083	2.63
7872	0.008 49	18 372	0.169
7282	0.008 67	16 047	0.0208
7207	0.008 69	15 761	0.0026
7198	0.008 70	15 727	4.45×10^{-4}
7196	0.008 70	15 719	6.36×10^{-6}

The relative error is calculated as

$$\frac{\text{calculated } \Delta P_f - \text{specified } \Delta P_f}{\text{specified } \Delta P_f}$$

This quantity gives a measure of convergence to the specified value. In this case, the calculation might be stopped after the fourth step, when the error is 0.0026, ie 0.26 per cent. The calculation converges to a value of Re (and hence u) very close to the value in Example 2.2. There is no point iterating beyond a discrepancy of about 1 per cent because the correlations are no better than this.

It should be borne in mind that frictional pressure drop calculations can be done to only limited accuracy because the roughness of the pipe will not be known accurately and will change during service.

2.4 Pressure drop in fittings and curved pipes

So far, only the frictional pressure drop in straight lengths of pipe of circular cross-section has been discussed. The pressure drop in pipelines containing valves and fittings can be calculated from equation 2.13 but with fittings represented by the length of plain pipe that causes the same pressure drop.

Equivalent lengths of various valves and fittings are readily available [Holland and Chapman (1966)] and a selection is given in Table 2.3.

Table 2.3

Fitting	*Number of equivalent pipe diameters* L_e/d_i	*Number of velocity heads* K
Sudden contraction	20	0.4
Sudden enlargement	40	0.8
90° elbow	30 – 40	0.6 – 0.8
Globe valve, fully open	60 – 300	1.2 – 6.0
Gate valve, fully open	7	0.15
Check valve	7	0.15

Equation 2.13 then becomes

$$\Delta P_f = 4f\left(\frac{\Sigma L_e}{d_i}\right)\frac{\rho u^2}{2} \qquad (2.22)$$

where ΣL_e is the sum of the equivalent lengths of the components, including the length of plain pipe.

If the frictional losses were expressed as the head loss, $h_f = \Delta P_f/\rho g$, then the quantity $4fL_e/d_i$ would multiply $u^2/2g$. Thus $4fL_e/d_i$ is the total number of velocity heads lost. Consequently, an alternative presentation of frictional losses for fittings is in terms of the number of velocity heads K lost for each fitting. In this case, the total frictional pressure drop may be calculated as

$$\Delta P_f = 4f\left(\frac{L}{d_i}\right)\frac{\rho u^2}{2} + \Sigma K\frac{\rho u^2}{2} \qquad (2.23)$$

In equation 2.23, the first term on the right hand side gives the frictional pressure drop for the plain pipe of length L and the second term represents the total loss for all the fittings. Values of K are given in Table 2.3.

Pipe entrance and exit pressure losses should also be calculated and added to obtain the overall pressure drop. The loss in pressure due to sudden expansion from a diameter d_{i1} to a larger diameter d_{i2} is given by the equation

$$\Delta P_e = \frac{\rho(u_1 - u_2)^2}{2} = \frac{\rho u_1^2}{2}\left[1 - \left(\frac{d_{i1}}{d_{i2}}\right)^2\right]^2 \qquad (2.24)$$

where u_1 and u_2 are the mean velocities in the smaller entrance pipe and the larger exit pipe respectively. When the expansion is very large, the pressure drop approaches $\rho u_1{}^2/2$, that is a head loss of one velocity head.

The loss in pressure due to sudden contraction from a diameter d_{i1} to a smaller diameter d_{i2} is given by the equation

$$\Delta P_c = K\left(\frac{\rho u_2^2}{2}\right) \qquad (2.25)$$

where

$$K = 0.4\left[1.25 - \left(\frac{d_{i2}}{d_{i1}}\right)^2\right] \quad \text{when} \quad \frac{d_{i2}^2}{d_{i1}^2} < 0.715$$

and

$$K = 0.75\left[1.0 - \left(\frac{d_{i2}}{d_{i1}}\right)^2\right] \quad \text{when} \quad \frac{d_{i2}^2}{d_{i1}^2} > 0.715$$

In these expressions u_2 is the mean velocity in the smaller exit pipe. When the contraction is very great, the pressure drop tends to $\frac{1}{2}(\rho u_2^2/2)$, or a head loss of half a velocity head, based on the smaller pipe.

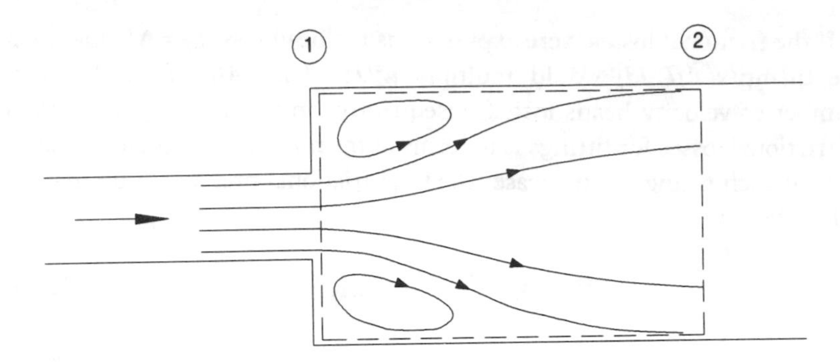

Figure 2.3
Flow through a sudden expansion

These expressions for the losses due to sudden expansion and sudden contraction can be derived by application of Bernoulli's equation and the momentum equation. Figure 2.3 shows a sudden expansion.

The control surface is taken as shown on the inside surface between planes 1 and 2. It might be thought that plane 1 should be placed upstream of the expansion but if this were done it would be necessary to include a term in the momentum equation representing the force exerted by the end wall on the fluid. In an expansion, the fluid flows as a jet and gradually expands to fill the larger pipe, there being a large recirculating flow outside the jet. The pressure must be fairly uniform in this recirculating fluid. (If this were not the case, that is if there were a significant pressure drop across that region, fluid would flow across it not circulate in it.) Consequently, the fluid jet at plane 1 does not flow radially and the pressure on the end wall must be uniform and equal to the pressure in the jet at plane 1. Owing to the low velocity gradients in the recirculating flow, there is very little frictional loss here: most of the losses occur where the expanding jet reattaches to the wall.

The momentum equation can be written as

$$P_1 S_2 - P_2 S_2 = M(u_2 - u_1) \tag{2.26}$$

Bernoulli's equation for this horizontal, turbulent flow is

$$\frac{P_2}{\rho g} + \frac{u_2^2}{2g} = \frac{P_1}{\rho g} + \frac{u_1^2}{2g} - h_f \tag{2.27}$$

where h_f is the loss due to the expansion. Substituting $M = \rho_2 u_2 S_2$ in equation 2.26, the pressure drop can be eliminated between equations 2.26 and 2.27 and the result rearranged to give

$$h_f = (u_1 - u_2)^2/2g \tag{2.28}$$

Now

$$\Delta P_e = \rho g h_f \tag{2.29}$$

so equation 2.24 is obtained.

Note that if plane 1 had been placed inside the pipe just upstream of the expansion, the pressure being the same as at the expansion, the momentum equation would be written as

$$P_1S_1 + P_1(S_2 - S_1) - P_2S_2 = M(u_2 - u_1) \tag{2.30}$$

where the first term is the pressure force at the left of the control volume and the second term is the force exerted on the fluid by the end wall: this is equal and opposite to the force exerted by the fluid on the wall. On cancelling the two P_1S_1 terms, equation 2.30 becomes identical to equation 2.26.

That the losses for sudden expansion and sudden contraction are markedly different may surprise the reader. The reason is that the flow pattern for a sudden contraction is different from that for an expansion as shown in Figure 2.4.

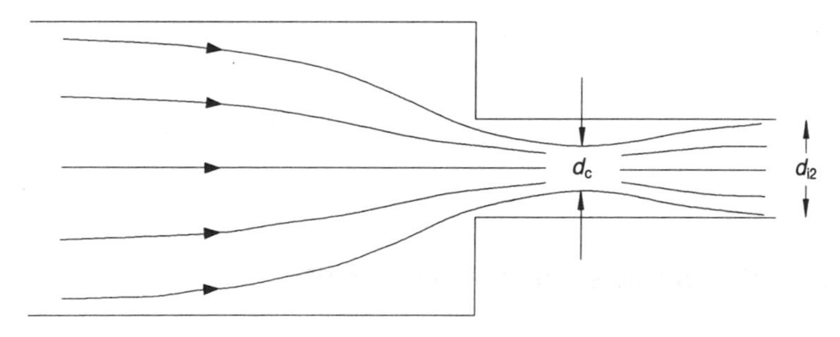

Figure 2.4
Flow through a sudden contraction

For a Newtonian liquid, there is negligible frictional loss due to the converging flow. However, a *vena contracta* forms in the smaller pipe and the expansion and reattachment downstream of it are similar to the flow in the sudden expansion. The difference is that it takes place from d_c to d_{i2}. The momentum equation cannot be applied over the whole flow because the pressure distribution is unknown in the region of the contraction. However, the analysis for the sudden expansion can be applied beyond the *vena contracta* but it is necessary to have experimental values of the contraction ratio.

2.4.1 *Pressure drop in coils*

A number of equations have been proposed for use in the calculation of pressure drop in coils of constant curvature [Srinivasan *et al* (1968)]. The latter are known as helices. For laminar flow, Kubair and Kuloor (1965) gave an equation for the Reynolds number range 170 to the critical value. In terms of the Fanning friction factor, their equation can be written as

$$f_c = \frac{16[2.8 + 12(d_i/D_c)]}{Re^{1.15}} \tag{2.31}$$

where d_i and D_c are the tube and coil diameters respectively. The critical Reynolds number may be defined as the highest Reynolds number for which the flow in a helix is still definitely in the viscous or laminar region. The critical Reynolds number can be calculated from the equation

$$Re_{\text{critical}} = 2100\left[1 + 12\left(\frac{d_i}{D_c}\right)^{1/2}\right] \tag{2.32}$$

For turbulent flow, White (1932) gave an equation for the Reynolds number range 15 000 to 100 000. In terms of the Fanning friction factor, White's equation can be written as

$$f = \frac{0.08}{Re^{1/4}} + 0.012\left(\frac{d_i}{D_c}\right)^{1/2} \tag{2.33}$$

2.5 Equivalent diameter for non-circular pipes

For pipes of non-circular cross section, the inside diameter d_i in equations 2.13 and 2.22 can be replaced by an equivalent diameter d_e defined as four times the cross-sectional flow area S divided by the appropriate flow perimeter.

For a circular cross section

$$d_e = \frac{4(\pi d_i^2/4)}{\pi d_i} = d_i \tag{2.34}$$

Consider a pipe of circular cross section with an inside and an outside diameter of d_i and d_o respectively. Let this pipe be placed symmetrically inside a larger pipe having an inside diameter of D_i and let a fluid flow through the annulus. Since the shear stress resisting the flow of fluid acts on both walls of the annulus, the appropriate flow perimeter required to calculate the equivalent diameter of the annulus d_e is $(\pi D_i + \pi d_o)$. Therefore

$$d_e = \frac{4[(\pi D_i^2/4) - (\pi d_o^2/4)]}{\pi D_i + \pi d_o} \tag{2.35}$$

$$= D_i - d_o$$

2.6 Velocity profile for laminar Newtonian flow in a pipe

The velocity profile and volumetric flow rate for steady, fully developed, laminar flow of an incompressible Newtonian fluid in a pipe were derived in Chapter 1. Here, the results are summarized for reference.

The velocity profile is related to the frictional pressure drop by equation 1.54

$$v_x = \frac{1}{4\mu}\left(\frac{\Delta P_f}{L}\right)(r_i^2 - r^2) \tag{1.54}$$

The volumetric flow rate Q is given by

$$Q = 2\pi \int_0^{r_i} r v_x \, \mathrm{d}r \tag{1.62}$$

On substituting for v_x in equation 1.62 using equation 1.54, and evaluating the integral, the volumetric flow rate is given by

$$Q = \frac{\pi r_i^4}{8\mu}\left(\frac{\Delta P_f}{L}\right) = \frac{\pi d_i^4}{128\mu}\left(\frac{\Delta P_f}{L}\right) \tag{1.64}$$

and consequently the volumetric average velocity u is given by

$$u = \frac{Q}{\pi r_i^2} = \frac{r_i^2}{8\mu}\left(\frac{\Delta P_f}{L}\right) = \frac{d_i^2}{32\mu}\left(\frac{\Delta P_f}{L}\right) \tag{1.65}$$

The velocity profile can therefore be written as

$$v = v_{\max}\left(1 - \frac{r^2}{r_i^2}\right) = 2u\left(1 - \frac{r^2}{r_i^2}\right) \tag{1.67}$$

Note that the volumetric average velocity u is exactly half the maximum velocity.

2.7 Kinetic energy in laminar flow

At a point where the velocity is v_x, the kinetic energy per unit volume is equal to $\rho v_x^2/2$. The volumetric flow rate through an element of area of infinitesimal width δr is equal to $2\pi r \delta r . v_x$. Thus the flow rate of kinetic energy $\delta(KE)$ through the element of area is given by

$$\delta(KE) = 2\pi r \delta r . v_x \rho v_x^2/2 = \pi \rho v_x^3 r \delta r \tag{2.36}$$

The total flow rate of kinetic energy through the whole cross section is obtained by integrating equation 2.36:

$$KE = \pi\rho \int_0^{r_i} v_x^3 r \, dr \tag{2.37}$$

Equation 2.37 is valid for any symmetric velocity profile. In the case of laminar flow of a Newtonian fluid, the velocity profile is given by equation 1.67, so the kinetic energy flow rate is found as

$$KE = \pi\rho \int_0^{r_i} 8u^3 \left(1 - \frac{r^2}{r_i^2}\right)^3 r \, dr = \pi\rho u^3 r_i^2 \tag{2.38}$$

The mass flow rate M is equal to $\rho\pi r_i^2 u$, so the average kinetic energy per unit mass is given by

$$\frac{KE}{M} = u^2 \tag{2.39}$$

Thus the kinetic energy per unit mass of a Newtonian fluid in steady laminar flow through a pipe of circular cross section is u^2. In terms of head this is u^2/g. Therefore for laminar flow, $\alpha = \frac{1}{2}$ in equation 1.14.

2.8 Velocity distribution for turbulent flow in a pipe

The theory for the turbulent flow of fluids through pipes is far less developed than that for laminar flow.

An approximate equation for the profile of the time-averaged velocity for steady turbulent flow of a Newtonian fluid through a pipe of circular cross section, corresponding to equation 1.67 for laminar flow, may be written as

$$\frac{\bar{v}_x}{\bar{v}_{max}} = \left(1 - \frac{r}{r_i}\right)^{1/7} = \left(1 - \frac{2r}{d_i}\right)^{1/7} \tag{2.40}$$

Equation 2.40 is an empirical equation known as the one-seventh power velocity distribution equation for turbulent flow. It fits the experimentally determined velocity distribution data with a fair degree of accuracy. In fact the value of the power decreases with increasing *Re* and at very high values of *Re* it falls as low as 1/10 [Schlichting (1968)]. Equation 2.40 is not valid in the viscous sublayer or in the buffer zone of the turbulent boundary layer and does not give the required zero velocity gradient at the centre-line. The 1/7th power law is commonly written in the form

$$\frac{\bar{v}_x}{\bar{v}_{max}} = \left(\frac{2y}{d_i}\right)^{1/7} \qquad y \leq d_i/2 \tag{2.41}$$

where y is the distance from the pipe wall:

$$y = \frac{d_i}{2} - r \tag{2.42}$$

For the steady turbulent flow of a Newtonian fluid at high values of *Re* in a pipe of circular cross section, the mean velocity u is related to the maximum velocity $\bar{v}_{max}$ by the equation

$$\frac{u}{\bar{v}_{max}} = 0.82 \tag{2.43}$$

Thus the turbulent flow velocity profile is flatter than the corresponding laminar flow profile, as shown in Figure 1.24.

Equation 2.43 can be derived from the velocity profile given in equation 2.40, using the same method as for laminar flow. The volumetric flow rate is given by

$$Q = 2\pi \int_0^{r_i} r\bar{v}_x \mathrm{d}r \tag{1.62}$$

Substituting for $\bar{v}_x$ from equation 2.40

$$Q = 2\pi \int_0^{r_i} r\bar{v}_{max}\left(1 - \frac{r}{r_i}\right)^{1/7} \mathrm{d}r$$

$$= \frac{49}{60}\pi r_i^2 \bar{v}_{max} \tag{2.44}$$

The volumetric average velocity is defined by $Q = \pi r_i^2 u$, so

$$u = \frac{49}{60}\bar{v}_{max} \approx 0.82\bar{v}_{max} \tag{2.45}$$

The average kinetic energy per unit mass can also be found as for laminar flow. The kinetic energy flow rate is given by

$$KE = \pi\rho \int_0^{r_i} (\bar{v}_x)^3 r\mathrm{d}r \tag{2.37}$$

Substituting for $\bar{v}_x$ from equation 2.40, the kinetic energy flow rate is given by

$$KE = \pi\rho \int_0^{r_i} (\bar{v}_{max})^3 \left(1 - \frac{r}{r_i}\right)^{3/7} r\mathrm{d}r \tag{2.46}$$

This integral can be evaluated conveniently by changing variables using equations 2.41 and 2.42. Thus

$$KE = \pi\rho(\bar{v}_{max})^3 \int_{r_i}^{0} \left(\frac{y}{r_i}\right)^{3/7} (r_i - y)(-\mathrm{d}y)$$

$$= \frac{\pi\rho(\bar{v}_{max})^3}{r_i^{3/7}} \int_0^{r_i} (y^{3/7} r_i - y^{10/7})\mathrm{d}y \tag{2.47}$$

Evaluating the integral gives

$$KE = \frac{49}{170}\pi r_i^2 \rho(\bar{v}_{max})^3 \tag{2.48}$$

From equation 2.45, the mass flow rate is given by

$$M = \frac{49}{60}\pi r_i^2 \bar{v}_{max}\rho \tag{2.49}$$

Thus, the average kinetic energy per unit mass is given by

$$\frac{KE}{M} = \frac{6}{17}(\bar{v}_{max})^2 \tag{2.50}$$

Substituting for the maximum velocity $\bar{v}_{max}$ from equation 2.45, the average kinetic energy per unit mass is equal to $0.52u^2$. This is so close to $u^2/2$, that the value of α in equation 1.14 can be taken as 1.0.

2.9 Universal velocity distribution for turbulent flow in a pipe

Consider a fully developed turbulent flow through a pipe of circular cross section. A turbulent boundary layer will exist with a thin viscous sublayer immediately adjacent to the wall, beyond which is the buffer or generation layer and finally the fully turbulent outer part of the boundary layer.

In the viscous sublayer, the magnitude of the time-averaged value of the shear stress $\bar{\tau}$ is given by Newton's law of viscosity which can be written in this case as

$$\bar{\tau} = -\mu \frac{d\bar{v}_x}{dr} = \mu \frac{d\bar{v}_x}{dy} \tag{2.51}$$

where y is the distance from the wall. The shear stress τ could be denoted by τ_{rx} or τ_{yx} but the subscripts will be omitted for brevity. The viscous sublayer is very thin compared with the pipe radius, so from equation 2.6 the shear stress is virtually equal to the wall shear stress τ_w throughout the sublayer. Thus, integrating equation 2.51,

$$\tau_w y = \mu \bar{v}_x + C \tag{2.52}$$

where C is a constant. Since the velocity $\bar{v}_x = 0$ at $y = 0$, then $C = 0$. Therefore, equation 2.52 can be rewritten as

$$v_x = \frac{\tau_w y}{\mu} = \frac{\tau_w y}{\rho \nu} \tag{2.53}$$

where ν is the kinematic viscosity.

The term τ_w/ρ in equation 2.53 is a constant and has the dimensions of velocity squared. Thus a quantity v_* can be defined by

$$v_* = \sqrt{\frac{\tau_w}{\rho}} \tag{2.54}$$

v_* is commonly known as the friction velocity or the shear stress velocity.

Combining equations 2.53 and 2.54 gives

$$\frac{\bar{v}_x}{v_*} = \frac{v_* y}{\nu} = \frac{\rho v_* y}{\mu} \tag{2.55}$$

A dimensionless velocity v^+ and dimensionless distance y^+ from the wall may be defined by

$$v^+ = \frac{\bar{v}_x}{v_*} \tag{2.56}$$

and

$$y^+ = \frac{v_* y}{\nu} \tag{2.57}$$

so equation 2.55 can also be written as

$$v^+ = y^+ \tag{2.58}$$

The dimensionless distance y^+ has the form of a Reynolds number. Equation 2.58 fits the experimental data in the range $0 \leq y^+ \leq 5$. In the viscous sublayer, the velocity increases linearly with distance from the wall.

Conditions in the fully turbulent outer part of the turbulent boundary layer are quite different. In a turbulent fluid, the shear stress $\bar{\tau}$ is given by equation 1.95. As illustrated in Example 1.10, outside the viscous sublayer and buffer zone the eddy kinematic viscosity ε is much greater than the molecular kinematic viscosity ν. Consequently equation 1.95 can be written as

$$\bar{\tau} = \varepsilon\rho\frac{\mathrm{d}\bar{v}_x}{\mathrm{d}y} \tag{2.59}$$

Prandtl assumed that in turbulent flow eddies move about in a similar manner to molecules in a gas. He defined a mixing length ℓ for turbulent flow analogous to the mean free path in the kinetic theory of gases. It is assumed that a turbulent fluctuation causes an element of fluid to travel a distance ℓ before losing its identity. The distance ℓ is known as the mixing length. Thus, a velocity fluctuation v'_y causes the element of fluid to travel from location y to location $y+\ell$ before losing its identity and the resulting momentum transfer produces a velocity fluctuation v'_x. As the mixing length ℓ is very small, the values of the mean velocity $\bar{v}_x$ at the two locations are related by

$$\bar{v}_x(y+\ell) - \bar{v}_x(y) \approx \ell\,\frac{\mathrm{d}\bar{v}_x}{\mathrm{d}y} \tag{2.60}$$

The velocity fluctuation must be of the same order of magnitude as this velocity difference, so that

$$\bar{v}'_x \approx -\ell\,\frac{\mathrm{d}\bar{v}_x}{\mathrm{d}y} \tag{2.61}$$

where the minus sign reflects the fact that, in a positive velocity gradient, a fluctuation in the positive y-direction will retard the faster fluid, thus producing a negative fluctuation.

Prandtl assumed further that the fluctuations in the x and y-directions are of the same order of magnitude, which is now known to be true for this type of flow. Consequently, the magnitude of the Reynolds stress is given by

$$\bar{\tau} = \rho \overline{v'_x v'_y} \approx \rho \ell^2 \left(\frac{\mathrm{d}\bar{v}_x}{\mathrm{d}y} \right)^2 \tag{2.62}$$

Comparing equations 2.59 and 2.62, it follows that the eddy kinematic viscosity can be expressed in terms of the mixing length:

$$\varepsilon \approx \ell^2 \left| \frac{\mathrm{d}\bar{v}_x}{\mathrm{d}y} \right| \tag{2.63}$$

The final assumption made by Prandtl was that ℓ is proportional to y, the distance from the solid wall. This is reasonable in that ℓ must be zero at the wall.

Writing

$$\ell = Ky \tag{2.64}$$

where K is a proportionality constant, equation 2.62 can be written as

$$\frac{\bar{\tau}}{\rho} = K^2 y^2 \left(\frac{\mathrm{d}\bar{v}_x}{\mathrm{d}y} \right)^2 \tag{2.65}$$

If the analysis is restricted to the region very close to the wall, then $\bar{\tau}$ remains nearly equal to τ_w. From the definition of v_* (equation 2.54), equation 2.65 can be written as

$$v_* = Ky \frac{\mathrm{d}\bar{v}_x}{\mathrm{d}y} \tag{2.66}$$

Integrating gives

$$\bar{v}_x = v_* \left(\frac{1}{K} \ln y + C_1 \right) \tag{2.67}$$

where C_1 is a constant.

Equation 2.67 can be written in modified form as

$$\bar{v}_x = v_* \left[\frac{1}{K} \ln \left(\frac{\rho v_* y}{\mu} \right) + C_2 \right] \tag{2.68}$$

where C_2 is another constant. Equation 2.68 is not applicable near the wall because it neglects the viscous shear stress and consequently gives $\bar{v}_x = -\infty$ instead of $\bar{v}_x = 0$ at $y = 0$.

Rewriting equation 2.68 in terms of $v^+ = \bar{v}_x/v_*$ and $y^+ = \rho v_* y/\mu$ gives

$$v^+ = \frac{1}{K} \ln y^+ + C \qquad (2.69)$$

where C is a constant.

Equation 2.69 fits the experimental data for turbulent flow in smooth pipes of circular cross section for $y^+ > 30$ when $1/K$ and C are given the values 2.5 and 5.5:

$$v^+ = 2.5 \ln y^+ + 5.5 \qquad (2.70)$$

It is interesting to note that this logarithmic velocity profile is followed over most of the cross section of the pipe, not just where $\bar{\tau} \approx \tau_w$.

For the buffer region, $5 < y^+ < 30$, viscous and turbulent stresses are of comparable magnitude. The data can be fitted by an equation of similar form:

$$v^+ = 5.0 \ln y^+ - 3.05 \qquad (2.71)$$

Equations 2.58, 2.70 and 2.71 enable the velocity distribution to be calculated for steady fully developed turbulent flow. These equations are only approximate and lead to a discontinuity of the gradient at $y^+ = 30$, which is where equations 2.70 and 2.71 intersect. The actual profile is, of course, smooth and the transition from the buffer zone to the fully turbulent outer zone is particularly gradual. As a result it is somewhat arbitrary where the limit of the buffer zone is taken: often the value $y^+ = 70$ rather than $y^+ = 30$ is used. The ability to represent the velocity profile in most turbulent boundary layers by the same $v^+ - y^+$ relationships (equations 2.58, 2.70 and 2.71) is the reason for calling this the universal velocity profile. The use of v_* in defining v^+ and y^+ demonstrates the fundamental importance of the wall shear stress.

It is the logarithmic profile of the outer region that can be approximated by Prandtl's 1/7th power law (equation 2.40 or 2.41).

The changing character of the flow in the different regions of the turbulent boundary layer explains certain aspects of the friction factor chart. If the absolute roughness of the pipe wall is smaller than the thickness of the viscous sublayer, flow disturbances caused by the roughness will be damped out by viscosity. The wall is subject to a viscous shear stress. Under these conditions, the line on the friction factor chart for smooth pipes is followed and the wall is said to be hydraulically smooth. If the roughness is large enough to protrude through the viscous sublayer into the buffer zone, the protuberances will be subject to form drag as well as the viscous drag. The proportion of form drag will increase as the roughness protrudes further and eventually, when it reaches right into the fully turbulent zone, form drag on the protuberances will be dominant. Under these conditions, the drag on the wall is proportional to u^2 and so the friction factor becomes independent of Re.

The viscous sublayer and the buffer zone both become thinner with increasing Re. (For example, it is easily shown that if the friction factor varies as $Re^{-1/4}$, as in the Blasius equation, then the thickness of each layer is inversely proportional to $Re^{7/8}$.) Consequently, a given pipe may be hydraulically smooth at intermediate values of Re but hydraulically rough at higher values of Re. This can be seen on the friction factor chart where the line for a given relative roughness continues along the line for a smooth pipe then curves away and eventually reaches a constant value at high values of Re.

As the wall becomes rougher, the velocity profile in the turbulent zone changes as shown in Figure 2.5, and the viscous sublayer and generation zone eventually disappear.

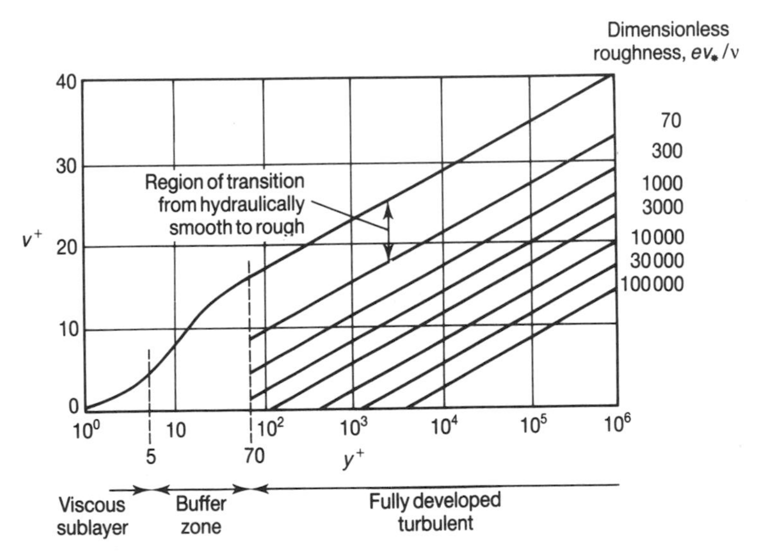

Figure 2.5

Universal velocity profile for turbulent flow

Source: N. Scholtz, *VDI–Beriche* **6**, 7–12 (1955)

2.10 **Flow in open channels**

Consider a liquid flowing in an open channel of uniform cross section under the influence of gravity. The liquid has a free surface subjected only to atmospheric pressure. If the flow is steady, the depth of the liquid is uniform and the hydraulic slope of the free liquid surface is parallel to the slope of the channel bed. Consider a length ΔL in Figure 2.6 in which the

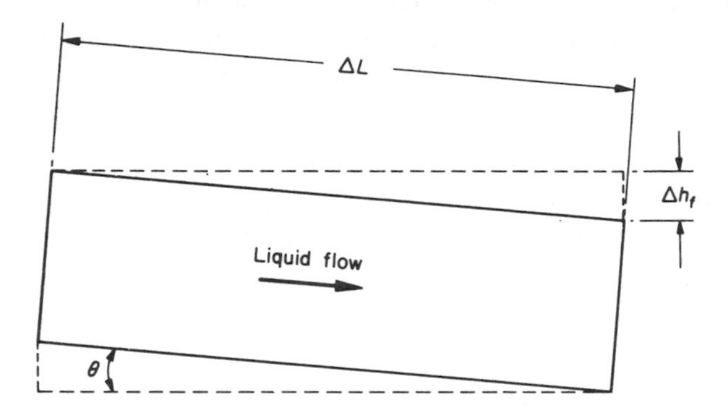

Figure 2.6
Flow in an open channel

frictional head loss is Δh_f. Let the channel slope at a small angle θ to the horizontal. The slope of the channel flow s is given by $s = \sin \theta = \Delta h_f / \Delta L$.

The frictional head loss is given by the equations

$$\Delta h_f = \Delta L s \tag{2.72}$$

and

$$\Delta h_f = 4f \left(\frac{\Delta L}{d_e} \right) \frac{u^2}{2g} \tag{2.73}$$

Equation 2.73 is another way of writing equation 2.13 where, in this case, the pressure drop is expressed in height of fluid instead of in force per unit area. In equation 2.73, d_e is the equivalent diameter defined as four times the cross-sectional flow area divided by the appropriate flow perimeter, f is the Fanning friction factor for flow in an open channel and u is the mean velocity. Combining equations 2.72 and 2.73, and solving for u gives

$$u = \sqrt{\frac{2g}{f}} \sqrt{\frac{d_e s}{4}} \tag{2.74}$$

where the mean velocity u is proportional to the square root of the channel slope s.

Equation 2.74 is frequently written in the form

$$u = C\sqrt{\frac{d_e s}{4}} \tag{2.75}$$

which is known as the Chezy formula.

The Chezy coefficient C is

$$C = \sqrt{\frac{2g}{f}} \tag{2.76}$$

Manning and others gave values of C for various types of surface roughness [Barna (1969)]. A typical value for C when water flows in a concrete channel is 100 $m^{1/2}/s$. In general, liquids such as water which commonly flow in open channels have a low viscosity and the flow is almost always turbulent.

References

Barna, P.S., *Fluid Mechanics for Engineers*, London, Butterworths, p. 85, (1969).

Haaland, S.E., Simple and explicit formulas for the friction factor in turbulent pipe flow, *Transactions of ASME, Journal of Fluids Engineering*, Series I **105**, pp. 89–90 (1983).

Holland, F.A. and Chapman, F.S., *Pumping of Liquids*, New York, Reinhold Publishing Corporation, p. 79, (1966).

Holland, F.A., Moores, R.M., Watson, F.A. and Wilkinson, J.K., *Heat Transfer*. London, Heinemann Educational Books Ltd, p. 461 (1970).

Kubair, V. and Kuloor, N.R., Non-isothermal pressure drop data for liquid flow in heated coils. *Indian Journal of Technology*, **3**, pp. 5–7 (1965).

Schlichting, H., *Boundary Layer Theory*, Sixth edition, New York: McGraw-Hill Book Company Inc. (1968).

Scholtz, N., Strömungsvorgänge in Grenzschichten, *VDI–Berichte*, **6**, pp. 7–12 (1955)

Srinivasan, P.S., Nandapurkar, S.S., and Holland, F.A., Pressure drop and heat transfer in coils. *The Chemical Engineer*, No. 218, pp. 113–119 (1968).

White, C.M., Fluid friction and its relation to heat transfer, *Transactions Institution of Chemical Engineers*, **10**, pp. 66–80 (1932).

3 Flow of incompressible non-Newtonian fluids in pipes

Among other characteristics, non-Newtonian fluids exhibit an apparent viscosity that varies with shear rate. Consequently, the determination of the shear stress–shear rate curve must be an initial consideration. Although the apparent viscosity of a thixotropic or a rheopectic fluid changes with the duration of shearing, meaningful measurements may be made if the change is relatively slow. Viscoelastic fluids also exhibit behaviour that is a function of time but their apparent viscosities can be measured provided conditions of steady shearing are obtained.

3.1 Elementary viscometry

There are two main types of viscometer: rotary instruments and tubular, often capillary, viscometers. When dealing with non-Newtonian fluids, it is desirable to use a viscometer that subjects the whole of the sample to the same shear rate and two such devices, the cone and plate viscometer and the narrow gap coaxial cylinders viscometer, will be considered first. With other instruments, which impose a non-uniform shear rate, the proper analysis of the measurements is more complicated.

With any viscometer the flow generated should ideally have only one non-zero velocity component, causing shearing in only one direction. The purpose of a viscometer is simultaneously to measure (or control) both the shear stress and the shear rate. Not only must the flow be laminar but viscous forces must be dominant, that is, inertial effects must be negligible.

3.3.1 *Cone and plate viscometer*

A schematic representation of a cone and plate viscometer is shown in Figure 3.1. In this case, the viscometer consists of a lower disc and an upper wide-angle cone. The sample fills the gap between the cone and the

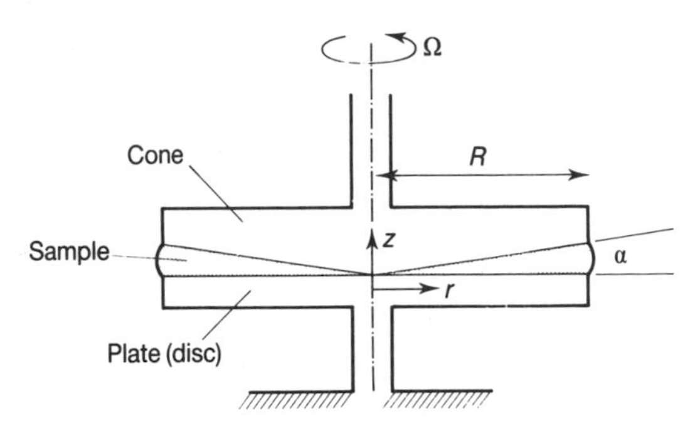

Figure 3.1
Cone and plate viscometer

plate. If the cone is rotated at a constant angular rate, the fluid is sheared with every element of fluid describing a horizontal circular path.

The tangential velocity component v_θ varies linearly from zero at the lower plate to the speed of the cone at the cone's surface. At a radial distance r, the cone's tangential speed is Ωr where Ω is in radians per second. At this location the height of the gap is αr where α is the angle of the gap in radians. Thus, the shear rate $\dot{\gamma}$ is given by

$$\dot{\gamma} = \frac{\partial v_\theta}{\partial z} = \frac{\Omega r}{\alpha r} = \frac{\Omega}{\alpha} \tag{3.1}$$

Equation 3.1 demonstrates the reason for using this geometry: both v_θ and the height of the gap are proportional to r so the shear rate is uniform throughout the sample. As a result, the shear stress $\tau_{z\theta}$ is uniform and can be determined by measuring the couple or torque required to maintain the steady shearing. Considering an element of area of the lower plate, forming an annular strip of width δr, the tangential force required is equal to $2\pi r \delta r . \tau_{z\theta}$ and the couple acting on this element of area at distance r is equal to $2\pi r^2 \delta r . \tau_{z\theta}$. The couple, C acting on the whole area is obtained by integrating over the area:

$$C = 2\pi\tau_{z\theta} \int_0^R r^2 \, \mathrm{d}r = \frac{2\pi R^3}{3} \tau_{z\theta} \tag{3.2}$$

Rearranging equation 3.2, the shear stress is given by

$$\tau_{z\theta} = \frac{3C}{2\pi R^3} \tag{3.3}$$

The couple acting on either the cone or the plate may be measured as they are equal but act in opposite directions. Thus $\tau_{z\theta}$ in equation 3.3 is strictly the magnitude of the shear stress. Dividing equation 3.3 by equation 3.1 gives an expression for the apparent viscosity:

$$\mu_a = \frac{3C}{2\pi R^3}\frac{\alpha}{\Omega} \tag{3.4}$$

Note that α must be in radians and Ω in radians per second. If the rotational speed is measured as N revolutions per minute (rpm), then the required conversion is

$$\Omega = \frac{2\pi N}{60} \tag{3.5}$$

The couple C is measured in N m. The angle between the cone and the plate is typically between $\frac{1}{2}°$ and 4° (0.0087 and 0.07 radians).

3.1.2 *Narrow gap coaxial cylinders viscometer*

A type of viscometer providing a nearly uniform shear rate in the sample is that shown in Figure 3.2. The sample fills the gap between the two cylinders, one of which is rotated steadily.

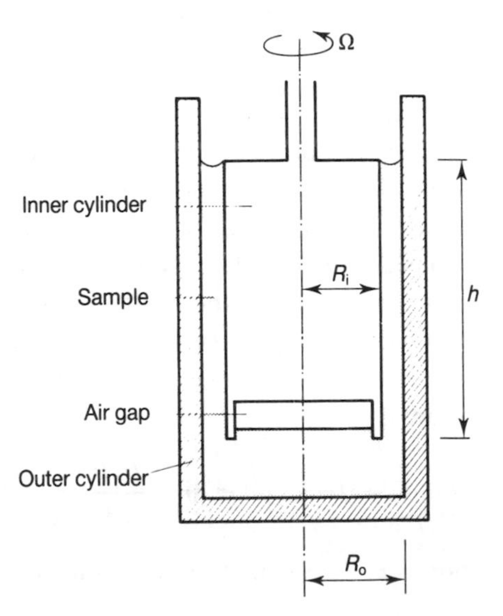

Figure 3.2
Narrow gap coaxial cylinders viscometer

Various arrangements at the bottom of the inner cylinder are available: in Figure 3.2 an indentation is provided so that an air gap is formed and shearing in the sample below the inner cylinder is negligible. Another arrangement is to make the bottom of the inner cylinder a cone. When one of the cylinders is rotated, a Couette flow is generated with fluid particles describing circular paths. The only non-zero velocity component is v_θ and it varies in the r–direction. In order to minimize secondary flow (Taylor vortices) it is preferable that the outer cylinder be rotated; however, in most commercial instruments it is the inner cylinder that rotates. In this case, the fluid's velocity is equal to ΩR_i at the surface of the inner cylinder and falls to zero at the surface of the outer cylinder. The shear stress is uniform over the curved surface of the inner cylinder and over the outer cylinder (to the bottom of the annular gap).

For steady conditions, the couple on each cylinder must be equal and opposite. Thus, denoting the magnitude of the shear stress on the inner and outer cylinders by τ_i and τ_o respectively, the couple C is given by

$$C = 2\pi R_i h \tau_i . R_i = 2\pi R_o h \tau_o . R_o \tag{3.6}$$

Thus

$$R_i^2 \tau_i = R_o^2 \tau_o \tag{3.7}$$

In fact, for any point in the sample, the product of the shear stress and the square of the distance from the axis of rotation is constant. Thus, in general for this type of flow between coaxial cylinders, the shear stress is inversely proportional to the square of the distance from the axis:

$$\tau_{r\theta} \propto \frac{1}{r^2} \tag{3.8}$$

When used with non-Newtonian liquids, this non-uniformity of the shear stress, and therefore the shear rate, is a limitation of the simple type of rotating shaft viscometer that can be placed in a vessel.

However, by making the width of the gap small compared with the radius of either cylinder, the magnitude of the shear stress on each cylinder, and throughout the sample, is nearly constant. As a result, the shear rate is virtually constant throughout the sample and can be written as

$$\dot{\gamma} = \frac{\Omega R_i}{R_o - R_i} \tag{3.9}$$

Again, Ω is in rad/s and may be calculated from equation 3.5. From equation 3.6, the magnitude of the shear stress can be calculated from the measured couple C on either cylinder:

$$\tau_{r\theta} \approx \frac{C}{2\pi R_i^2 h} \approx \frac{C}{2\pi R_o^2 h} \tag{3.10}$$

Assuming that the shear stress is calculated for the inner cylinder, the apparent viscosity is given by

$$\mu_a \approx \frac{C(R_o - R_i)}{2\pi R_i^3 h \Omega} \tag{3.11}$$

Equations 3.4 and 3.11 are unnecessary for the calculation of μ_a because both τ and $\dot{\gamma}$ will be evaluated and μ_a calculated from the ratio. However, the expressions for μ_a do indicate what changes of instrument parameters may be necessary to accommodate materials of different viscosities. Note, in particular, that the measured couple is proportional to R^3 so large diameter viscometers will allow measurable couples to be obtained with low viscosity materials, while smaller instruments will be more suitable with very viscous materials.

A general limitation of rotational instruments is that the high shear rates typical of most engineering applications cannot be achieved because, at the high speeds needed, secondary flow occurs and compromises the measurements.

3.1.3 *Tubular viscometer*

A tubular viscometer is shown schematically in Figure 3.3. The fluid

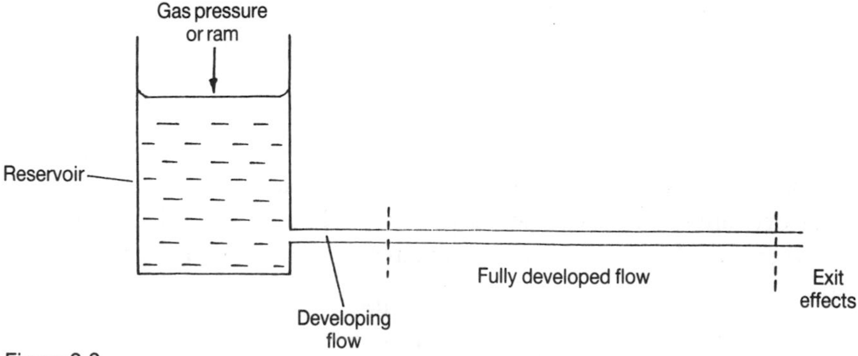

Figure 3.3
Tubular viscometer

under test is forced through the tube either by driving a ram through the reservoir or by applying gas pressure. In the case of testing a molten polymer, the reservoir might be a vertical heated barrel with a capillary tube fitted into its end. The capillary tube may be only about 50 mm long and have a diameter of 0.5 mm or less. A piston driven down the barrel extrudes the viscous molten polymer through the capillary tube. In contrast, relatively large diameter tubing is required when studying suspensions. In this case, an arrangement like that in Figure 3.3 might be used with the suspension being forced through very long tubes by the application of pressurized nitrogen to the upper part of the reservoir.

It is important to appreciate that fully-developed flow occurs along only part of the tube's length. There will be an entrance length in which the velocity profile develops and where the pressure gradient is larger than for fully-developed flow. This entrance length will be at least 50 tube diameters and may be as long as 400 diameters with some viscoelastic fluids. There will also be end effects near the tube exit but these are usually small. Ideally, pressure measurements would be made at several locations along the tube so that the constant, fully developed, pressure gradient could be determined. However, this is difficult, particularly with dispersed materials and the pressure difference between the reservoir and the tube exit has to be used. In this case, measurements should be made over a series of flow rates with at least two tubes of the same diameter but different lengths. By comparing the pressure drop measurements at the same flow rate in the two tubes, the differential pressure drop for the differential length can be calculated, thus eliminating end effects. Often measurements are simply made with long tubes in the hope that end effects are negligible.

If the frictional component of the pressure drop over a length L of steady, fully–developed flow is ΔP_f, then the shear stress τ_{rx} at distance r from the tube axis is given by

$$\tau_{rx} = \frac{r}{2}\left(\frac{\Delta P_f}{L}\right) \tag{2.4}$$

It was stressed in Chapters 1 and 2 that this relationship is valid for any steady, fully-developed flow and, in particular, does not depend on the type of fluid. The value of the shear stress at the wall is given by

$$\tau_w = \frac{r_i}{2}\left(\frac{\Delta P_f}{L}\right) = \frac{d_i}{4}\left(\frac{\Delta P_f}{L}\right) \tag{2.5}$$

so that by measuring the pressure drop ΔP_f the value of τ_w can be determined. In the case of flow in a vertical tube, the measured pressure drop must be corrected for the static head. In order to find the shear stress–shear rate relationship, ie to be able to plot a τ–$\dot{\gamma}$ curve, it is necessary to know the shear rate at the wall. Thus, τ_w can be plotted against corresponding values of $\dot{\gamma}_w$. The difficulty that arises is that $\dot{\gamma}_w$ is equal to the gradient of the velocity profile at the wall and the shape of the velocity profile is different for different types of fluid. Thus it appears that in order to calculate $\dot{\gamma}_w$ it is necessary to known the fluid's shear stress–shear rate behaviour but this is the very point of making the measurements. This difficulty can be overcome but it will be helpful first to consider the case of a Newtonian fluid.

It was shown in Section 1.10 that the velocity profile for a Newtonian fluid in laminar flow is parabolic and may be expressed as

$$v_x = 2u\left(1 - \frac{r^2}{r_i^2}\right) \tag{1.67}$$

Differentiating wrt r and putting $r = r_i$, the velocity gradient at the wall, which is equal to the shear rate, is given by

$$\dot{\gamma}_{wN} = \left.\frac{\mathrm{d}v_x}{\mathrm{d}r}\right|_{r=r_i} = -\frac{4u}{r_i} = -\frac{8u}{d_i} \tag{3.12}$$

or, in terms of the volumetric flow rate Q

$$\dot{\gamma}_{wN} = -\frac{4Q}{\pi r_i^3} \tag{3.13}$$

The second subscript N is a reminder that this is the wall shear rate for a Newtonian fluid. The quantity $(8u/d_i)$, or the equivalent form in equation 3.13, is known as the flow characteristic. It is a quantity that can be calculated for the flow of any fluid in a pipe or tube but it is only in the case of a Newtonian fluid in laminar flow that it is equal to the magnitude of the shear rate at the wall.

Owing to the different relationship between τ and $\dot{\gamma}$ for a non-Newtonian fluid, the shear rate at the wall is not given by equation 3.13 but can be expressed as the flow characteristic multiplied by a correction factor as shown in Section 3.2.

3.2 Rabinowitsch–Mooney equation

The solution to the problem of determining the wall shear rate for a non-Newtonian fluid in laminar flow in a tube relies on equation 2.6.

The volumetric flow rate through an annular element of area perpendicular to the flow and of width δr is given by

$$\delta Q = 2\pi r \delta r . v_x \tag{1.61}$$

and, consequently, the flow rate through the whole tube is

$$Q = 2\pi \int_0^{r_i} r v_x \, \mathrm{d}r \tag{1.62}$$

Integrating by parts gives

$$Q = 2\pi \left\{ \left[\frac{r^2 v_x}{2} \right]_0^{r_i} + \int_0^{r_i} \frac{r^2}{2} \left(-\frac{\mathrm{d}v_x}{\mathrm{d}r} \right) \mathrm{d}r \right\} \tag{3.14}$$

Provided there is no slip at the tube wall, the first term in equation 3.14 vanishes.

Now, the velocity gradient is equal to the shear rate $\dot{\gamma}$ so equation 3.14 can be written as

$$Q = \pi \int_0^{r_i} r^2 (-\dot{\gamma}) \, \mathrm{d}r \tag{3.15}$$

Just as the variation of v_x with r was unknown, so is the variation of $\dot{\gamma}$. However, if the fluid is time–independent and homogeneous, the shear stress is a function of shear rate only. The inverse is that the shear rate $\dot{\gamma}$ is a function of shear stress τ_{rx} only and the variation of τ_{rx} with r is known:

$$\frac{\tau_{rx}}{\tau_w} = \frac{r}{r_i} \tag{2.6}$$

Changing variables in equation 3.15, using equation 2.6, and dropping the subscripts rx, equation 3.15 can be written as

$$Q = \pi \int_0^{\tau_w} \frac{\tau^2 r_i^2}{\tau_w^2} (-\dot{\gamma}) \frac{r_i}{\tau_w} \, \mathrm{d}\tau = \frac{\pi r_i^3}{\tau_w^3} \int_0^{\tau_w} \tau^2 (-\dot{\gamma}) \, \mathrm{d}\tau \tag{3.16}$$

where $\dot{\gamma}$ is interpreted as a function of τ instead of r. Writing equation 3.16 in terms of the flow characteristic gives

$$\frac{8u}{d_i} = \frac{4Q}{\pi r_i^3} = \frac{4}{\tau_w^3} \int_0^{\tau_w} \tau^2 (-\dot{\gamma}) \, \mathrm{d}\tau \tag{3.17}$$

For flow in a pipe or tube the shear rate is negative so the integral in equation 3.17 is positive. For a given relationship between τ and $\dot{\gamma}$, the value of the integral depends only on the value of τ_w. Thus, for a non-Newtonian fluid, as well as for a Newtonian fluid, the flow characteristic $8u/d_i$ is a unique function of the wall shear stress τ_w.

The shear rate $\dot{\gamma}$ can be extracted from equation 3.17 by differentiating with respect to τ. Moreover, if a definite integral is differentiated wrt the upper limit (here τ_w), the result is the integrand evaluated at the upper limit. It is convenient first to multiply equation 3.17 by τ_w^3 throughout, then differentiating wrt τ_w gives

$$3\tau_w^2\left(\frac{8u}{d_i}\right) + \tau_w^3 \frac{\mathrm{d}(8u/d_i)}{\mathrm{d}\tau_w} = 4\tau_w^2(-\dot{\gamma})_w \tag{3.18}$$

Rearranging equation 3.18 gives the wall shear rate $\dot{\gamma}_w$ as

$$-\dot{\gamma}_w = \frac{8u}{d_i}\left[\frac{3}{4} + \frac{1}{4}\frac{\tau_w}{(8u/d_i)}\frac{\mathrm{d}(8u/d_i)}{\mathrm{d}\tau_w}\right] \tag{3.19}$$

Making use of the relationship $\mathrm{d}x/x = \mathrm{d}\ln x$, equation 3.19 can be written as

$$-\dot{\gamma}_w = \frac{8u}{d_i}\left[\frac{3}{4} + \frac{1}{4}\frac{\mathrm{d}\ln(8u/d_i)}{\mathrm{d}\ln\tau_w}\right] \tag{3.20}$$

As the wall shear rate $\dot{\gamma}_{wN}$ for a Newtonian fluid in laminar flow is equal to $(-8u/d_i)$, equation 3.20 can be expressed as

$$\dot{\gamma}_w = \dot{\gamma}_{wN}\left[\frac{3}{4} + \frac{1}{4}\frac{\mathrm{d}\ln(8u/d_i)}{\mathrm{d}\ln\tau_w}\right] \tag{3.21}$$

Equations 3.20 and 3.21 are forms of the Rabinowitsch–Mooney equation. It shows that the wall shear rate for a non-Newtonian fluid can be calculated from the value for a Newtonian fluid having the same flow rate in the same pipe, the correction factor being the quantity in the square brackets. The derivative can be estimated by plotting $\ln(8u/d_i)$ against $\ln\tau_w$ and measuring the gradient. Alternatively the gradient may be calculated from the (finite) differences between values of $\ln(8u/d_i)$ and $\ln\tau_w$. Thus the flow curve τ_w against $\dot{\gamma}_w$ can be determined. The measurements required and the calculation procedure are as follows.

1 Measure Q at various values of $\Delta P_f/L$, preferably eliminating end effects.
2 Calculate τ_w from the pressure drop measurements using equation 2.5 and the corresponding values of the flow characteristic ($8u/d_i = 4Q/\pi r_i^3$) from the flow rate measurements.

3 Plot $\ln(8u/d_i)$ against $\ln\tau_w$ and measure the gradient at various points on the curve. [$\log(8u/d_i)$ and $\log\tau_w$ may be used if more convenient.] Alternatively, calculate the gradient from the differences between the successive values of these quantities.
4 Calculate the true wall shear rate from equation 3.20 with the derivative determined in 3. In general, the plot of $\ln(8u/d_i)$ against $\ln\tau_w$ will not be a straight line and the gradient must be evaluated at the appropriate points on the curve.

Example 3.1
The flow rate–pressure drop measurements shown in Table 3.1 were made in a horizontal tube having an internal diameter $d_i = 6$ mm, the pressure drop being measured between two tappings 2.00 m apart. The density of the fluid, ρ, was 870 kg/m^3. Determine the wall shear stress–flow characteristic curve and the shear stress–true shear rate curve for this material.

Table 3.1

Pressure drop (bar)	*Mass flow rate* $\times 10^3$ (kg/s)
0.384	0.0864
0.519	0.463
0.716	1.37
0.965	2.76
1.16	4.13
1.29	5.20
1.46	6.78
1.60	8.15

Calculations
The shear stress at the wall is given by

$$\tau_w = \frac{d_i \Delta P_f}{4L} = \frac{(6 \times 10^{-3}\ \text{m})}{4(2\ \text{m})} \Delta P_f = 7.50 \times 10^{-4} \Delta P_f \ \text{Pa} \qquad (2.5)$$

where ΔP_f is in Pa. Note: 1 bar = 10^5 Pa. The flow characteristic is

$$\frac{8u}{d_i} = \frac{4Q}{\pi r_i^3} = \frac{32Q}{\pi d_i^3} = \frac{32M}{\pi d_i^3 \rho}$$

Using the given values

$$\frac{8u}{d_i} = \frac{32M}{\pi(6 \times 10^{-3}\ \mathrm{m})^3(870\ \mathrm{kg/m^3})} = 54200M\ \mathrm{s^{-1}}$$

where the mass flow rate M is in kg/s.

Using these expressions for τ_w and $8u/d_i$ enables the values in the first two columns of Table 3.2 to be calculated. This provides the shear stress–flow characteristic curve.

In order to determine the true shear rate at the wall it is necessary to use the Rabinowitsch–Mooney equation:

$$-\dot{\gamma}_w = \frac{8u}{d_i}\left[\frac{3}{4} + \frac{1}{4}\frac{\mathrm{d}\ln(8u/d_i)}{\mathrm{d}\ln\tau_w}\right] \tag{3.20}$$

By plotting the calculated values of τ_w against $8u/d_i$ on logarithmic axes, the gradient of the curve is equal to the reciprocal of the derivative in equation 3.20. Denoting the gradient by n', equation 3.20 can be written as

$$-\dot{\gamma}_w = \frac{8u}{d_i}\left(\frac{3}{4} + \frac{1}{4n'}\right) = \frac{8u}{d_i}\left(\frac{3n'+1}{4n'}\right)$$

On plotting the graph and estimating the gradient at each point, the values of the gradient n' shown in column 3 of Table 3.2 are found and the corresponding values of the correction factor are shown in column 4. The value of the shear rate at the wall is then given by multiplying the corresponding value of $8u/d_i$ by the correction factor.

Table 3.2

τ_w (Pa)	$8u/d_i$ ($\mathrm{s^{-1}}$)	n'	$(3n'+1)/4n'$	$-\dot{\gamma}_w$ ($\mathrm{s^{-1}}$)
28.8	4.68	0.157	2.34	11.0
38.9	25.1	0.232	1.83	45.9
53.7	74.3	0.375	1.42	106
72.4	150	0.439	1.32	197
87.0	224	0.475	1.28	286
96.8	282	0.475	1.28	360
110	367	0.475	1.28	469
120	442	0.475	1.28	564

Both curves are shown in Figure 3.4.

Table 3.2 shows values of n' calculated by measuring the gradient of the $\ln\tau_w - \ln(8u/d_i)$ curve. This is the simplest way of determining the gradient

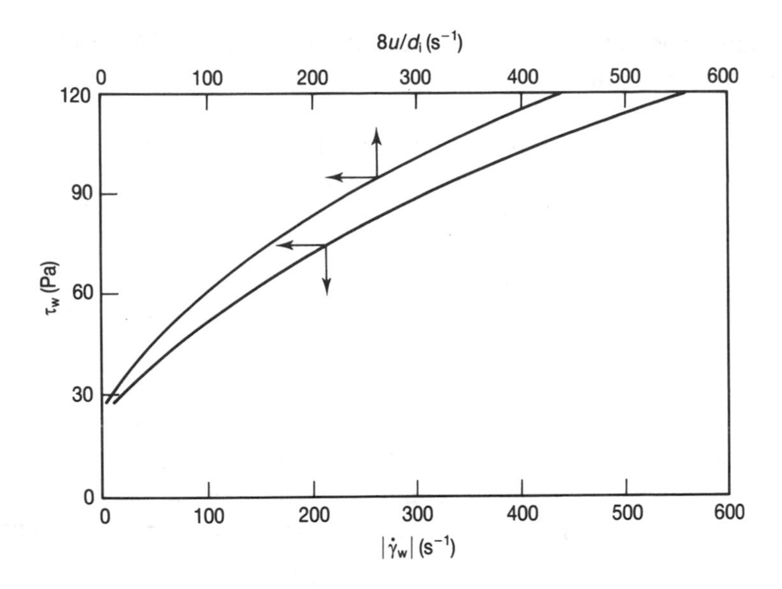

Figure 3.4
$\tau_w - 8u/d_i$ and $\tau_w - \dot{\gamma}$ curves for Example 3.1

and hence the derivative in equation 3.20; however, the reader should be aware that graphical differentiation is notoriously inaccurate and consequently requires great care. Other methods include fitting a low order polynomial equation to the data and hence finding the derivative algebraically. Another method is to calculate the (finite) differences between successive values of $\ln \tau_w$ and $\ln(8u/d_i)$ and hence calculate the derivative as their ratio. The approximation is then made that this is the derivative at the *middle of the interval* (on the logarithmic scale).

The graphical method has the advantage of smoothing the data: if the other methods are used it may be necessary to smooth the data first. Application of the difference method directly to the data in the above table will show good agreement in the lower part of the range but oscillations will be observed towards the top of the range.

This material is seen to be shear thinning. It is possible that it may exhibit a yield stress but confirmation of this would require measurements at lower shear rates. Note that the Rabinowitsch–Mooney equation is still valid when a non-zero yield stress occurs.

3.3 Calculation of flow rate–pressure drop relationship for laminar flow using τ–$\dot{\gamma}$ data

Flow rate–pressure drop calculations for laminar non-Newtonian flow in pipes may be made in various ways depending on the type of flow information available. When the flow data are in the form of flow rate and pressure gradient measured in a tubular viscometer or in a pilot scale pipeline, direct scale-up can be done as described in Section 3.4. When the data are in the form of shear stress–shear rate values (tabular or graphical), the flow rate can be calculated directly using equation 3.17, where d_i is the diameter of the pipe to be used and τ_w the wall shear stress corresponding to the specified pressure gradient. Whether obtained with a rotational instrument or with a tubular viscometer, the data provide the relationship between τ and $\dot{\gamma}$. Numerical evaluation of the integral in equation 3.17 can be done using selected pairs of values of τ and $\dot{\gamma}$ ranging from 0 to τ_w. The value of τ_w is given by equation 2.5.

If the τ–$\dot{\gamma}$ relationship can be accurately represented by a simple algebraic expression, such as the power law, over the required range then this may be used to substitute for $\dot{\gamma}$ in equation 3.17, allowing the integral to be evaluated analytically. Both these methods are illustrated in Example 3.2.

Example 3.2

Using the viscometric data given in Table 3.3 calculate the average velocity for the material flowing through a pipe of diameter 37 mm when the pressure gradient is 1.1 kPa/m.

Table 3.3

$\dot{\gamma}$ (s^{-1})	τ (Pa)	μ_a (Pa s)
0.00911	0.0417	4.58
0.0911	0.178	1.95
0.911	0.708	0.777
9.111	2.82	0.310
91.11	11.22	0.123
102.3	12.03	0.118

Note that the above table gives absolute values of $\dot{\gamma}$ and τ.

Calculations

The wall shear stress is given by

$$\tau_w = \frac{d_i \Delta P}{4L} \tag{2.5}$$

$$= \frac{(37 \times 10^{-3}\ \text{m})(1100\ \text{Pa/m})}{4}$$

$$= 10.18\ \text{Pa}$$

The volumetric average velocity u is given by equation 3.17 as

$$\frac{8u}{d_i} = \frac{4}{\tau_w^3} \int_0^{\tau_w} \tau^2(-\dot{\gamma})\,\mathrm{d}r \tag{3.17}$$

It is necessary to evaluate the integral from $\tau = 0$ to $\tau = 10.18$ Pa. This can be done by calculating $\tau^2\dot{\gamma}$ for each of the values given in the table and plotting $\tau^2\dot{\gamma}$ against τ. The area under the curve between $\tau = 0$ and $\tau = 10.18$ Pa can then be measured. An alternative, which will be used here, is to use a numerical method such as Simpson's rule. This requires values at equal intervals of τ.

Dividing the range of integration into six strips and interpolating the data allows Table 3.4 to be constructed.

Table 3.4

τ (Pa)	$\dot{\gamma}$ (s^{-1})	$\tau^2\dot{\gamma}$ (Pa^2s^{-1})	
0.00	0.00	0.00	Centre-line
1.70	3.91	11.24	
3.39	12.41	142.8	
5.09	24.38	631.0	
6.78	39.39	1812	
8.48	57.14	4108	
10.18	77.43	8016	Pipe wall

By Simpson's rule

$$\int_0^{10.18} \tau^2\dot{\gamma}\,\mathrm{d}\tau \approx \frac{10.18/6}{3}\,[0 + 8016 + 4(11.24 + 631 + 4108) + 2(142.8 + 1812)]$$

$$= 17490\ \text{Pa}^3\text{s}^{-1}$$

From equation 3.17

$$u = \frac{(37 \times 10^{-3}\ \text{m})(17\,490\ \text{Pa}^3\ \text{s}^{-1})}{2(10.18\ \text{Pa})^3} = \underline{0.307\ \text{m/s}}$$

The above is the general method but in this case the viscometric data can be well represented by

$$\tau = 0.749\dot{\gamma}^{0.60}\ \mathrm{Pa}$$

Thus

$$\dot{\gamma} = 1.62\tau^{1.667}\ \mathrm{s}^{-1}$$

This allows the integral in equation 3.17 to be evaluated analytically.

$$\int_0^{\tau_w} \tau^2\dot{\gamma}\,\mathrm{d}\tau = 1.62\int_0^{10.18} \tau^{3.667}\,\mathrm{d}\tau = 17510\ \mathrm{Pa^3s^{-1}}$$

This agrees with the value found by numerical integration and would give the same value for u.

Note that the values of the apparent viscosity μ_a were not used; they were provided to show that the fluid is strongly shear thinning. If the data were available as values of μ_a at corresponding values of $\dot{\gamma}$, then τ should be calculated as their product. The table of values of $\tau^2\dot{\gamma}$ (Table 3.4) illustrates the fact that flow in the centre makes a small contribution to the total flow: flow in the outer parts of the pipe is most significant.

As mentioned previously, the minus sign in equation 3.17 reflects the fact that the shear rate is negative for flow in a pipe. In the above calculations, the absolute values of $\dot{\gamma}$ and τ have been used and the minus sign has therefore been dropped.

3.4 Wall shear stress–flow characteristic curves and scale-up for laminar flow

When data are available in the form of the flow rate–pressure gradient relationship obtained in a small diameter tube, direct scale-up for flow in larger pipes can be done. It is not necessary to determine the τ–$\dot{\gamma}$ curve with the true value of $\dot{\gamma}$ calculated from the Rabinowitsch–Mooney equation (equation 3.20).

Equation 3.17 shows that the flow characteristic is a unique function of the wall shear stress for a particular fluid:

$$\frac{8u}{d_i} = \frac{4}{\tau_w^3}\int_0^{\tau_w} \tau^2(-\dot{\gamma})\,\mathrm{d}\tau \qquad (3.17)$$

In the case of a Newtonian fluid, substituting $\dot{\gamma} = -\tau/\mu$ in equation 3.17 and evaluating the integral gives

$$\frac{8u}{d_i} = \frac{\tau_w}{\mu} \tag{3.22}$$

Recall that the wall shear rate for a Newtonian fluid in laminar flow in a tube is equal to $-8u/d_i$. In the case of a non-Newtonian fluid in laminar flow, the flow characteristic is no longer equal to the magnitude of the wall shear rate. However, the flow characteristic is still related uniquely to τ_w because the value of the integral, and hence the right hand side of equation 3.17, is determined by the value of τ_w.

If the fluid flows in two pipes having internal diameters d_{i1} and d_{i2} with the same value of the wall shear stress in both pipes, then from equation 3.17 the values of the flow characteristic are equal in both pipes:

$$\frac{8u_1}{d_{i1}} = \frac{8u_2}{d_{i2}} \quad \text{at same } \tau_w \tag{3.23}$$

So the average velocities are related by

$$\frac{u_1}{u_2} = \frac{d_{i1}}{d_{i2}} \tag{3.24}$$

By substituting for u or by writing the flow characteristic as $4Q/\pi r_i^3$, the volumetric flow rates are related by

$$\frac{Q_1}{Q_2} = \left(\frac{d_{i1}}{d_{i2}}\right)^3 \tag{3.25}$$

It is important to appreciate that the same value of τ_w requires different values of the pressure gradient in the two pipes:

$$\tau_w = \frac{d_i \Delta P_f}{4L} \tag{2.5}$$

It is convenient to represent the flow behaviour as a graph of τ_w plotted against $8u/d_i$, as shown in Figure 3.5. In accordance with the above discussion, all data fit a single line for laminar flow. The graph is steeper for turbulent flow and different lines are found for different pipe diameters. (Note that the same would be found for Newtonian flow if the data were plotted in this way. The laminar flow line would be a straight line of gradient μ passing through the origin. For a given value of Re, such as the laminar–turbulent transition value, $8u/d_i$ increases with decreasing d_i.) The plot in Figure 3.5 is not a true flow curve because the flow characteristic is equal to the magnitude of the wall shear rate only in the case of Newtonian laminar flow.

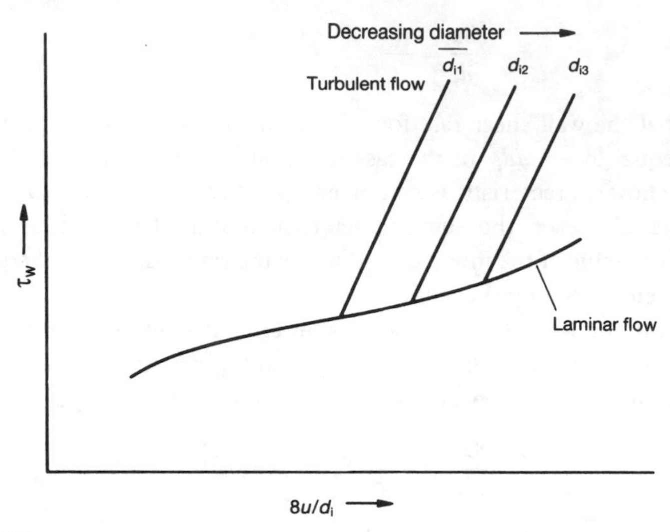

Figure 3.5
Shear stress at the pipe wall against flow characteristic for a non-Newtonian fluid flowing in a pipe

Given a wall shear stress–flow characteristic curve such as that in Figure 3.5, the flow rate–pressure drop relationship can be found for any diameter of pipe provided the flow remains laminar and is within the range of the graph. For example, if it is required to calculate the pressure drop for flow in a pipe of given diameter at a specified volumetric flow rate, the value of the flow characteristic ($8u/d_i = 4Q/\pi r_i^3$) is calculated and the corresponding value of the wall shear stress τ_w read from the graph. The pressure gradient, and hence the pressure drop for a given pipe length, can then be calculated using equation 2.7.

It is found useful to define two quantities K' and n' in order to describe the τ_w–flow characteristic curve. If the laminar flow data are plotted on logarithmic axes as in Figure 3.6, then the gradient of the curve defines the value of n':

$$n' = \frac{\mathrm{d}\ln\tau_w}{\mathrm{d}\ln(8u/d_i)} \tag{3.26}$$

The equation of the tangent can be written as

$$\tau_w = K'\left(\frac{8u}{d_i}\right)^{n'} \tag{3.27}$$

and this defines K'.

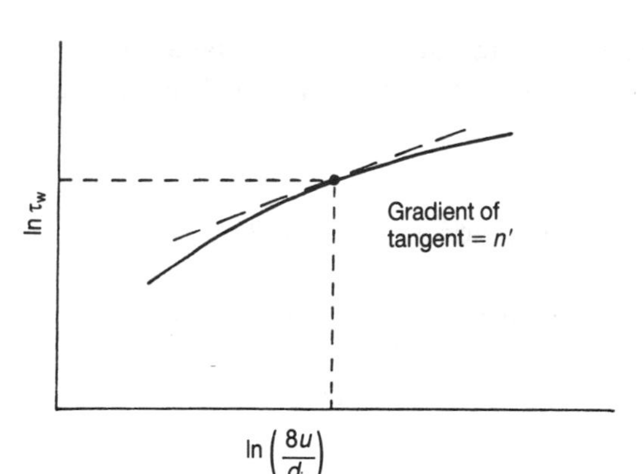

Figure 3.6
Logarithmic plot of wall shear stress against flow characteristic: the gradient at a point defines n′

In general, both K' and n' have different values at different points along the curve. The values should be found at the point corresponding to the required value of τ_w. In some cases, the curve in Figure 3.6 will be virtually straight over the range required and a single value may be used for each of K' and n'. Although equation 3.27 is similar to the equation of a power law fluid, the two must not be confused.

The reason for defining n' in this way can be seen from equation 3.21 where the inverse of the derivative occurs in the correction factor.

Equation 3.20 can be written in terms of n' as

$$-\dot{\gamma}_w = \frac{8u}{d_i}\left(\frac{3}{4} + \frac{1}{4n'}\right) = \frac{8u}{d_i}\left(\frac{3n'+1}{4n'}\right) \tag{3.28}$$

and equation 3.21 as

$$\dot{\gamma}_w = \dot{\gamma}_{wN}\left(\frac{3n'+1}{4n'}\right) \tag{3.29}$$

Equation 3.29 is helpful in showing how the value of the correction factor in the Rabinowitsch–Mooney equation corresponds to different types of flow behaviour. For a Newtonian fluid, $n' = 1$ and therefore the correction factor has the value unity. Shear thinning behaviour corresponds to $n' < 1$ and consequently the correction factor has values greater than unity, showing that the wall shear rate $\dot{\gamma}_w$ is of greater magnitude than the value for Newtonian flow. Similarly, for shear thickening behaviour, $\dot{\gamma}_w$ is of a smaller magnitude than the Newtonian value $\dot{\gamma}_{wN}$. The value of the correction factor varies from 2.0 for $n' = 0.2$ to 0.94 for $n' = 1.3$.

3.5 Generalized Reynolds number for flow in pipes

For Newtonian flow in a pipe, the Reynolds number is defined by

$$Re = \frac{\rho u d_i}{\mu} \tag{1.3}$$

In the case of non-Newtonian flow, it is necessary to use an appropriate apparent viscosity. Although the apparent viscosity μ_a is defined by equation 1.71 in the same way as for a Newtonian fluid, it no longer has the same fundamental significance and other, equally valid, definitions of apparent viscosities may be made. In flow in a pipe, where the shear stress varies with radial location, the value of μ_a varies. As pointed out in Example 3.1, it is the conditions near the pipe wall that are most important. The value of μ_a evaluated at the wall is given by

$$\mu_a = -\frac{\text{shear stress at wall}}{\text{shear rate at wall}} = \frac{\tau_w}{(-\mathrm{d}v_x/\mathrm{d}r)_w} \tag{3.30}$$

Another definition is based, not on the true shear rate at the wall, but on the flow characteristic. This quantity, which may be called the apparent viscosity for pipe flow, is given by

$$\mu_{ap} = \frac{\text{shear stress at wall}}{\text{flow characteristic}} = \frac{\tau_w}{8u/d_i} \tag{3.31}$$

For laminar flow, μ_{ap} has the property that it is the viscosity of a Newtonian fluid having the same flow characteristic as the non-Newtonian fluid when subjected to the same value of wall shear stress. In particular, this corresponds to the same volumetric flow rate for the same pressure gradient in the same pipe. This suggests that μ_{ap} might be a useful quantity for correlating flow rate–pressure gradient data for non-Newtonian flow in pipes. This is found to be the case and it is on μ_{ap} that a generalized Reynolds number Re' is based:

$$Re' = \frac{\rho u d_i}{\mu_{ap}} \tag{3.32}$$

Representing the fluid's laminar flow behaviour in terms of K' and n'

$$\tau_w = K'\left(\frac{8u}{d_i}\right)^{n'} \tag{3.27}$$

The pipe flow apparent viscosity, defined by equation 3.31, is given by

$$\mu_{ap} = \frac{\tau_w}{8u/d_i} = K'\left(\frac{8u}{d_i}\right)^{n'-1} \tag{3.33}$$

When this equation for μ_{ap} is substituted into equation 3.32, the generalized Reynolds number takes the form

$$Re' = \frac{\rho u^{2-n'} d_i^{n'}}{8^{n'-1} K'} \tag{3.34}$$

Use of this generalized Reynolds number was suggested by Metzner and Reed (1955). For Newtonian behaviour, $K' = \mu$ and $n' = 1$ so that the generalized Reynolds number reduces to the normal Reynolds number.

3.6 Turbulent flow of inelastic non-Newtonian fluids in pipes

Turbulent flow of Newtonian fluids is described in terms of the Fanning friction factor, which is correlated against the Reynolds number with the relative roughness of the pipe wall as a parameter. The same approach is adopted for non-Newtonian flow but the generalized Reynolds number is used.

The Fanning friction factor is defined by

$$f = \frac{\tau_w}{\frac{1}{2}\rho u^2} \tag{2.10}$$

For laminar flow of a non-Newtonian fluid, the wall shear stress can be expressed in terms of K' and n' as

$$\tau_w = K'\left(\frac{8u}{d_i}\right)^{n'} \tag{3.27}$$

On substituting for τ_w in equation 2.10, the Fanning friction factor for laminar non-Newtonian flow becomes

$$f = \frac{16}{Re'} \tag{3.35}$$

This is of the same form as equation 2.14 for Newtonian flow and is one reason for using this form of generalized Reynolds number. Equation 3.35

provides another way of calculating the pressure gradient for a given flow rate for laminar non-Newtonian flow, instead of using the methods of Sections 3.3 and 3.4.

3.6.1 *Laminar–turbulent transition*

A stability analysis made by Ryan and Johnson (1959) suggests that the transition from laminar to turbulent flow for inelastic non-Newtonian fluids occurs at a critical value of the generalized Reynolds number that depends on the value of n'. The results of this analysis are shown in Figure 3.7. This relationship has been tested for shear thinning and for Bingham

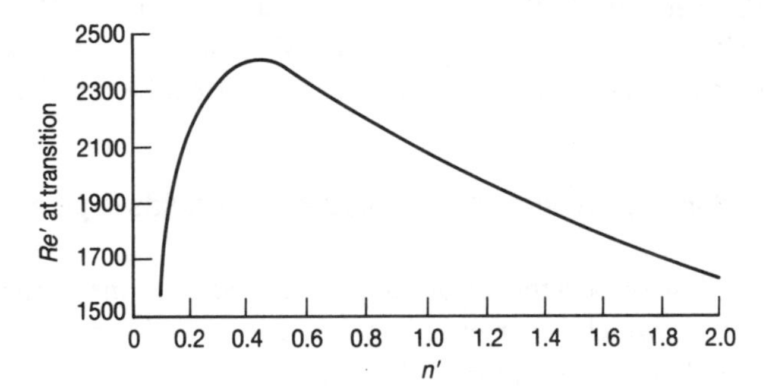

Figure 3.7
Variation of the critical value of the Reynolds number with n'

plastic fluids and has been found to be accurate. Over the range of shear thinning behaviour encountered in practice, $0.2 \leq n' < 1.0$, the critical value of Re' is in the range $2100 \leq n' \leq 2400$.

3.6.2 *Friction factors for turbulent flow in smooth pipes*

Experimental results for the Fanning friction factor for turbulent flow of shear thinning fluids in smooth pipes have been correlated by Dodge and Metzner (1959) as a generalized form of the von Kármán equation:

$$\frac{1}{f^{1/2}} = \frac{4.0}{(n')^{0.75}} \log\left[f^{(1-n'/2)} Re'\right] - \frac{0.40}{(n')^{1.2}} \tag{3.36}$$

This correlation is shown in Figure 3.8. The broken lines represent extrapolation of equation 3.36 for values of n' and Re' beyond those of the measurements made by Dodge and Metzner. More recent studies tend to

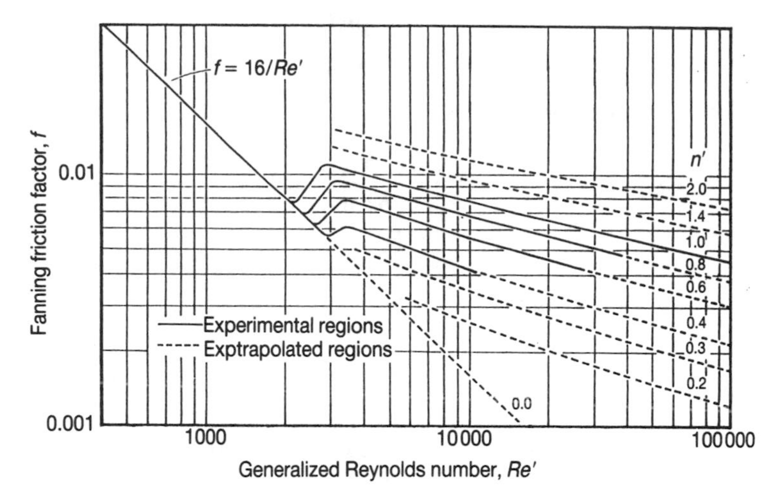

Figure 3.8
Friction factor chart for purely viscous non-Newtonian fluids. (See Friction Factor Charts on page 349.)
Source: D. W. Dodge and A. B. Metzner, *AIChE Journal* **5**, pp. 189–204 (1959)

confirm the findings of Dodge and Metzner but do not significantly extend the range of applicability.

Having determined the value of the friction factor f for a specified flow rate and hence Re', the pressure gradient can be calculated in the normal way using equation 2.13.

Example 3.3

A general time-independent non-Newtonian liquid of density 961 kg/m^3 flows steadily with an average velocity of 2.0 m/s through a tube 3.048 m long with an inside diameter of 0.0762 m. For these conditions, the pipe flow consistency coefficient K' has a value of 1.48 Pa s$^{0.3}$ and n' a value of 0.3. Calculate the values of the apparent viscosity for pipe flow μ_{ap}, the generalized Reynolds number Re' and the pressure drop across the tube, neglecting end effects.

Calculations

Apparent viscosity is given by

$$\mu_{ap} = K'\left(\frac{8u}{d_i}\right)^{n'-1} \tag{3.33}$$

and generalized Reynolds number by

$$Re' = \frac{\rho u d_i}{\mu_{ap}} \tag{3.32}$$

The flow characteristic is given by

$$\frac{8u}{d_i} = \frac{8(2.0 \text{ m/s})}{0.0762 \text{ m}} = 210 \text{ s}^{-1}$$

and

$$\left(\frac{8u}{d_i}\right)^{n'-1} = 210^{(0.3-1.0)} = 0.0237 \text{ s}^{0.7}$$

Hence

$$\mu_{ap} = (1.48 \text{ Pa s}^{0.3})(0.0237 \text{ s}^{0.7}) = \underline{0.0351 \text{ Pa s}}$$

and

$$Re' = \frac{(0.0762 \text{ m})(2.0 \text{ m/s})(961 \text{ kg/m}^3)}{(0.0351 \text{ Pa s})} = \underline{4178}$$

From Figure 3.8, the Fanning friction factor f has a value 0.0047. Therefore the pressure drop is given by

$$\Delta P_f = 4f\left(\frac{L}{d_i}\right)\frac{\rho u^2}{2} = \frac{2fL\rho u^2}{d_i} \tag{2.13}$$

$$= \frac{2(0.0047)(3.048 \text{ m})(961 \text{ kg/m}^3)(2.0 \text{ m/s})^2}{(0.0762 \text{ m})}$$

$$= \underline{1445 \text{ Pa}}$$

3.7 Power law fluids

The methods presented in Sections 3.1 to 3.6 are general and do not require the assumption of any particular flow model. While the flow of power law fluids and Bingham plastics can be treated by those methods, some results specific to these materials will be considered in this and the next sections.

It was mentioned in Section 3.4 that the equation of the tangent to the $\ln\tau_w$–$\ln(8u/d_i)$ curve, ie

$$\tau_w = K'\left(\frac{8u}{d_i}\right)^{n'} \tag{3.27}$$

must not be confused with equation 1.72a defining a power law fluid. The relationship between n' and n, and K' and K will now be demonstrated.

For the conditions at the pipe wall, denoted by the subscript w, the equation of the power law fluid can be written as

$$\tau_w = K(-\dot{\gamma}_w)^n = K\left(-\frac{dv_x}{dr}\right)_w^n \tag{3.37}$$

The minus sign has been placed inside the parentheses recognizing the fact that the shear rate $\dot{\gamma}$ (equal to dv_x/dr) is negative. $\dot{\gamma}_w$ is the *true* shear rate at the wall and is related to the flow characteristic $(8u/d_i)$ by the Rabinowitsch–Mooney equation:

$$-\dot{\gamma}_w = \frac{8u}{d_i}\left(\frac{3n'+1}{4n'}\right) \tag{3.28}$$

Therefore the behaviour of a power law fluid, evaluated at the wall conditions, is given by

$$\tau_w = K\left(\frac{8u}{d_i}\right)^n \left(\frac{3n'+1}{4n'}\right)^n \tag{3.38}$$

Equation 3.38 shows that a plot of $\ln\tau_w$ against $\ln(8u/d_i)$ has a constant gradient of n. Consequently, for a power law fluid

$$n' = n \tag{3.39}$$

Comparing equation 3.38 with equation 3.27 (both with $n' = n$), shows that

$$K' = K\left(\frac{3n+1}{4n}\right) \tag{3.40}$$

3.7.1 *Velocity profile for laminar flow in a pipe*

The velocity profile for steady, fully developed, laminar flow in a pipe can be determined easily by the same method as that used in Example 1.9 but using the equation of a power law fluid instead of Newton's law of viscosity. The shear stress distribution is given by

$$\tau_{rx} = \frac{r}{2}\left(\frac{\Delta P_f}{L}\right) \tag{2.4}$$

For a power law fluid

$$\tau_{rx} = K\left(-\frac{dv_x}{dr}\right)^n \tag{3.41}$$

Combining equations 2.4 and 3.41 gives the velocity gradient:

$$\frac{dv_x}{dr} = -\left(\frac{\Delta P_f}{2LK}\right)^{1/n} r^{1/n} \tag{3.42}$$

On integrating equation 3.42 with the boundary conditions $v_x = 0$ at $r = r_i$, the velocity profile is found as

$$v_x = \left(\frac{\Delta P_f}{4KL/d_i}\right)^{1/n}\left(\frac{n}{n+1}\right) r_i \left[1 - \left(\frac{r}{r_i}\right)^{(n+1)/n}\right] \tag{3.43}$$

The volumetric flow rate is readily calculated from

$$Q = 2\pi \int_0^{r_i} r v_x \, dr \tag{1.62}$$

with v_x given by equation 3.43. The result is

$$Q = \left(\frac{n}{3n+1}\right)\left(\frac{\Delta P_f}{4KL/d_i}\right)^{1/n} \pi r_i^3 \tag{3.44}$$

and the volumetric average velocity u is equal to $Q/\pi r_i^2$. Consequently, the velocity profile can be expressed as

$$\frac{v_x}{u} = \left(\frac{3n+1}{n+1}\right)\left[1 - \left(\frac{r}{r_i}\right)^{(n+1)/n}\right] \tag{3.45}$$

Figure 3.9 shows velocity profiles for various values of n. The profiles for the limiting values $n = 0$ and $n = \infty$ are of interest but it should be remembered that the behaviour of real fluids lies in the approximate range $0.2 < n < 1.3$.

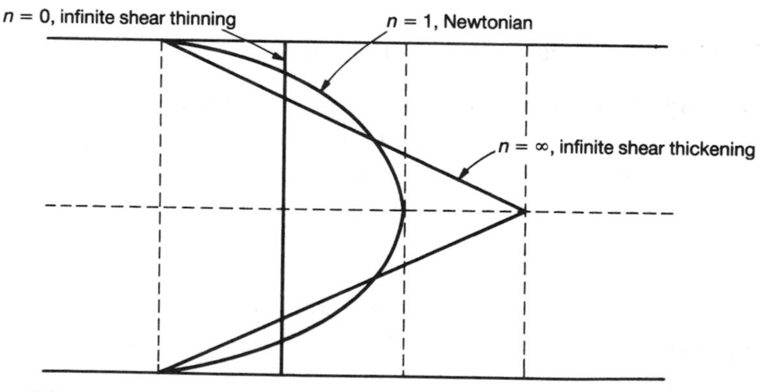

Figure 3.9
Velocity profiles for power law fluids showing the effect of the power law index, n

It can be seen from equaton 3.45 that

$$\frac{v_{max}}{u} = \frac{3n+1}{n+1} \tag{3.46}$$

All the results for a power law fluid reduce to the corresponding ones for a Newtonian fluid on putting $n = 1$ and $K = \mu$.

3.7.2 *Velocity profile for turbulent flow in a pipe*

Dodge and Metzner (1959) deduced the velocity profile from their measurements of flow rate and pressure gradient for turbulent flow of power law fluids in pipes. For the turbulent core, the appropriate equation is

$$v^+ = \frac{5.66}{n^{0.75}} \log y^+ - \frac{0.566}{n^{1.2}} + \frac{3.475}{n^{0.75}} \times \left[1.960 + 0.815n - 1.628n \log\left(\frac{3n+1}{n}\right)\right] \tag{3.47}$$

where v^+ and y^+ are respectively a dimensionless velocity and distance from the wall, defined as follows:

$$v^+ = \frac{\bar{v}_x}{v_*} \tag{2.56}$$

and

$$y^+ = \frac{\rho v_*^{(2-n)} y^n}{K} \tag{3.48}$$

where

$$v_* = \sqrt{\frac{\tau_w}{\rho}} \tag{2.54}$$

The coefficients in equation 3.47 are the corrected values given by Skelland (1967).

For Newtonian fluids, $n = 1$ and $K = \mu$ and equation 3.47 reduces to

$$v^+ = 5.66 \log y^+ + 5.7 \tag{3.49a}$$

or

$$v^+ = 2.46 \ln y^+ + 5.7 \tag{3.49b}$$

This is in excellent agreement with equation 2.70, considering that equation 3.47 was derived from flow rate and pressure gradient measurements rather than from direct measurement of the velocity profiles.

For Newtonian fluids the velocity profile in the viscous sublayer adjacent to the wall is

$$v^+ = y^+ \tag{2.58}$$

The corresponding equation for power law fluids is [Dodge and Metzner (1959)]

$$v^+ = (y^+)^{1/n} \tag{3.50}$$

3.7.3 *Expansion and contraction losses for power law fluids*

For laminar flow of a power law fluid through a sudden expansion in a pipe of circular cross section, it is easily shown [Skelland (1967)] that the pressure drop is given by

$$\Delta P_e = \rho u_1^2 \left(\frac{3n+1}{2n+1}\right)\left[\frac{n+3}{2(5n+3)}\left(\frac{S_1}{S_2}\right)^2 - \left(\frac{S_1}{S_2}\right) + \frac{3(3n+1)}{2(5n+3)}\right] \tag{3.51}$$

In equation 3.51, ρ is the density and n the power law flow behaviour index of the fluid. S_1 and S_2 are the cross-sectional areas of the smaller and larger pipes respectively and u_1 the volumetric average velocity in the smaller pipe.

On putting $n = 0$, equation 3.51 reduces to

$$\Delta P_e = \frac{\rho u_1^2}{2}\left[1 - \left(\frac{S_1}{S_2}\right)\right]^2 = \frac{\rho u_1^2}{2}\left[1 - \left(\frac{d_{i1}}{d_{i2}}\right)^2\right]^2 \tag{3.52}$$

This is identical to equation 2.24 for turbulent flow of a Newtonian fluid because $n = 0$ corresponds to a flat velocity profile and this is a good approximation for turbulent flow.

Measurements suggest that the pressure loss for laminar flow of power law fluids through a sudden contraction is not significantly different from that for Newtonian flow [Skelland (1967)]. This statement applies to inelastic power law fluids: in the case of elastic liquids, very high contraction pressure losses occur as discussed in Section 3.10.

As would be expected from equation 3.52, expansion and contraction losses for turbulent flow of power law fluids are similar to those for turbulent flow of Newtonian fluids.

3.8 Pressure drop for Bingham plastics in laminar flow

The behaviour of a Bingham plastic is described by

$$\tau - \tau_y = -\beta\dot{\gamma} \qquad \text{for } \tau \geq \tau_y$$

and (1.73)

$$\dot{\gamma} = 0 \qquad \text{for } \tau < \tau_y$$

As the shear stress for flow in a pipe varies from zero at the centre-line to a maximum at the wall, genuine flow, ie deformation, of a Bingham plastic occurs only in that part of the cross section where the shear stress is greater than the yield stress τ_y. In the part where $\tau < \tau_y$ the material remains as a solid plug and is transported by the genuinely flowing outer material.

As part of the Rabinowitsch–Mooney analysis, it was shown that the volumetric flow rate can be written in terms of the shear stress distribution:

$$Q = \frac{\pi r_i^3}{\tau_w^3} \int_0^{\tau_w} \tau^2(-\dot{\gamma})\,d\tau \tag{3.16}$$

For a Bingham plastic, there is a change in the flow behaviour at $\tau = \tau_y$ and the range of integration must be split into two parts:

$$Q = \frac{\pi r_i^3}{\tau_w^3}\left[\int_0^{\tau_y} \tau^2(-\dot{\gamma})\,d\tau + \int_{\tau_y}^{\tau_w} \tau^2(-\dot{\gamma})\,d\tau\right] \tag{3.53}$$

The first integral vanishes because $\dot{\gamma} = 0$ for $0 \leq \tau \leq \tau_y$.

Substituting for $\dot{\gamma}$ from equation 1.73, equation 3.53 becomes

$$Q = \frac{\pi r_i^3}{\tau_w^3} \int_{\tau_y}^{\tau_w} \frac{\tau^2(\tau - \tau_y)}{\beta}\,d\tau \tag{3.54}$$

Evaluating this integral and writing the result in terms of the flow characteristic gives the well-known Buckingham equation:

$$\frac{8u}{d_i} = \frac{4Q}{\pi r_i^3} = \frac{\tau_w}{\beta}\left[1 - \frac{4}{3}\left(\frac{\tau_y}{\tau_w}\right) + \frac{1}{3}\left(\frac{\tau_y}{\tau_w}\right)^4\right] \tag{3.55}$$

The value of τ_w is found in the usual way:

$$\tau_w = \frac{\Delta P_f}{4L/d_i} \tag{2.5}$$

These equations allow u or Q to be calculated if τ_w (and therefore d_i) are specified.

Govier (1959) developed a method for solving equation 3.55 for τ_w and the pressure gradient for a given value of the Reynolds number. He defined a modified Reynolds number Re_B in terms of β:

$$Re_B = \frac{\rho u d_i}{\beta} \tag{3.56}$$

It was pointed out in Section 1.12 that the coefficient of rigidity β is equal to the apparent viscosity at infinite shear rate. Govier also defined a dimensionless yield number Y by

$$Y = \frac{\tau_y d_i}{\beta u} \tag{3.57}$$

Dividing equation 3.57 by equation 3.56 and using the definition of the Fanning friction factor (equation 2.10) gives

$$\frac{Y}{Re_B} = \frac{\tau_y}{\rho u^2} = \frac{f\tau_y}{2\tau_w} \tag{3.58}$$

allowing τ_y/τ_w to be replaced by $2Y/fRe_B$ in equation 3.55.

In addition

$$\frac{8u}{d_i} \bigg/ \frac{\tau_w}{\beta} = \frac{8u\beta}{d_i\tau_w} = \frac{8u\beta}{d_i(\frac{1}{2}\rho u^2)f} = \frac{16\beta}{\rho u d_i f} = \frac{16}{fRe_B} \tag{3.59}$$

Using these results, equation 3.55 can be written as

$$\frac{1}{fRe_B} = \frac{1}{16} - \frac{Y}{6(fRe_B)} + \frac{Y^4}{3(fRe_B)^4} \tag{3.60}$$

Note that equation 3.60 reduces to $f = 16/Re$ for $Y = 0$.
The product fRe_B is a unique function of the yield number [Hedström (1952)] and Govier (1959) has tabulated corresponding values of fRe_B and Y. A slightly different presentation is in terms of the Hedström number He, which is given by

$$He = Re_B Y = \frac{d_i^2 \rho \tau_y}{\beta^2} \tag{3.61}$$

Figure 3.10 shows a friction factor – Reynolds number chart for Bingham plastics at various values of the Hedström number. The turbulent flow line is that for Newtonian behaviour and is followed by some Bingham plastics with low values of the yield stress [Thomas (1962)].

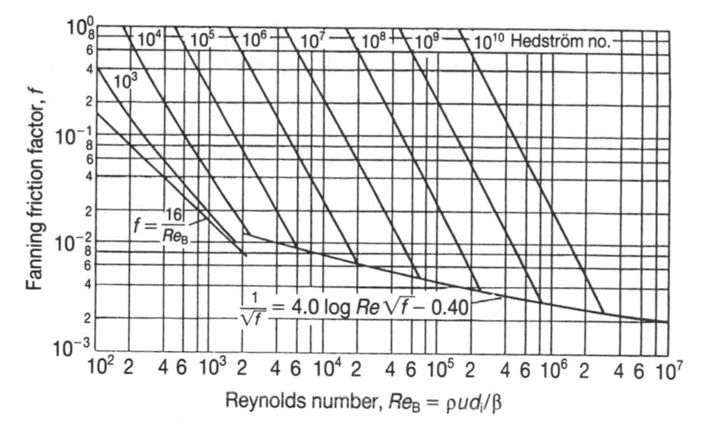

Figure 3.10

Friction factor chart for laminar flow of Bingham plastic materials. (See Friction Factor Charts on page 349.)

Source: D. G. Thomas, *AIChE Journal* **6**, pp. 631–9 (1960)

3.9 **Laminar flow of concentrated suspensions and apparent slip at the pipe wall**

The flow properties of suspensions are complex. The apparent viscosity at a given shear rate increases with increasing solids concentration and rises extremely rapidly when the volume fraction of solids reaches about 50 per cent. The flow properties also depend on the particle size distribution and the particle shape, as well as the flow properties of the suspending liquid.

With some concentrated suspensions of solid particles, particularly those in which the liquid has a relatively low viscosity, the suspension appears to slip at the pipe wall or at the solid surfaces of a viscometer. Slip occurs because the suspension is depleted of particles in the vicinity of the solid surface. In the case of concentrated suspensions, the main reason is probably that of physical exclusion: if the suspension at the solid surface were to have the same spatial distribution of particles as that in the bulk, some particles would have to overlap the wall. As a result of the lower concentration of particles in the immediate vicinity of the wall, the effective viscosity of the suspension near the wall may be significantly lower than that of the bulk and consequently this wall layer may have an extremely high shear rate. If this happens, the bulk material appears to slip on this lubricating layer of low viscosity material.

Although this type of flow can be treated by a multiple layer model, the usual approach is to treat the suspension as being uniform up to the wall but assuming that slip occurs at the wall. In reality there must be a very high velocity gradient but the thickness of the layer over which this occurs is unknown, although it may be expected to be of the order of the particle diameter. By supposing slip to occur, this high velocity gradient over a small distance is replaced by a discontinuity in the velocity at the wall.

3.9.1 *Modification of the Rabinowitsch–Mooney analysis*

The occurrence of slip invalidates all normal analyses because they assume that the velocity is zero at the wall. Returning to the Rabinowitsch–Mooney analysis, the total volumetric flow rate for laminar flow in a pipe is given by

$$Q = 2\pi\left\{\left[\frac{r^2 v_x}{2}\right]_0^{r_i} + \int_0^{r_i} \frac{r^2}{2}\left(-\frac{\mathrm{d}v_x}{\mathrm{d}r}\right)\mathrm{d}r\right\} \tag{3.14}$$

If slip occurs, with a slip velocity v_s at the wall, then the first term on the right hand side of equation 3.14 does not vanish as before:

$$\left[\frac{r^2 v_x}{2}\right]_0^{r_i} = \frac{r_i^2 v_s}{2} \tag{3.62}$$

The total measured flow rate Q is therefore given by

$$Q = \pi r_i^2 v_s + \pi \int_0^{r_i} r^2\left(-\frac{\mathrm{d}v_x}{\mathrm{d}r}\right)\mathrm{d}r \tag{3.63}$$

As before, the variable of integration can be changed and equation 3.63 can be written as

$$Q = \pi r_i^2 v_s + \frac{\pi r_i^3}{\tau_w^3} \int_0^{\tau_w} \tau^2(-\dot{\gamma})\,\mathrm{d}\tau \tag{3.64}$$

Equation 3.64 shows that the total discharge rate Q that is measured consists of normal, genuine flow given by the integral term, plus the extra discharge due to slip. The slip term is simply the slip velocity v_s multiplied by the cross-sectional area of the pipe.

Writing equation 3.64 in terms of the flow characteristic gives

$$\frac{4Q}{\pi r_i^3} = \frac{4v_s}{r_i} + \frac{4}{\tau_w^3} \int_0^{\tau_w} \tau^2(-\dot{\gamma})\,\mathrm{d}\tau \tag{3.65}$$

or

$$\frac{8u}{d_i} = \frac{8v_s}{d_i} + \frac{4}{\tau_w^3} \int_0^{\tau_w} \tau^2(-\dot{\gamma})\,d\tau \qquad (3.66)$$

Equations 3.65 and 3.66 reduce to equation 3.17 when $v_s = 0$.

It is important to remember that in these equations Q is the measured total flow rate and u is calculated from Q.

When trying to determine the flow behaviour of a material suspected of exhibiting wall slip, the procedure is first to establish whether slip occurs and how significant it is. The magnitude of slip is then determined and by subtracting the 'flow' due to slip from the measured flow rate, the genuine flow rate can be determined. The standard Rabinowitsch–Mooney equation can then be used with the corrected flow rates to determine the τ_w-$\dot{\gamma}_w$ curve. Alternatively, the results can be presented as a plot of τ_w against the corrected flow characteristic, where the latter is calculated from the corrected value of the flow rate.

In order to determine whether slip occurs with a particular material, it is essential to make measurements with tubes of various diameters. In equation 3.66, the value of the integral term is a function of the wall shear stress only. Thus, in the absence of wall slip, the flow characteristic $8u/d_i$ is a unique function of τ_w. However, if slip occurs, the term $8v_s/d_i$ will be different for different values of d_i at the same value of τ_w, as shown in Figure 3.11. It is clear from equation 3.66 that for a given value of the slip velocity v_s, the effect of slip is greater in tubes of smaller diameter. If the effect of slip is dominant, that is the bulk of the material experiences negligible shearing, then it can be seen from equation 3.66 that on a plot of

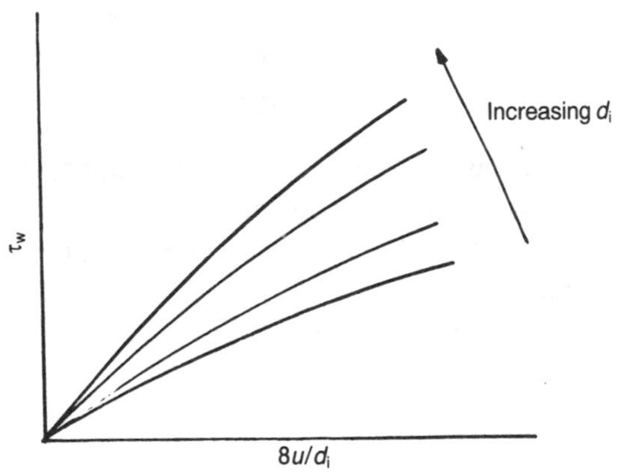

Figure 3.11

The varying effect of wall slip in tubes of different diameters

u against τ_w, all the results for different tube diameters will fall on, or close to, a single line. If this is the case, the genuine flow behaviour cannot be determined but for pipe flow calculations the u–τ_w plot gives the information that is required for scale-up, assuming that the slip behaviour is the same in the large pipe.

Having established that wall slip occurs but is not dominant, the procedure is to estimate the value of v_s and hence calculate a corrected flow rate by subtracting the slip 'flow' from the measured flow rate. In general it is found that the slip velocity increases with τ_w and decreases with d_i, although in some cases v_s is independent of d_i. Consequently, it can be seen from equation 3.66 that the effect of slip decreases as d_i increases, and becomes negligible at very large diameters.

It is convenient to divide equation 3.66 by τ_w throughout:

$$\left(\frac{8u}{d_i}\right)\Big/\tau_w = \frac{8v_s}{d_i\tau_w} + \frac{4}{\tau_w^4}\int_0^{\tau_i} \tau^2(-\dot{\gamma})\,d\tau \tag{3.67}$$

The quantity on the left hand side of equation 3.67 is the reciprocal of the apparent viscosity for pipe flow μ_{ap} and is often called the apparent fluidity.

Schofield and Scott Blair (1930) assumed that the slip velocity was a linear function of the wall shear stress τ_w but independent of d_i. Similarly, Mooney (1931) assumed that the slip velocity can be written as

$$v_s = C_s\tau_w \tag{3.68}$$

where C_s is called the slip coefficient. The slip term in equation 3.67 can then be written as

$$\frac{8v_s}{d_i\tau_w} = \frac{8C_s}{d_i} \tag{3.69}$$

If this type of behaviour is followed, a plot of apparent fluidity against $1/d_i$ at constant τ_w will be linear as shown in Figure 3.12. The gradient of a line is equal to $8C_s$ for the corresponding value of τ_w; hence the value of v_s can be calculated for that value of τ_w.

The flow Q_s due to slip is then given by

$$Q_s = \frac{\pi d_i^2}{4} v_s \tag{3.70}$$

and the corrected flow rate Q_c can be calculated from the measured flow rate Q:

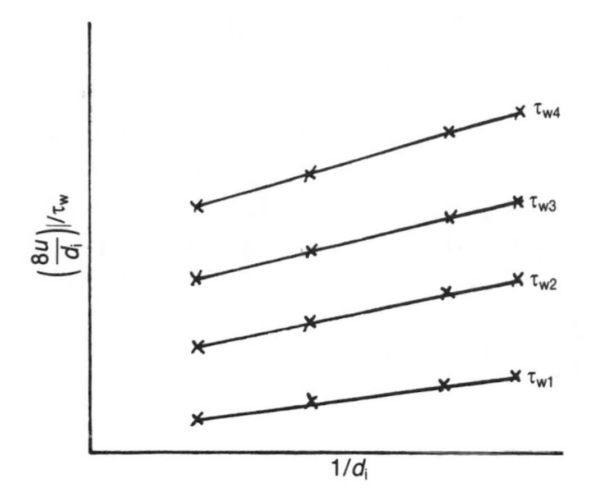

Figure 3.12
Plot of apparent fluidity against $1/d_i$ to determine the slip velocity (Mooney's method)

$$Q_c = Q - Q_s \tag{3.71}$$

The corrected volumetric average velocity u_c is given by

$$u_c = \frac{Q_c}{\pi d_i^2/4} \tag{3.72}$$

This must be done for each of a range of values of the wall shear stress τ_w. The standard Rabinowitsch–Mooney equation can then be used with the corrected values of u_c:

$$-\dot{\gamma}_w = \frac{8u_c}{d_i}\left[\frac{3}{4} + \frac{1}{4}\frac{\mathrm{d}\ln(8u_c/d_i)}{\mathrm{d}\ln\tau_w}\right] \tag{3.73}$$

This enables the true τ_w–$\dot{\gamma}_w$ curve to be determined.

Alternatively, the results may be presented as a plot of τ_w against the corrected flow characteristic $8u_c/d_i$.

Mooney's method has been modified in various ways to allow for the observation that, with many suspensions, the slip velocity depends on the tube diameter as well as the wall shear stress. Jastrzebski (1967) deduced that, for certain kaolinite–water suspensions, v_s was inversely proportional to d_i. Thus a modified slip coefficient C_J may be defined by

$$v_s = C_J\tau_w/d_i \tag{3.74}$$

Thus, the slip term in equation 3.67 can be written as

$$\frac{8v_s}{d_i\tau_w} = \frac{8C_J}{d_i^2} \tag{3.75}$$

If this type of behaviour occurs, a plot of apparent fluidity against $1/d_i^2$ at constant τ_w will be a straight line. The gradient of the line is equal to $8C_J$ for the corresponding value of τ_w. Hence, v_s can be calculated from equation 3.74 and then the corrected flow rate as before.

Several workers [see for example Cheng (1984)] have generalized Jastrzebski's method, writing the slip term in equation 3.67 in the form C/d_i^m. A suitable value of m is sought so that a plot of apparent fluidity against $1/d_i^m$ is a straight line.

3.9.2 *Scale up*

In principle, scale up to larger pipes is straightforward provided the basic tube flow viscometric measurements are available. If Mooney's method has been found satisfactory, the plot of apparent fluidity against $1/d_i$ can be used to extrapolate the measurements to the value of $1/d_i$ corresponding to the pipe diameter. Similarly, if Jastrzebski's method or its generalization has been used, the corresponding plot can be used in the same way. The value of $(8u/d_i)/\tau_w$ is read off the graph for the pipe diameter and a selected value of τ_w. In this way, a curve of u against τ_w can be constructed for the specified pipe diameter. This method simultaneously scales the genuine flow and that due to slip.

The other approach is to scale up the genuine flow, then add the slip flow for the appropriate pipe diameter. Scale up of the genuine flow can be done as described in Section 3.3 or Section 3.4. In order to assess the flow due to wall slip in the pipe, it is necessary to have information about the variation of v_s with τ_w and d_i unless it is assumed that the pipe is large enough for the effect of slip to be negligible. If slip velocity data are available, implying that the apparent fluidity plots are also available, then it would be easier to use these plots directly.

It is implicit in these methods that the wall slip behaviour in the pipe is similar to that in the tubes. There is evidence [Fitzgerald (1990)] that the wall roughness can have a dramatic effect on the flow of some suspensions, making the assumption of similar slip behaviour very dangerous.

Similar problems occur when using rotational viscometers and it is usually impracticable to vary the gap sufficiently to determine the effect of slip. An approach often adopted is to roughen the surfaces or even to use vanes in the hope of eliminating slip. It is possible that these methods may allow a good estimate of the genuine bulk flow behaviour to be obtained but use of these methods for calculating the flow rate in a pipe requires great caution, with the possible exception of very large pipes in which the effect of slip is negligible. The safest procedure is to make measurements in pipes as large as possible, made of the same material and having the same wall characteristics as the full-scale pipe.

Not all suspensions will exhibit wall slip. Concentrated suspensions of finely ground coal in water have been found to exhibit wall slip [Fitzgerald (1990)]. This is to be expected because the coal suspension has a much higher apparent viscosity than the water. In contrast, when the liquid is a very viscous gum, the addition of solids may have a relatively small effect. In this case, the layer at the wall will behave only marginally differently from the material in the bulk.

3.10 Viscoelasticity

As noted in Chapter 1, viscoelastic fluids exhibit a combination of solid-like and liquid-like behaviour. Even a simple analysis of viscoelastic effects in process plant is beyond the scope of this book. This section is restricted to an outline of practical implications of elastic effects and a demonstration of the fact that viscoelastic liquids exhibit stronger elastic behaviour as the deformation rate is increased.

3.10.1 *Practical manifestations of viscoelastic behaviour*

When a viscoelastic liquid is sheared, for example between two parallel plates as in Figure 1.12, large unequal normal stress components are generated. This is in contrast to the behaviour of a purely viscous fluid where the three normal stress components τ_{xx}, τ_{yy} and τ_{zz} are all equal to the pressure. For the viscoelastic liquid, the differences among the normal stress components are equivalent to a tension along the streamlines (the x-direction in Figure 1.12) and the opposite, an outward directed force, in the direction of the velocity gradient (the y-direction in Figure 1.12). This is a manifestation of elastic behaviour; the solid in Figure 1.10 would generate similar normal stress components. Some startling effects occur when the normal stress differences are not negligible compared with the shear stress components.

When a viscoelastic liquid flows through a tube, the normal stress differences cause the liquid to be under an axial tension while a normal stress pushes radially outwards. Consequently, when the liquid emerges from the end of the tube it swells under the influence of the normal stresses. This phenomenon, known as die swell, is in contrast to the contraction of the Newtonian liquid under similar conditions (see example 1.5).

Another well-known phenomenon is the Weissenberg effect, which occurs when a long vertical rod is rotated in a viscoelastic liquid. Again, the shearing generates a tension along the streamlines, which are circles centred on the axis of the rod. The only way in which the liquid can respond is to flow inwards and it therefore climbs up the rod until the hydrostatic head balances the force due to the normal stresses.

A less well known effect occurs in open channel flow of a viscoelastic liquid, when the normal stress differences cause the free surface to bow upwards in the centre [Tanner (1985)].

Of greater engineering importance are the various consequences of the extremely high resistance that viscoelastic liquids exhibit to being stretched. In addition to fibre spinning, in which the fibres are deliberately stretched by the take-up spool, stretching occurs in converging flow such as that into a die or a pipe, and in calendering. For flow of a viscoelastic liquid into a tube, this resistance to stretching is manifest as an excessively high entrance pressure drop, which must be allowed for when determining the apparent viscosity using a tubular viscometer. The entrance pressure drop, after subtracting the known pressure drop for an inelastic liquid, may be used as a rough measure of elasticity.

When a viscoelastic fluid flows into a sudden contraction, it flows only from a central conical region, which is surrounded by a large toroidal vortex [Oliver and Bragg (1973)]. This is in contrast to the behaviour of a Newtonian fluid where the flow is from all points upstream of the contraction. In flowing only from the slender cone, the viscoelastic fluid adopts a flow that reduces the rate of stretching. This is important in the extrusion of molten polymers where, when the flow rate is too high, the axial tension generated is so great that it momentarily interrupts the flow causing the phenomenon called melt fracture.

Consider a stream of liquid that is subject to a purely elongational flow in the x-direction. The elongational rate of strain may be defined as the velocity gradient in the direction of flow, ie $\mathrm{d}v_x/\mathrm{d}x$. Now consider the case in which the elongational strain rate is constant:

$$\frac{\mathrm{d}v_x}{\mathrm{d}x} = k \tag{3.76}$$

Integrating

$$v_x = kx + C \tag{3.77}$$

If v_x is negligibly small at $x = 0$, then $C = 0$. Noting that $v_x = \mathrm{d}x/\mathrm{d}t$, equation 3.77 may be integrated to give the position x of a material point:

$$x = \mathrm{e}^{kt} \tag{3.78}$$

Thus, even when the elongation rate, as defined by equation 3.76, is constant, the separation of two material points increases exponentially with time. As stress relaxation occurs exponentially, it is clear that at high elongation rates the stress will increase very rapidly. In a purely viscous liquid the stress relaxes instantaneously and consequently this high resistance to stretching does not occur.

Elongational flow is a very severe form of deformation, partly because of the high rate of stretching of elements of the material but also because fluid elements do not rotate in pure elongation. In shearing, the velocity gradient causes fluid elements to rotate thus 'evening out' the force on molecules. By superimposing shearing on stretching, the effect of the stretching can be made less severe: this can be done by using a tapered die.

The high resistance to stretching is exhibited even by dilute polymer solutions [Metzner and Metzner (1970), Bragg and Oliver (1973)] and is believed to play an important role in the phenomenon of drag reduction. For example, if a polymer such as poly(ethylene oxide) or polyacrylamide is dissolved in water at a concentration of only a few parts per million by weight, it is found that the friction factor for turbulent flow of the solution is lower than that of water under comparable conditions, as shown in Figure 3.13. At these low concentrations it is impossible to detect any change in viscosity from that of water. An elongational viscosity μ_e can be defined using the normal stress and the elongation rate $\mathrm{d}v_x/\mathrm{d}x$ and it can be shown that for a Newtonian liquid $\mu_e = 3\mu$. However, for dilute drag reducing solutions, the elongational viscosity can be as high as 1000 times the shear viscosity, and for concentrated solutions the ratio may reach 10000. It has been found that with these drag reducing solutions the flow in the buffer zone is modified, both in terms of the mean velocity profile and the distribution of velocity fluctuations. The viscous sublayer is not modified ($v^+ = y^+$ but note that τ_w on which these quantities are based is different for a given flow rate). In Chapter 1 it was explained that the buffer zone, or generation zone, is where turbulence originates and that bursts appear to be important. It is thought that the stretching flow in the burst process is inhibited by the elastic nature of the solution.

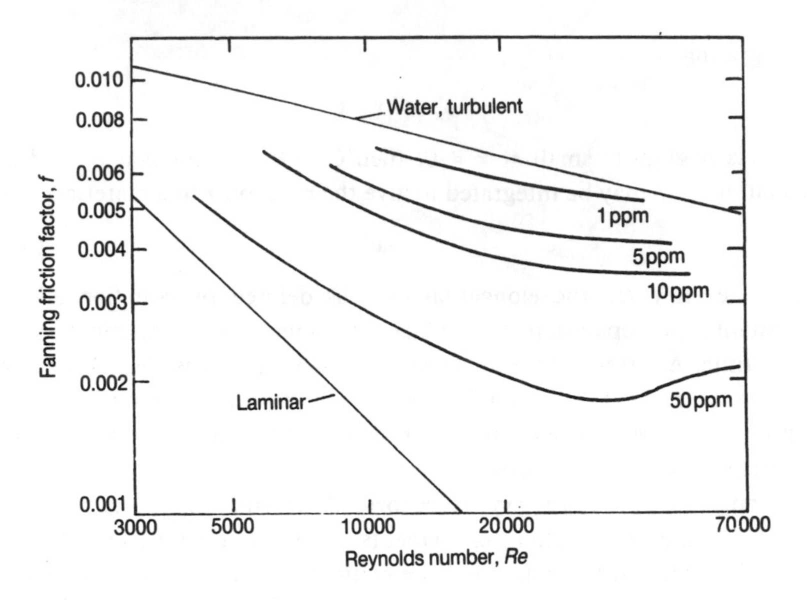

Figure 3.13
Friction factors for turbulent flow of dilute solutions of poly ethylene oxide
Source: R. W. Patterson and F. H. Abernathy, *Journal of Fluid Mechanics*, **51**, pp. 177–85 (1972)

Drag reducing polymers are susceptible to degradation and consequently find only limited application. However, suspensions of fibres, particularly of asbestos, exhibit drag reducing properties and may be more suitable for prolonged use.

An excellent discussion of the effects of drag reducing materials on turbulence has been given by McComb (1990).

3.10.2 *Response of a Maxwell fluid to oscillatory shearing*

In Chapter 1 it was pointed out that the Maxwell fluid is a very simple model of the first order effects observed with viscoelastic liquids. The equation of a Maxwell fluid is

$$\tau + \lambda\dot{\tau} = -\mu\dot{\gamma} \tag{1.80}$$

where $\lambda = \mu/G$ is known as the relaxation time. The Maxwell fluid is a combination of the behaviour of a Newtonian liquid of viscosity μ and a Hookean elastic solid with modulus G.

As noted in Chapter 1, purely viscous behaviour corresponds to $\lambda = 0$, while purely elastic behaviour is approached as $\lambda \to \infty$. It will now be shown that the response of a viscoelastic fluid to unsteady shearing depends on how rapidly it is deformed compared with the rate at which it can relax.

Imagine a Maxwell liquid placed between two parallel plates and sheared by moving the upper plate in its own plane. However, instead of moving the plate at a constant velocity as discussed in Chapter 1, let the displacement of the plate vary sinusoidally with time, ie the plate undergoes simple harmonic motion. If the maximum displacement of the upper plate is X and the distance between the plates is h, then the amplitude A of the shear strain in the liquid is given by

$$A = \frac{X}{h} \tag{3.79}$$

At time t the shear strain γ is therefore

$$\gamma = A \sin \omega t \tag{3.80}$$

where ω is the angular frequency of oscillation. The rate of strain $\dot{\gamma}$ is then

$$\dot{\gamma} = A\omega \cos \omega t \tag{3.81}$$

Note that the strain and rate of strain are out of phase.

Provided ω is not too high, inertia is negligible and the fluid's motion is dominated by viscous stresses. Substituting for $\dot{\gamma}$ in equation 1.80 gives

$$\tau + \lambda\dot{\tau} = -\mu A\omega \cos \omega t \tag{3.82}$$

The shear stress component is τ_{yx} but the subscripts have been omitted for brevity.

Equation 3.82 can be integrated easily by introducing the integrating factor $e^{t/\lambda}$. Thus

$$\lambda e^{t/\lambda}\tau = \mu A\omega \int e^{t/\lambda} \cos \omega t \, dt \tag{3.83}$$

This is a standard integral of the form [Dwight (1961)]

$$\int e^{ax} \cos nx \, dx = \frac{e^{ax}}{a^2 + n^2}(a \cos nx + n \sin nx) + C \tag{3.84}$$

Thus, integration of equation 3.82 leads to

$$\tau = -\frac{\mu A\omega}{1 + \lambda^2\omega^2}(\cos \omega t + \lambda\omega \sin \omega t) \tag{3.85}$$

Equation 3.85 is valid after initial transients have died away.

Comparing equation 3.85 with equations 3.80 and 3.81 shows that the first term in parentheses is in phase with the rate of strain, while the second term is in phase with the strain. Thus the first term represents the viscous part of the fluid's response and the second term the elastic part. It can be seen from equation 3.85 that

$$\tau \to -\mu A \omega \cos \omega t \qquad \text{as } \lambda\omega \to 0 \tag{3.86}$$

and

$$\tau \to -\frac{\mu A}{\lambda} \sin \omega t = -GA \sin \omega t \qquad \text{as } \lambda\omega \to \infty \tag{3.87}$$

Thus purely viscous behaviour is approached as $\lambda\omega \to 0$ and purely elastic behaviour as $\lambda\omega \to \infty$.

It is helpful here to introduce the Deborah number De defined by

$$De = \frac{\text{characteristic time of fluid}}{\text{characteristic time of flow}} \tag{3.88}$$

The fluid's relaxation time λ is the characteristic time of the fluid and, for oscillatory shearing, ω^{-1} can be taken as a measure of the characteristic time of the flow process, so $De = \lambda\omega$. Thus, viscous behaviour occurs when the Deborah number is low, reflecting the fact that the fluid is able to relax. When the Deborah number is high, elastic behaviour is observed because the fluid is unable to relax sufficiently quickly.

In practical flow processes, the characteristic time of the flow will be of the order of a characteristic distance in the direction of flow divided by a characteristic velocity. For example, in flow through a die or into a tube, the characteristic distance will be a small multiple of the diameter d_i of the die or tube and the characteristic velocity can be taken as the volumetric average velocity u based on d_i. Thus the Deborah number can be taken as $\lambda/(d_i/u)$. This will usually be a high value, for a molten polymer for example. In contrast, flow through a long tube or in the barrel of an extruder will have a low value of the Deborah number. Elastic effects will be important in the former case but not in the latter. As a result it is possible to measure the apparent viscosity of viscoelastic fluids in long tubes, provided the flow rate is not too high. Relaxation times of molten polymers are generally in the range 10^{-2} to 10 seconds.

Oscillatory shearing is used to characterize viscoelastic fluids using coaxial cylinders or cone and plate instruments.

It is interesting to consider the response of a Maxwell fluid to an arbitrary shear rate history. Denoting the shear rate as $\dot{\gamma}(t)$, an arbitrary function of time, the equivalent of equation 3.83 is

$$\lambda e^{t/\lambda}\tau = -\mu \int_{-\infty}^{t} e^{t/\lambda}\dot{\gamma}(t)\,dt \tag{3.89}$$

where the shearing at all times in the past is considered. In the integral of equation 3.89, t is a dummy variable so it can be changed to another variable t' without changing the value of the integral provided that the upper limit remains unchanged:

$$\lambda e^{t/\lambda}\tau = -\mu \int_{-\infty}^{t} e^{t'/\lambda}\dot{\gamma}(t')\,dt' \tag{3.90}$$

Thus

$$\tau = -G \int_{-\infty}^{t} e^{-(t-t')/\lambda}\dot{\gamma}(t')\,dt' \tag{3.91}$$

Equation 3.91 gives the shear stress at time t arising from the complete shear history over all earlier times t'.

This is a particular case of the general form

$$\tau = -G \int_{-\infty}^{t} \psi(t-t')\dot{\gamma}(t')\,dt' \tag{3.92}$$

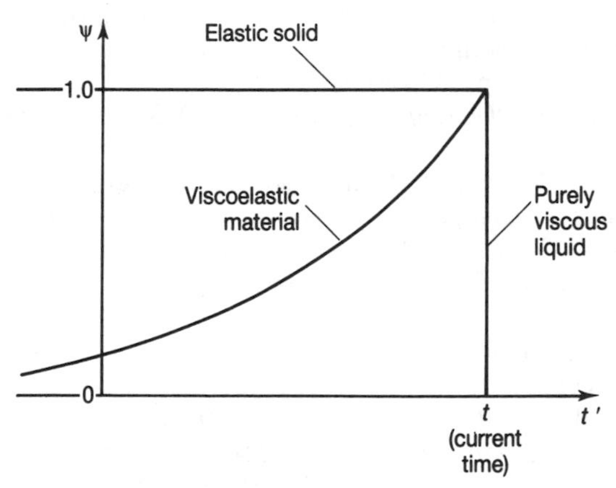

Figure 3.14
Forms of the memory function for different types of material

The function $\psi(t-t')$ may be interpreted as a memory function having a form as shown in Figure 3.14. For an elastic solid, ψ has the value unity at all times, while for a purely viscous liquid ψ has the value unity at the current time but zero at all other times. Thus, a solid behaves as if it 'remembers' the whole of its deformation history, while a purely viscous liquid responds only to its instantaneous deformation rate and is uninfluenced by its history. The viscoelastic fluid is intermediate, behaving as if it had a memory that fades exponentially with time. The purely elastic solid and the purely viscous fluid are just extreme cases of viscoelastic behaviour.

References

Bragg, R. and Oliver, D.R., The triple jet: a new method for extensional viscosity measurement, *Nature Physical Science*, **241**, No. 111, pp. 131–2 (1973).

Cheng, D.C.-H., Further observations on the rheological behaviour of dense suspension, *Powder Technology*, **37**, pp. 255–73 (1984).

Dodge, D.W. and Metzner, A.B., Turbulent flow of non-Newtonian systems, *AIChE Journal*, **5**, pp. 189–204 (1959).

Dwight, H.B., *Tables of Integrals and Other Mathematical Data*, Fourth edition, New York, The Macmillan Company (1961).

Fitzgerald, F.D., *Viscometric properties of an ultra-fine coal–water suspension*, Ph.D. thesis, UMIST (1990).

Govier, G.W., Use chart to find friction factors, *Chemical Engineering*, **66**, pp. 139–42 (24 Aug 1959).

Hedström, B.O.A., Flow of plastics materials in pipes, *Industrial and Engineering Chemistry*, **44**, No. 3, pp. 651–6 (1952).

Jastrzebski, Z.D., Entrance effects and wall effects in an extrusion rheometer during the flow of concentrated suspensions, *Industrial and Engineering Chemistry*, **6**, No. 3, pp. 445–54 (1967).

McComb, W.D., *The Physics of Fluid Turbulence*, Oxford, Oxford University Press, 1990.

Metzner, A.B. and Metzner, A.P., Stress levels in rapid extensional flow of polymeric fluids, *Rheologica Acta* **9**, No. 2, pp. 174–81 (1970).

Metzner, A.B. and Reed, J.C., Flow of non-Newtonian fluids – correlation of the laminar, transition, and turbulent-flow regimes, *AIChE Journal*, **1**, pp. 434–40 (1955).

Mooney, M., Explicit formulas for slip and fluidity, *Journal of Rheology*, **2**, No. 2, pp. 210–22 (1931).

Oliver, D.R. and Bragg, R., Flow patterns in viscoelastic liquids upstream of orifices, *Canadian Journal of Chemical Engineering*, **51**, pp. 287–90 (1973).

Patterson, R.W. and Abernathy, F.H., Transition to turbulence in pipe flow for water and dilute solutions of polyethylene oxide, *Journal of Fluid Mechanics*, **51**, pp. 177–85 (1972).

Ryan, N.W. and Johnson, M.M., Transition from laminar to turbulent flow in pipes, *AIChE Journal*, 5, pp. 433–5 (1959).

Schofield, R.K. and Scott Blair, G.W., The influence of the proximity of a solid wall on the consistency of viscous and plastic materials, *Journal of Physical Chemistry*, **34**, No. 1, pp. 248–62 (1930).

Skelland, A.H.P., *Non-Newtonian Flow and Heat Transfer*, New York, John Wiley and Sons Inc, 1967.

Tanner, R., *Engineering Rheology*, Oxford, Oxford University Press, 1985.

Thomas, D.G., Heat and momentum transport characteristics of non-Newtonian aqueous thorium oxide suspensions, *AIChE Journal*, **6**, pp. 631–9 (1960).

Thomas, D.G., Transport characteristics of suspensions: Part IV. Friction loss of concentrated-flocculated suspensions in turbulent flow, *AIChE Journal*, **8**, pp. 266–71 (1962).

4 Pumping of liquids

4.1 Pumps and pumping

Pumps are devices for supplying energy or head to a flowing liquid in order to overcome head losses due to friction and also, if necessary, to raise the liquid to a higher level. The head imparted to a flowing liquid by a pump is known as the total head Δh. If a pump is placed between points 1 and 2 in a pipeline, the heads for steady flow are related by equation 1.14

$$\left(z_2+\frac{P_2}{\rho_2 g}+\frac{u_2^2}{2g\alpha_2}\right)-\left(z_1+\frac{P_1}{\rho_1 g}+\frac{u_1^2}{2g\alpha_1}\right)=\Delta h-h_f \quad (1.14)$$

In equation 1.14, z, $P/(\rho g)$, and $u^2/(2g\alpha)$ are the static, pressure and velocity heads respectively and h_f is the head loss due to friction. The dimensionless velocity distribution factor α is $\frac{1}{2}$ for laminar flow and approximately 1 for turbulent flow.

For a liquid of density ρ flowing with a constant mean velocity u through a pipeline of circular cross section and constant diameter between points 1 and 2 separated by a pump, equation 1.14 can be written as

$$\left(z_2+\frac{P_2}{\rho g}+\frac{u^2}{2g\alpha}\right)-\left(z_1+\frac{P_1}{\rho g}+\frac{u^2}{2g\alpha}\right)=\Delta h-h_f \quad (4.1)$$

For the most part, pumps can be classified into centrifugal and positive displacement pumps.

4.2 System heads

The important heads to consider in a pumping system are the suction, discharge, total and available net positive suction heads. The following definitions are given in reference to the typical pumping system shown in Figure 4.1 where the arbitrarily chosen base line is the centre-line of the pump.

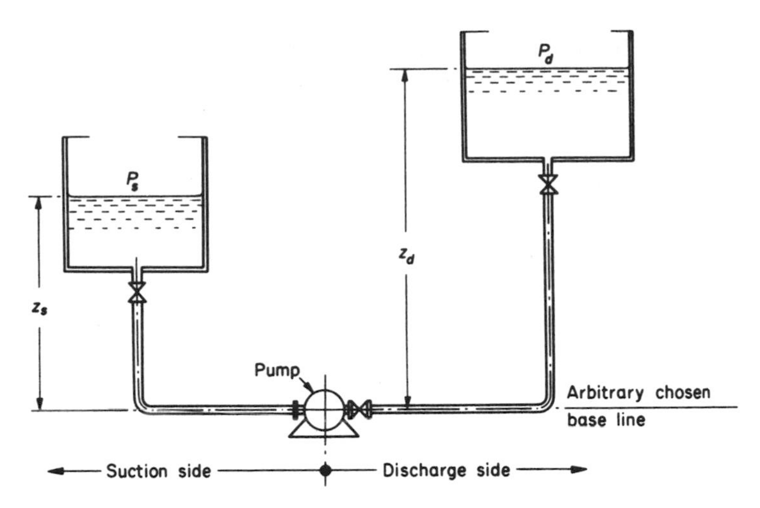

Figure 4.1
Typical pumping system

Suction head:

$$h_s = z_s + \frac{P_s}{\rho g} - h_{fs} \tag{4.2}$$

Discharge head:

$$h_d = z_d + \frac{P_d}{\rho g} + h_{fd} \tag{4.3}$$

In equation 4.2, h_{fs} is the head loss due to friction, z_s is the static head and P_s is the gas pressure above the liquid in the tank on the suction side of the pump. If the liquid level on the suction side is below the centre-line of the pump, z_s is negative.

In equation 4.3, h_{fd} is the head loss due to friction, z_d is the static head and P_d is the gas pressure above the liquid in the tank on the discharge side of the pump.

h_s and h_d are the values of $(P/\rho g + u^2/2g\alpha + z)$ at the suction flange and at the discharge flange respectively. Equations 4.2 and 4.3 are obtained by applying Bernoulli's equation between the supply tank and the suction flange, and between the discharge flange and the receiving tank, respectively. On the suction side, the frictional loss h_{fs} reduces the total head at the suction flange but on the discharge side, h_{fd} increases the head at the discharge flange.

The total head Δh which the pump is required to impart to the flowing liquid is the difference between the discharge and suction heads:

$$\Delta h = h_d - h_s \tag{4.4}$$

Equation 4.4 can be written in terms of equations 4.2 and 4.3 as

$$\Delta h = (z_d - z_s) + \frac{(P_d - P_s)}{\rho g} + (h_{fd} + h_{fs}) \tag{4.5}$$

The head losses due to friction are given by the equations

$$h_{fs} = 4f\left(\frac{\Sigma L_{es}}{d_i}\right)\frac{u^2}{2g} \tag{4.6}$$

and

$$h_{fd} = 4f\left(\frac{\Sigma L_{ed}}{d_i}\right)\frac{u^2}{2g} \tag{4.7}$$

where ΣL_{es} and ΣL_{ed} are the total equivalent lengths on the suction and discharge sides of the pump respectively.

The suction head h_s decreases and the discharge head h_d increases with increasing liquid flow rate because of the increasing value of the friction head loss terms h_{fs} and h_{fd}. Thus the total head Δh which the pump is required to impart to the flowing liquid increases with the liquid pumping rate.

It is clear from equation 4.2 that the suction head h_s can fall to a very low value, for example when the suction frictional head loss is high and the static head z_s is low. If the absolute pressure in the liquid at the suction flange falls to, or below, the absolute vapour pressure P_v of the liquid, bubbles of vapour will be formed at the pump inlet. Worse still, even if the pressure at the suction flange is slightly higher than the vapour pressure, cavitation—the formation and subsequent collapse of vapour bubbles—will occur within the body of the pump because the pressure in the pump falls further as the liquid is accelerated.

In order that cavitation may be avoided, pump manufacturers specify a minimum value by which the total head at the suction flange must exceed the head corresponding to the liquid's vapour pressure.

The difference between the suction head and the vapour pressure head is known as the Net Positive Suction Head, NPSH:

$$\text{NPSH} = h_s - \frac{P_v}{\rho g} \tag{4.8}$$

Substituting for h_s from equation 4.2, the available NPSH is given by

$$\text{NPSH} = z_s + \frac{P_s - P_v}{\rho g} - h_{fs} \tag{4.9}$$

The available NPSH given by equations 4.8 and 4.9 must exceed the value required by the pump and specified by the manufacturer. The required NPSH increases with increasing flow rate as discussed below.

4.3 Centrifugal pumps

In centrifugal pumps, energy or head is imparted to a flowing liquid by centrifugal action. The most common type of centrifugal pump is the volute pump. In volute pumps, liquid enters near the axis of a high speed impeller and is thrown radially outward into a progressively widening spiral casing as shown in Figure 4.2. The impeller vanes are curved to

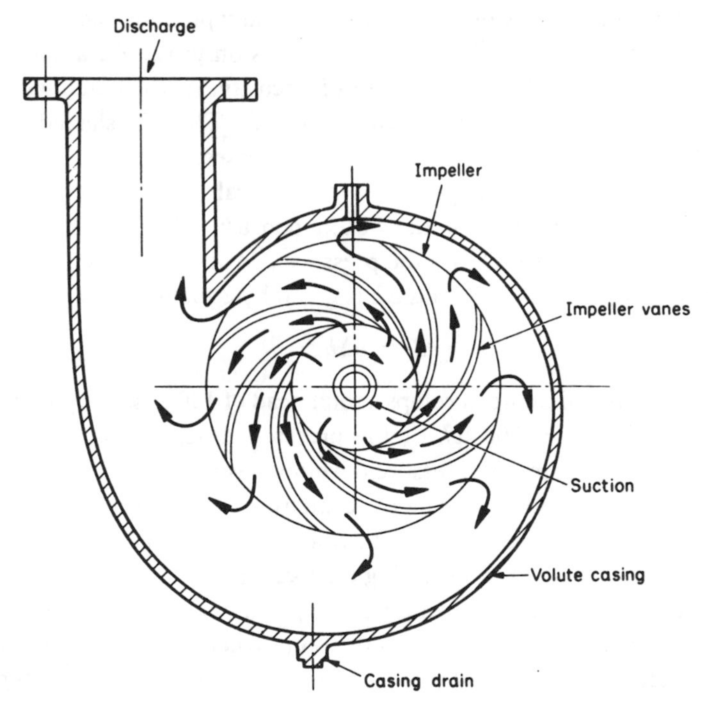

Figure 4.2
Volute centrifugal pump casing design

ensure a smooth flow of liquid. The velocity head imparted to the liquid is gradually converted into pressure head as the velocity of the liquid is reduced. The efficiency of this conversion is a function of the design of the impeller and casing and the physical properties of the liquid.

The performance of a centrifugal pump for a particular rotational speed of the impeller and liquid viscosity is represented by plots of total head against capacity, power against capacity, and required NPSH against capacity. These are known as characteristic curves of the pump. Characteristic curves have a variety of shapes depending on the geometry of the impeller and pump casing. Pump manufacturers normally supply these curves only for operation with water. However, methods are available for plotting curves for other viscosities from the water curves [Holland and Chapman (1966)].

The most common shape of a total head against capacity curve for a conventional volute centrifugal pump is shown in Figure 4.3, where Δh is the total head developed by the pump and Q is the volumetric flow rate of liquid or capacity. The maximum total head developed by the pump is at zero capacity. As the liquid throughput is increased, the total head developed decreases. The pump can operate at any point on the Δh against Q curve. Any individual Δh against Q curve is only true for a particular rotational speed of the impeller and liquid viscosity. As the liquid viscosity increases the Δh against Q curve becomes steeper. Thus the shaded area in Figure 4.3 increases as the liquid viscosity increases.

The total head Δh developed by a centrifugal pump at a particular capacity Q is independent of the liquid density. Thus the higher the density of the liquid, the higher the pressure ΔP developed by the pump. The relationship between ΔP and Δh is given by equation 4.10

$$\Delta P = \rho \Delta h g \tag{4.10}$$

Thus if a centrifugal pump develops a total head of 100 m when pumping a liquid of density $\rho = 1000$ kg/m^3, the pressure developed is 981 000 Pa; while for $\rho = 917$ kg/m^3 the pressure developed is 900 000 Pa.

Equation 4.10 shows that when a centrifugal pump runs on air, the pressure developed is very small. In fact, a conventional centrifugal pump can never prime itself when operating on a suction lift.

In a particular system, a centrifugal pump can only operate at one point on the Δh against Q curve and that is the point where the pump Δh against Q curve intersects with the system Δh against Q curve as shown in Figure 4.4.

Equation 4.5 gives the system total head at a particular liquid flow rate.

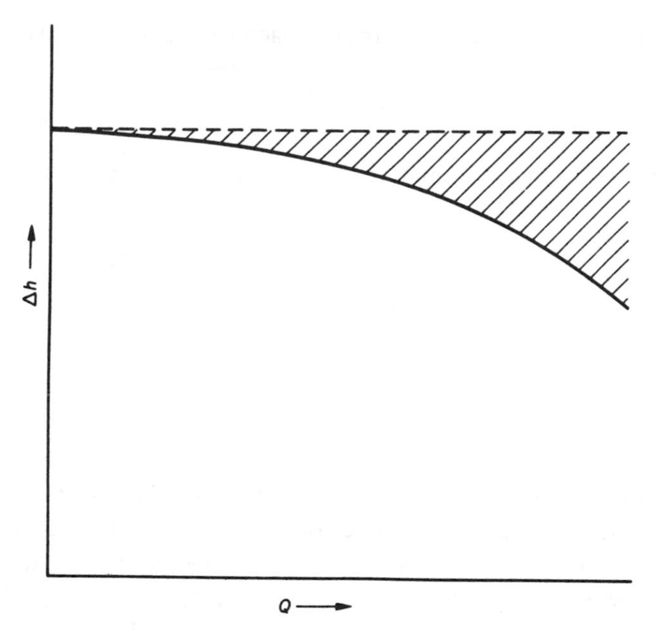

Figure 4.3
Total head against capacity characteristic curve for a volute centrifugal pump

$$\Delta h = (z_d - z_s) + \frac{(P_d - P_s)}{\rho g} + (h_{fd} + h_{fs}) \tag{4.5}$$

Combining equation 4.5 with equations 4.6 and 4.7, which give the frictional head losses h_{fs} and h_{fd} respectively, allows the total head to be written as

$$\Delta h = (z_d - z_s) + \frac{(P_d - P_s)}{\rho g} + 4f\left[\frac{(\Sigma L_{es} + \Sigma L_{ed})}{d_i}\right]\frac{u^2}{2g} \tag{4.11}$$

The mean velocity u of the liquid is related to the volumetric flow rate or capacity Q by

$$u = \frac{Q}{\pi d_i^2/4} \qquad \text{from (1.6)}$$

Substituting for u in equation 4.11 gives

$$\Delta h = (z_d - z_s) + \frac{(P_d - P_s)}{\rho g} + \frac{2f}{g}\left[\frac{(\Sigma L_{es} + \Sigma L_{ed})}{d_i}\right]\left(\frac{Q}{\pi d_i^2/4}\right)^2 \tag{4.12}$$

For laminar flow, the Fanning friction factor f is given by equation 2.15

$$f = \frac{16}{Re} \tag{2.15}$$

Substituting for f in equation 4.12, the total head for laminar flow can be written as

$$\Delta h = (z_d - z_s) + \frac{(P_d - P_s)}{\rho g} + \left(\frac{32\mu}{\rho d_i g}\right)\left[\frac{(\Sigma L_{es} + \Sigma L_{ed})}{d_i}\right] u \tag{4.13}$$

or as

$$\Delta h = (z_d - z_s) + \frac{(P_d - P_s)}{\rho g} + \left(\frac{32\mu}{\rho d_i g}\right)\left[\frac{(\Sigma L_{es} + \Sigma L_{ed})}{d_i}\right]\left(\frac{Q}{\pi d_i^2/4}\right) \tag{4.14}$$

The system Δh against Q curve shown in Figure 4.4 can be plotted using equation 4.12 to calculate the values of the system total head Δh at each volumetric flow rate of liquid or capacity Q. Equation 4.14 shows that for laminar flow the total head Δh increases linearly with capacity Q. Thus for laminar flow, the system Δh against Q curve is a straight line.

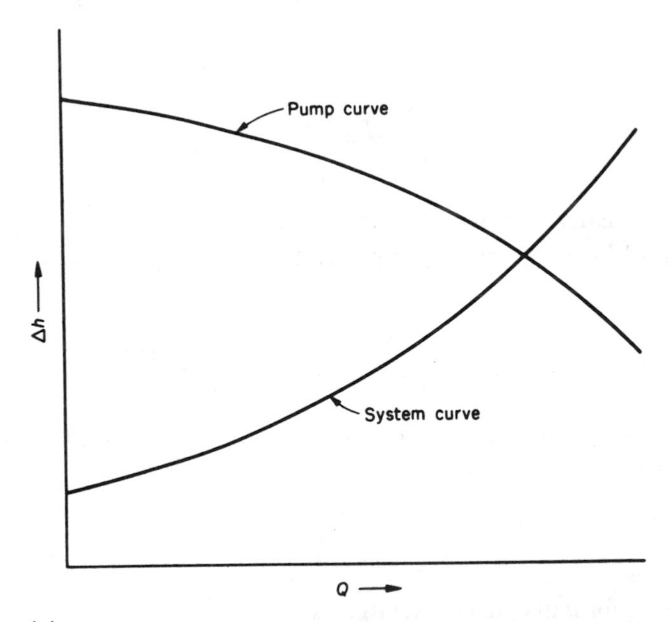

Figure 4.4

System and pump total head against capacity curves. The intersection of the two curves defines the operating point

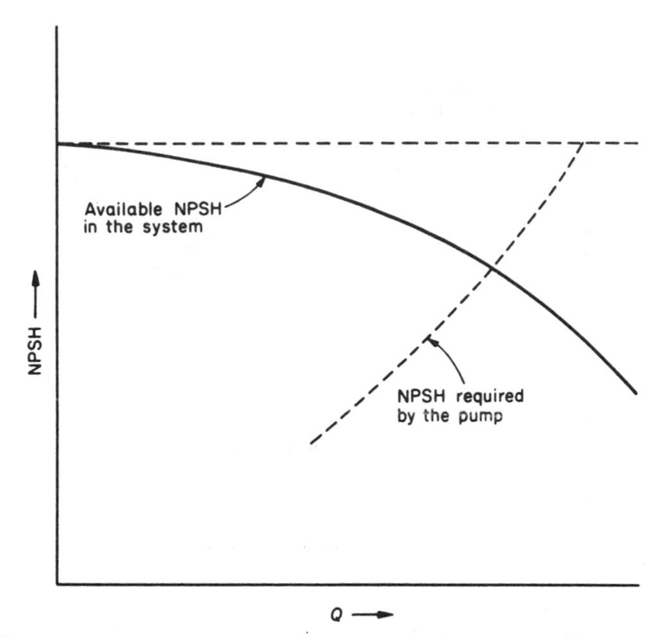

Figure 4.5
Available and required net positive suction heads against capacity in a pumping system

In the above discussion it is assumed that the available NPSH in the system is adequate to support the flow rate of liquid into the suction side of the pump. If the available NPSH is less than that required by the pump, cavitation occurs and the normal curves do not apply. In cavitation, some of the liquid vaporizes as it flows into the pump. As the vapour bubbles are carried into higher pressure regions of the pump they collapse, resulting in noise and vibration. High speed pumps are more prone to cavitation than low speed pumps.

Figure 4.5 shows a typical relationship between the available NPSH in the system and the NPSH required by the pump as the volumetric flow rate of liquid or capacity Q is varied. The NPSH required by a centrifugal pump increases approximately with the square of the liquid throughput. The available NPSH in a system can be calculated from equation 4.9 having substituted for h_{fs}

$$\text{NPSH} = z_s + \frac{(P_s - P_v)}{\rho g} - \frac{2f}{g}\left(\frac{\Sigma L_{es}}{d_i}\right)\left(\frac{Q}{\pi d_i^2/4}\right)^2 \qquad (4.15)$$

Equation 4.15 shows that the available NPSH in a system decreases as the liquid throughput increases because of the greater frictional head losses.

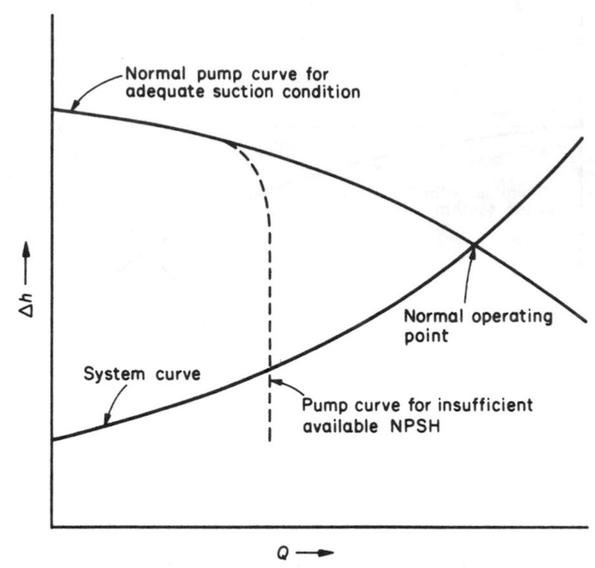

Figure 4.6
Effect of insufficient NPSH on the performance of a centrifugal pump

A centrifugal pump will operate normally at a point on its total head against capacity characteristic curve until the available NPSH falls below the required NPSH curve. Beyond this point, the total head generated by a centrifugal pump falls drastically as shown in Figure 4.6 as the pump begins to operate in cavitation conditions.

In centrifugal pump systems, a throttling valve is located on the discharge side of the pump. When this valve is throttled, the system Δh against Q curve is altered to incorporate the increased frictional head loss. The effect of throttling is illustrated in Figure 4.7. Throttling can be used to decrease cavitation. A flow regulating valve or other constriction must not be placed on the *suction* side of the pump.

System total heads should be estimated as accurately as possible. Safety factors should never be added to these estimated total head values. This is illustrated by Figure 4.8. Suppose that OA_1 is the correct curve and that the centrifugal pump is required to operate at point A_1. Let a safety factor be added to the total head values to give a system curve OA_2. On the basis of curve OA_2, the manufacturer will supply a pump to operate at point A_2. However, since the true system curve is OA_1, the pump will operate at point A_3. Not only is the capacity higher than that specified, but the pump motor may be overloaded.

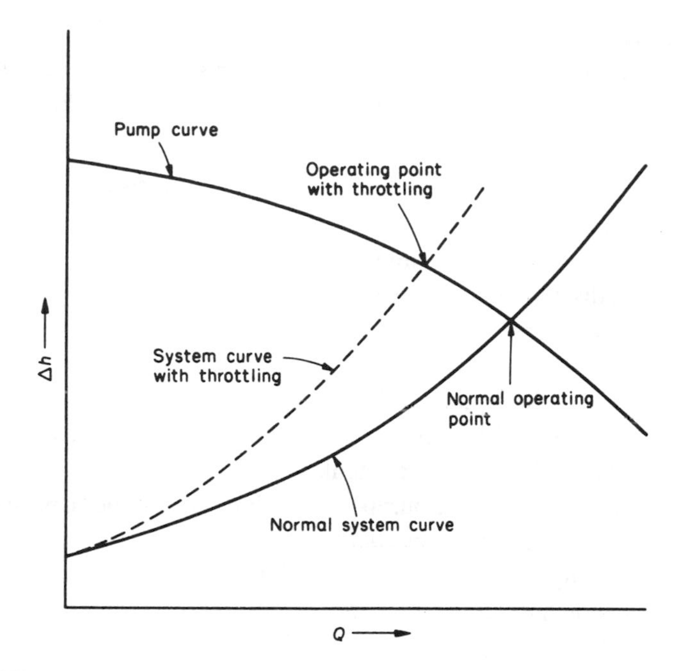

Figure 4.7

Effect of throttling the discharge valve on the operating point of a centrifugal pump

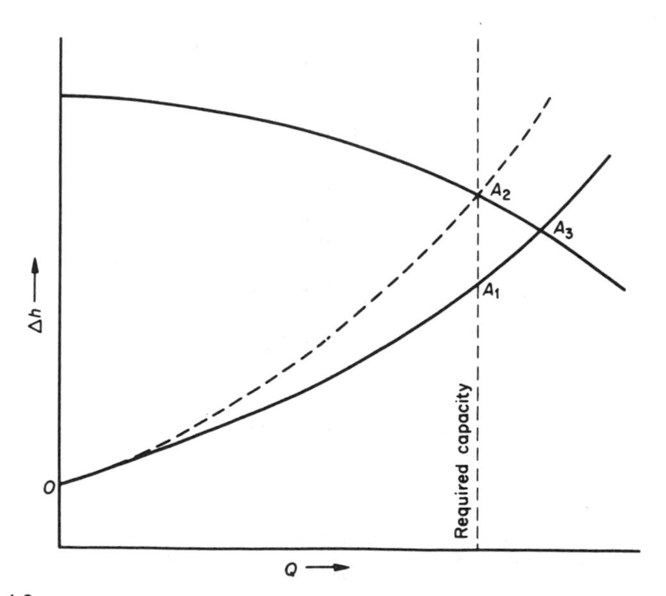

Figure 4.8

Effect of adding a safety factor to the system total head against capacity curve

Example 4.1

Calculate the values for a system total head against capacity curve for the initial conditions of the system shown in Figure 4.1 given the following data:

dynamic viscosity of liquid	$\mu = 0.04$ Pa s
density of liquid	$\rho = 1200$ kg/m^3
static head on suction side of pump	$z_s = 3$ m
static head on discharge side of pump	$z_d = 7$ m
inside diameter of pipe	$d_i = 0.0526$ m
pipe roughness	$e = 0.000045$ m
gas pressure above the liquid in the tank on the suction side of the pump	P_s = atmospheric pressure
gas pressure above the liquid in the tank on the discharge side of the pump	P_d = atmospheric pressure
total equivalent length on the suction side of the pump	$\Sigma L_{es} = 4.9$ m
total equivalent length on the discharge side of the pump	$\Sigma L_{ed} = 63.2$ m

Calculations

$$\text{Reynolds number } Re = \frac{\rho u d_i}{\mu} \tag{1.3}$$

$$\rho = 1200 \text{ kg/m}^3$$

initially take

$$u = 1.0 \text{ m/s}$$

$$d_i = 0.0526 \text{ m}$$

$$\mu = 0.04 \text{ Pa s}$$

$$Re = \frac{(1200 \text{ kg/m}^3)(1.0 \text{ m/s})(0.0526 \text{ m})}{0.04 \text{ Pa s}} = 1578$$

$$\text{pipe roughness } e = 0.000045 \text{ m}$$

$$d_i = 0.0526 \text{ m}$$

$$\text{relative roughness } \frac{e}{d_i} = \frac{0.000045 \text{ m}}{0.0526 \text{ m}} = 0.000856$$

from f against Re graph in Figure 2.1

$$f = 0.0112 \quad \text{for} \quad Re = 1578 \quad \text{and} \quad \frac{e}{d_i} = 0.000856$$

$$\frac{(\Sigma L_{es} + \Sigma L_{ed})}{d_i} = \frac{4.9\text{ m} + 63.2\text{ m}}{0.0526\text{ m}} = 1294.7$$

$$\frac{u^2}{2g} = \frac{(1.0\text{ m/s})^2}{(2)(9.81\text{ m/s}^2)} = 0.05097\text{ m}$$

$$\text{total head } \Delta h = (z_d - z_s) + \frac{(P_d - P_s)}{\rho g} + 4f\left[\frac{(\Sigma L_{es} + \Sigma L_{ed})}{d_i}\right]\frac{u^2}{2g} \quad (4.11)$$

$$= 4\text{ m} + 4(0.0112)(1294.7)(0.05097\text{ m})$$

$$= \underline{6.671\text{ m}}$$

$$\text{mean velocity } u = \frac{Q}{\pi d_i^2/4}$$

$$\frac{\pi d_i^2}{4} = \frac{(3.142)(0.0526\text{ m})^2}{4} = 0.002173\text{ m}^2$$

$$\text{capacity } Q = u\frac{\pi d_i^2}{4}$$

$$= (1.0\text{ m/s})(0.002173\text{ m}^2)$$

$$= 0.002173\text{ m}^3\text{/s} = \underline{0.00217\text{ m}^3\text{/s}}$$

Repeating the calculations for other values of u gives the following results:

Table 4.1

u m/s	Re	$f/2$	$u^2/2g$ m	$4f\left[\frac{(\Sigma L_{es} + \Sigma L_{es})}{d_i}\right]\frac{u^2}{2g}$ m	Δh m	Q $\text{m}^3\text{/s}$
0.5	789	0.01014	0.01274	1.338	5.3	0.00109
1.0	1 578	0.00506	0.05097	2.671	6.7	0.00217
1.5	2 367	—	0.1149	—	—	0.00326
2.0	3 156	0.0052	0.2039	10.98	15.0	0.00435
2.5	3 945	0.0050	0.3186	16.50	20.5	0.00544
3.0	4 734	0.0048	0.4487	22.31	26.3	0.00653

At $u = 1.5$ m/s the flow is transitional and no reliable value can be given to f.

4.4 Centrifugal pump relations

The power P_E required in an ideal centrifugal pump can be expected to be a function of the liquid density ρ, the impeller diameter D and the rotational speed of the impeller N. If the relationship is assumed to be given by the equation

$$P_E = C\rho^a N^b D^c \tag{4.16}$$

then it can be shown by dimensional analysis [Holland and Chapman (1966)] that

$$P_E = C_1 \rho N^3 D^5 \tag{4.17}$$

where C_1 is a constant which depends on the geometry of the system.

The power P_E is also proportional to the product of the volumetric flow rate Q and the total head Δh developed by the pump.

$$P_E = C_2 Q \Delta h \tag{4.18}$$

where C_2 is a constant.

The volumetric flow rate Q and the total head Δh developed by the pump are related to the rotational speed of the impeller N and the impeller diameter D by equations 4.19 and 4.20 respectively:

$$Q = C_3 N D^3 \tag{4.19}$$

$$\Delta h = C_4 N^2 D^2 \tag{4.20}$$

where C_3 and C_4 are constants.

Note that equation 4.20 is dimensionally consistent only if C_4 has the dimensions T^2/L and consequently the numerical value of C_4 is different for different sets of units.

Eliminating D between equations 4.19 and 4.20 allows the following result to be obtained:

$$\frac{N\sqrt{Q}}{\Delta h^{3/4}} = \text{a constant} \tag{4.21}$$

The constant in equation 4.21 is known as the Specific Speed N_s of the pump. Although commonly used, this definition of the specific speed is unsatisfactory because, following from equation 4.20, the value of N_s depends on the units used. Moreover, manufacturers sometimes mix the units. When using specific speed data it is essential to know the definition of N_s and the units of N, Q and h employed.

A more satisfactory definition is that of a dimensionless specific speed N'_s. The above deficiency can be removed by replacing equation 4.20 by equation 4.22:

$$g\Delta h = C'_4 N^2 D^2 \tag{4.22}$$

where C'_4 is a dimensionless constant. The dimensionless specific speed N'_s is given by

$$N'_s = \frac{N\sqrt{Q}}{(g\Delta h)^{3/4}} \tag{4.23}$$

The value of N'_s is a unique number provided that consistent units are used. In SI units, the units are N in rev/s, Q in m^3/s, h in m, and g has the value 9.81 m/s^2. The specific speed is used as an index of pump types and is always evaluated at the best efficiency point (bep) of the pump.

Two different size pumps are said to be geometrically similar when the ratios of corresponding dimensions in one pump are equal to those of the other pump [Holland and Chapman (1966)]. Geometrically similar pumps are said to be homologous. A set of equations known as the affinity laws govern the performance of homologous centrifugal pumps at various impeller speeds.

Consider a centrifugal pump with an impeller diameter D_1 operating at a rotational speed N_1 and developing a total head Δh_1. Consider an homologous pump with an impeller diameter D_2 operating at a rotational speed N_2 and developing a total head Δh_2.

Equations 4.19 and 4.22 (or 4.20) for this case can be rewritten respectively in the form

$$\frac{Q_1}{Q_2} = \left(\frac{N_1}{N_2}\right)\left(\frac{D_1}{D_2}\right)^3 \tag{4.24}$$

and

$$\frac{\Delta h_1}{\Delta h_2} = \left(\frac{N_1}{N_2}\right)^2\left(\frac{D_1}{D_2}\right)^2 \tag{4.25}$$

Similarly equation 4.17 can be rewritten in the form

$$\frac{P_{E1}}{P_{E2}} = \left(\frac{N_1}{N_2}\right)^3\left(\frac{D_1}{D_2}\right)^5 \tag{4.26}$$

and by analogy with equation 4.25 the net positive suction heads for the two homologous pumps can be related by the equation

$$\frac{\text{NPSH}_1}{\text{NPSH}_2} = \left(\frac{N_1}{N_2}\right)^2 \left(\frac{D_1}{D_2}\right)^2 \tag{4.27}$$

Equations 4.24 to 4.27 are the affinity laws for homologous centrifugal pumps.

For a particular pump where the impeller of diameter D_1 is replaced by an impeller with a slightly different diameter D_2, the following equations hold [Holland and Chapman (1966)]:

$$\frac{Q_1}{Q_2} = \left(\frac{N_1}{N_2}\right) \left(\frac{D_1}{D_2}\right) \tag{4.28}$$

$$\frac{\Delta h_1}{\Delta h_2} = \left(\frac{N_1}{N_2}\right)^2 \left(\frac{D_1}{D_2}\right)^2 \tag{4.29}$$

and

$$\frac{P_{E1}}{P_{E2}} = \left(\frac{N_1}{N_2}\right)^3 \left(\frac{D_1}{D_2}\right)^3 \tag{4.30}$$

If the characteristic performance curves are available for a centrifugal pump operating at a given rotation speed, equations 4.28 to 4.30 enable the characteristic performance curves to be plotted for other operating speeds and for other slightly different impeller diameters.

Example 4.2

A volute centrifugal pump with an impeller diameter of 0.02 m has the following performance data when pumping water at the best efficiency point:

impeller speed N	= 58.3 rev/s
capacity Q	= 0.012 m^3/s
total head Δh	= 70 m
required net positive suction head NPSH	= 18 m
brake power P_B	= 12 000 W

Evaluate the performance characteristics of a homologous pump with twice the impeller diameter operating at half the impeller speed.

Calculations

Let subscripts 1 and 2 refer to the first and second pumps respectively.

The ratio of impeller spreeds $N_1/N_2 = 2$ and the ratio of impeller diameters $D_1/D_2 = 1/2$. The ratio of capacities is given by

$$\frac{Q_1}{Q_2} = \left(\frac{N_1}{N_2}\right)\left(\frac{D_1}{D_2}\right)^3 \tag{4.24}$$

$$= (2)(\tfrac{1}{8}) = \tfrac{1}{4}$$

The capacity of the second pump is

$$Q_2 = 4Q_1 = (4)(0.012 \text{ m}^3/\text{s})$$

$$= \underline{0.048 \text{ m}^3/\text{s}}$$

The ratio of total heads is

$$\frac{\Delta h_1}{\Delta h_2} = \left(\frac{N_1}{N_2}\right)^2\left(\frac{D_1}{D_2}\right)^2 \tag{4.25}$$

$$= (4)(\tfrac{1}{4}) = 1$$

The total head of the second pump is

$$\Delta h_2 = \Delta h_1 = \underline{70 \text{ m}}$$

The ratio of powers is

$$\frac{P_{E1}}{P_{E2}} = \left(\frac{N_1}{N_2}\right)^3\left(\frac{D_1}{D_2}\right)^5 \tag{4.26}$$

$$= (8)(\tfrac{1}{32}) = \tfrac{1}{4}$$

Assume

$$\frac{P_{B1}}{P_{B2}} = \frac{P_{E1}}{P_{E2}}$$

Then

$$\frac{P_{B1}}{P_{B2}} = \frac{1}{4}$$

(This is equivalent to assuming that the two pumps operate at the same efficiency.)

The power for the second pump is given by

$$P_{B2} = 4P_{B1} = (4)(12\,000 \text{ W})$$

$$= \underline{48\,000 \text{ W}}$$

The ratio of required net positive suction heads is

$$\frac{\text{NPSH}_1}{\text{NPSH}_2} = \left(\frac{N_1}{N_2}\right)^2 \left(\frac{D_1}{D_2}\right)^2 \tag{4.27}$$

$$= (4)(\tfrac{1}{4}) = 1$$

Therefore net positive suction head of second pump

$$\text{NPSH}_2 = \text{NPSH}_1 = \underline{18 \text{ m}}$$

4.5 **Centrifugal pumps in series and in parallel**

Diskind (1959) determined the operating characteristics for centrifugal pumps in parallel and in series using a simple graphical method.

Consider two centrifugal pumps in parallel as shown in Figure 4.9. The

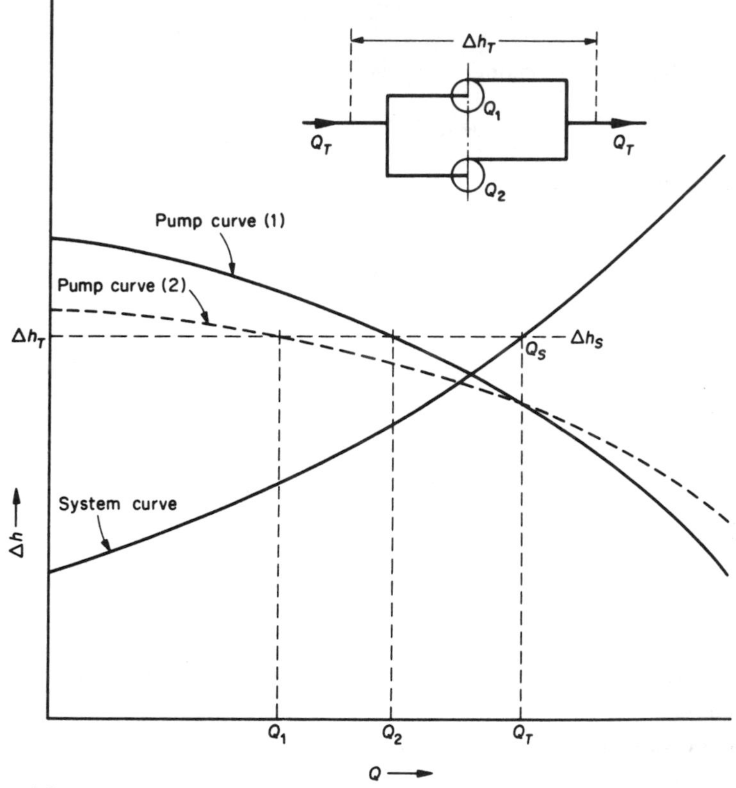

Figure 4.9

Operating point for centrifugal pumps in parallel

total head for the pump combination Δh_T is the same as the total head for each pump, ie

$$\Delta h_T = \Delta h_1 = \Delta h_2 \tag{4.31}$$

The volumetric flow rate or capacity for the pump combination Q_T is the sum of the capacities for the two pumps, ie

$$Q_T = Q_1 + Q_2 \tag{4.32}$$

The operating characteristics for two pumps in parallel are obtained as follows.

1 Draw the Δh against Q characteristic curves for each pump together with the system Δh_s against Q_s curve on the same plot as shown in Figure 4.9.
2 Draw a horizontal constant total head line in Figure 4.9 which intersects the two pump curves at capacities Q_1 and Q_2 respectively, and the system curve at capacity Q_s.
3 Add the values of Q_1 and Q_2 obtained in Step 2 to give

$$Q_T = Q_1 + Q_2 \tag{4.32}$$

4 Compare Q_T from Step 3 with Q_s from Step 2. If they are not equal repeat Steps 2, 3 and 4 until $Q_T = Q_s$. This is the operating point of the two pumps in parallel.

An alternative to this trial and error procedure for two pumps in parallel is to calculate Q_T from equation 4.32 for various values of the total head from known values of Q_1 and Q_2 at these total heads. The operating point for stable operation is at the intersection of the Δh_T against Q_T curve with the Δh_s against Q_s curve.

Consider two centrifugal pumps in series as shown in Figure 4.10. The total head for the pump combination Δh_T is the sum of the total heads for the two pumps, ie

$$\Delta h_T = \Delta h_1 + \Delta h_2 \tag{4.33}$$

The volumetric flow rate or capacity for the pump combination Q_T is the same as the capacity for each pump, ie

$$Q_T = Q_1 = Q_2 \tag{4.34}$$

The operating characteristics for two pumps in series are obtained as follows.

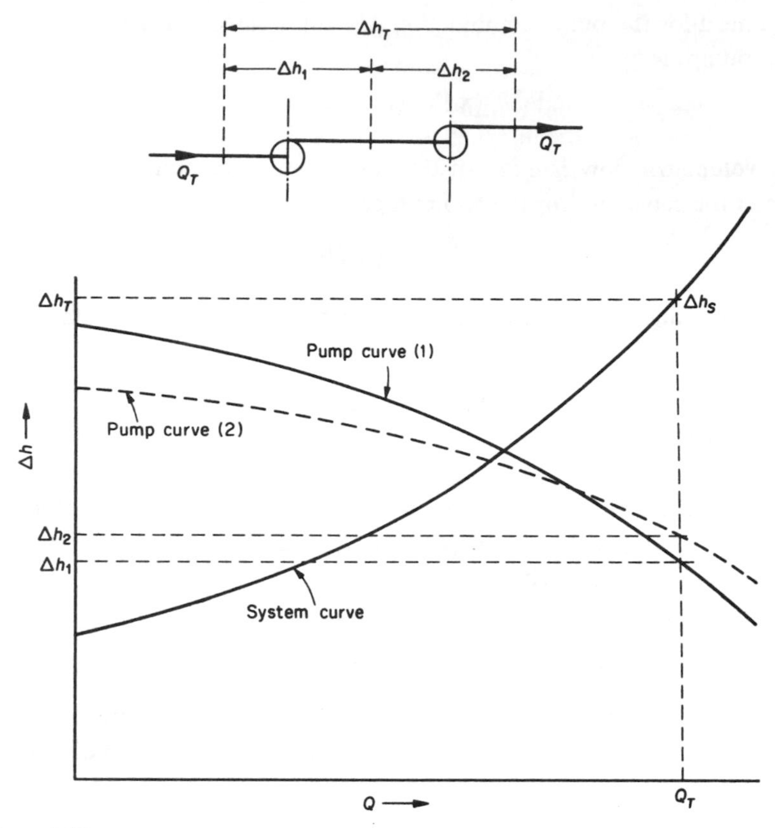

Figure 4.10
Operating point for centrifugal pumps in series

1 Draw the Δh against Q characteristic curves for each pump together with the system Δh_s against Q_s curve on the same plot as shown in Figure 4.10.
2 Draw a vertical constant capacity line in Figure 4.10 which intersects the two pump curves at total heads Δh_1 and Δh_2 respectively, and the system curve at total head Δh_s.
3 Add the values of Δh_1 and Δh_2 obtained in Step 2 to give

$$\Delta h_T = \Delta h_1 + \Delta h_2 \tag{4.33}$$

4 Compare Δh_T from Step 3 with Δh_s from Step 2. If they are not equal repeat Steps 2, 3 and 4 until $\Delta h_T = \Delta h_s$. This is the operating point of the two pumps in series.

An alternative to this trial and error procedure for two pumps in series is to calculate Δh_T from equation 4.33 for various values of the capacity from known values of Δh_1 and Δh_2 at these capacities. The operating point for stable operation is at the intersection of the Δh_T against Q_T curve with the Δh_s against Q_s curve.

The piping and valves may be arranged to enable two centrifugal pumps to be operated either in series or in parallel. For two identical pumps, series operation gives a total head of $2\Delta h$ at a capacity Q and parallel operation gives a capacity of $2Q$ at a total head Δh. The efficiency of either the series or parallel combination is practically the same as for a single pump.

4.6 Positive displacement pumps

For the most part, positive displacement pumps can be classified either as rotary pumps or as reciprocating pumps. However, pumps do exist which exhibit some of the characteristics of both types.

Rotary pumps forcibly transfer liquid through the action of rotating gears, lobes, vanes, screws etc, which operate inside a rigid container. Normally, pumping rates are varied by changing the rotational speed of the rotor. Rotary pumps do not require valves in order to operate.

Reciprocating pumps forcibly transfer liquid by changing the internal volume of the pump. Pumping rates are varied by altering either the frequency or the length of the stroke. Valves are required on both the suction and discharge sides of the pump.

One of the most common rotary pumps is the external gear pump illustrated in Figure 4.11. The fixed casing contains two meshing gears of equal size. The driving gear is coupled to the drive shaft which transmits

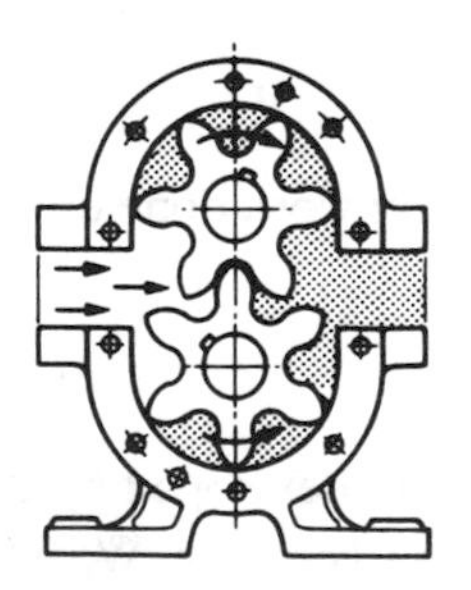

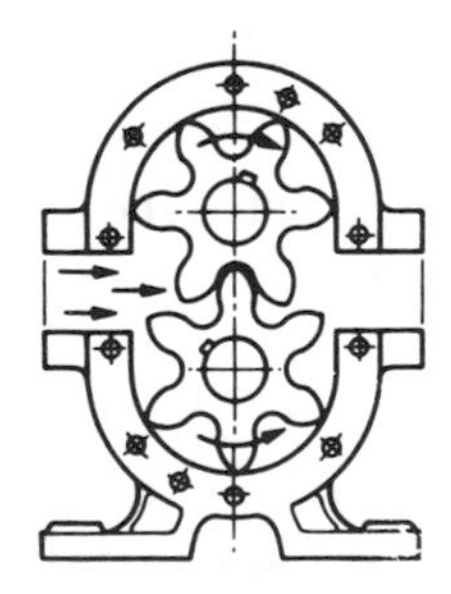

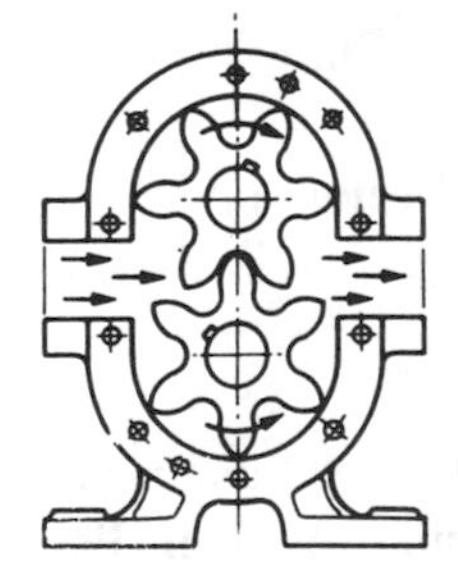

Figure 4.11
Operation of an external gear pump

the power from the motor. The idler gear runs free. As the rotating gears unmesh they create a partial vacuum which causes liquid from the suction line to flow into the pump. Liquid is carried through the pump between the rotating gear teeth and the fixed casing. The meshing of the rotating gears generates an increase in pressure which forces the liquid into the outlet line. In principle an external gear pump can discharge liquid either way depending on the direction of the gear rotation. In practice, external gear pumps are equipped with relief valves to limit the discharge pressures generated since they cannot be operated against a closed discharge without damage to the pump. In this case the direction of gear rotation is fixed and is clearly marked on the pump.

External gear pumps are self-priming since the rotating gears are capable of pumping air. They give a constant delivery of liquid for a set rotor speed with negligible pulsations. Changes in capacity are small with variations in discharge pressure and liquid viscosity. External gear pumps depend on the liquid pumped to lubricate the internal moving parts. They can be damaged if run dry. In order to provide for changes in pumping rate, variable speed drives are required.

Since close clearances are essential between the moving parts, alignment is critical. Some leakage occurs between the discharge and suction sides of a pump through the clearances. This is known as slip. Slip increases with pressure difference across the pump and decreases with increasing liquid viscosity. Since slip is independent of pump speed, it is an advantage to pump low viscosity liquids at high speeds. Slip is negligible for high viscosity liquids and in fact external gear pumps are often used as metering pumps.

Rotary pumps can normally be divided into two classes: small liquid cavity high speed pumps and large liquid cavity low speed pumps.

4.7 Pumping efficiencies

The liquid power P_E can be defined as the rate of useful work done on the liquid. It is given by the equation

$$P_E = Q\Delta P \tag{4.35}$$

If the volumetric flow rate Q is in m^3/s and the pressure developed by the pump ΔP is in Pa or N/m^2, the liquid power P_E is in N m/s or W. The pressure developed by the pump ΔP is related to the total head developed by the pump Δh by equation 4.10.

$$\Delta P = \rho \Delta h g \tag{4.10}$$

Substituting equation 4.10 into equation 4.35 gives

$$P_E = \rho Q \Delta h g \tag{4.36}$$

which can also be written as

$$P_E = M \Delta h g \tag{4.37}$$

since $M = \rho Q$ is the mass flow rate. If M is in kg/s, Δh is in m and the gravitational accelerating $g = 9.81$ m/s^2, P_E is in W.

The brake power P_B can be defined as the actual power delivered to the pump by the prime mover. It is the sum of liquid power and power lost due to friction and is given by the equation

$$P_B = P_E\left(\frac{100}{\eta}\right) \tag{4.38}$$

where η is the mechanical efficiency expressed in per cent.

The mechanical efficiency decreases as the liquid viscosity and hence the frictional losses increase. The mechanical efficiency is also decreased by power losses in gears, bearings, seals etc. In rotary pumps contact between the rotor and the fixed casing increases power losses and decreases the mechanical efficiency. These losses are not proportional to pump size. Relatively large pumps tend to have the best efficiencies whilst small pumps usually have low efficiencies. Furthermore, high speed pumps tend to be more efficient than low speed pumps. In general, high efficiency pumps have high NPSH requirements. Sometimes a compromise may have to be made between efficiency and NPSH.

Another efficiency which is important for positive displacement pumps is the volumetric efficiency. This is the delivered capacity per cycle as a percentage of the true displacement per cycle. If no slip occurs, the volumetric efficiency of the pump is 100 per cent. For zero pressure difference across the pump, there is no slip and the delivered capacity is the true displacement. The volumetric efficiency of a pump is reduced by the presence of entrained air or gas in the pumped liquid. It is important to know the volumetric efficiency of a positive displacement pump when it is to be used for metering.

4.8 Factors in pump selection

The selection of a pump depends on many factors which include the required rate and properties of the pumped liquid and the desired location of the pump.

In general, high viscosity liquids are pumped with positive displacement pumps. Centrifugal pumps are not only very inefficient when pumping high viscosity liquids but their performance is very sensitive to changes in liquid viscosity. A high viscosity also leads to high frictional head losses and hence a reduced available NPSH. Since the latter must always be greater than the NPSH required by the pump, a low available NPSH imposes a severe limitation on the choice of a pump. Liquids with a high vapour pressure also reduce the available NPSH. If these liquids are pumped at a high temperature, this may cause the gears to seize in a close clearance gear pump.

If the pumped liquid is shear thinning, its apparent viscosity will decrease with an increase in shear rate and hence pumping rate. It is therefore an advantage to use high speed pumps to pump shear thinning liquids and in fact centrifugal pumps are frequently used. In contrast, the apparent viscosity of a shear thickening liquid will increase with an increase in shear rate and hence pumping rate. It is therefore an advantage to use large cavity positive displacement pumps with a low cycle speed to pump shear thickening liquids.

Some liquids can be permanently damaged by subjecting them to high shear in a high speed pump. For example, certain liquid detergents can be broken down into two phases if subject to too much shear. Even though these detergents may exhibit shear thinning characteristics they should be pumped with relatively low speed pumps.

Wear is a more serious problem with positive displacement pumps than with centrifugal pumps. Liquids with poor lubricating qualities increase the wear on a pump. Wear is also caused by corrosion and by the pumping of liquids containing suspended solids which are abrasive.

In general, centrifugal pumps are less expensive, last longer and are more robust than positive displacement pumps. However, they are unsuitable for pumping high viscosity liquids and when changes in viscosity occur.

References

Diskind, T., Solve muliple hookups graphically, *Chemical Engineering*, **66** p. 102 (2 Nov 1959).

Holland, F.A. and Chapman, F.S., *Pumping of Liquids*, New York, Reinhold Publishing Corporation (1966).

5 Mixing of liquids in tanks

5.1 Mixers and mixing

Mixing may be defined as the 'intermingling of two or more dissimilar portions of a material, resulting in the attainment of a desired level of uniformity, either physical or chemical, in the final product' [Quillen (1954)]. Since natural diffusion in liquids is relatively slow, liquid mixing is most commonly accomplished by rotating an agitator in the liquid confined in a tank. It is possible to waste much of this input of mechanical energy if the wrong kind of agitator is used. Parker (1964) defined agitation as 'the creation of a state of activity such as flow or turbulence, apart from any mixing accomplished'.

A rotating agitator generates high speed streams of liquid which in turn entrain stagnant or slower moving regions of liquid resulting in uniform mixing by momentum transfer. As the viscosity of the liquid is increased, the mixing process becomes more difficult since frictional drag retards the high speed streams and confines them to the immediate vicinity of the rotating agitator.

In general, agitators can be classified into the following two groups.

1 Agitators with a small blade area which rotate at high speeds. These include turbines and marine type propellers.
2 Agitators with a large blade area which rotate at low speeds. These include anchors, paddles and helical screws.

The second group is more effective than the first in the mixing of high viscosity liquids.

The mean shear rate produced by an agitator in a mixing tank $\dot{\gamma}_m$ is proportional to the rotational speed of the agitator N [Metzner and Otto (1957)].

Thus

$$\dot{\gamma}_m = kN \tag{5.1}$$

where k is a dimensionless proportionality constant for a particular system.

For a liquid mixed in a tank with a rotating agitator, the shear rate is greatest in the immediate vicinity of the agitator. In fact the shear rate decreases exponentially with distance from the agitator [Norwood and Metzner (1960)]. Thus the shear stresses and strain rates vary greatly throughout an agitated liquid in a tank. Since the dynamic viscosity of a Newtonian liquid is independent of shear at a given temperature, its viscosity will be the same at all points in the tank. In contrast the apparent viscosity of a non-Newtonian liquid varies throughout the tank. This in turn significantly influences the mixing process. For shear thinning liquids, the apparent viscosity is at a minimum in the immediate vicinity of the agitator. The progressive increase in the apparent viscosity of a shear thinning liquid with distance away from the agitator tends to dampen eddy currents in the mixing tank. In contrast, for shear thickening liquids, the apparent viscosity is at a maximum in the immediate vicinity of the agitator. In general shear thinning and shear thickening liquids should be mixed using high and low speed agitators respectively.

It is desirable to produce a particular mixing result in the minimum time t and with the minimum input of power per unit volume P_A/V. Thus an efficiency function E can be defined as

$$E = \left(\frac{1}{P_A/V}\right)\left(\frac{1}{t}\right) \tag{5.2}$$

5.2 Small blade high speed agitators

Small blade high speed agitators are used to mix low to medium viscosity liquids. Two of the most common types are the six-blade flat blade turbine and the marine type propeller shown in Figures 5.1 and 5.2 respectively. Flat blade turbines used to mix liquids in baffled tanks produce radial flow patterns primarily perpendicular to the vessel wall as shown in Figure 5.3. In contrast marine type propellers used to mix liquids in baffled tanks produce axial flow patterns primarily parallel to the vessel wall as shown in Figure 5.4. Marine type propellers and flat blade turbines are suitable to mix liquids with dynamic viscosities up to 10 and 50 Pa s, respectively.

Figure 5.5 shows a turbine agitator of diameter D_A in a cylindical tank of diameter D_T filled with liquid to a height H_L. The agitator is located at a height H_A from the bottom of the tank and the baffles which are located

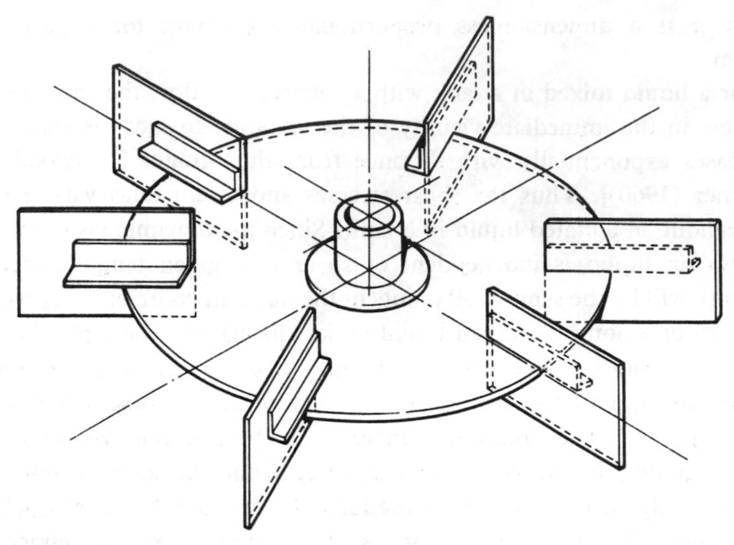

Figure 5.1
Six-blade flat blade turbine

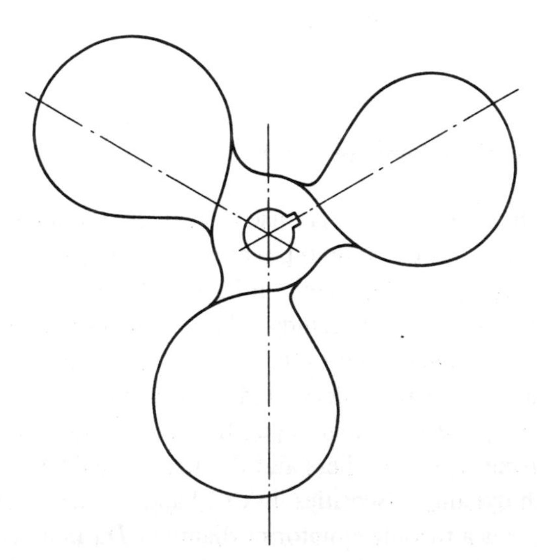

Figure 5.2
Marine propeller

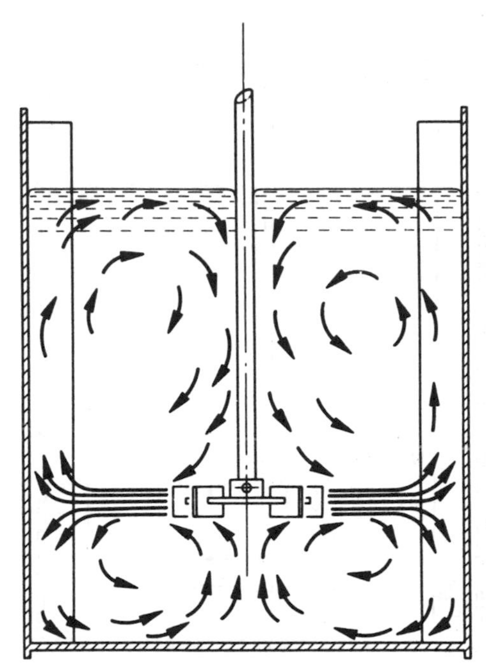

Figure 5.3

Radial flow pattern produced by a flat blade turbine

Figure 5.4

Axial flow pattern produced by a marine propeller

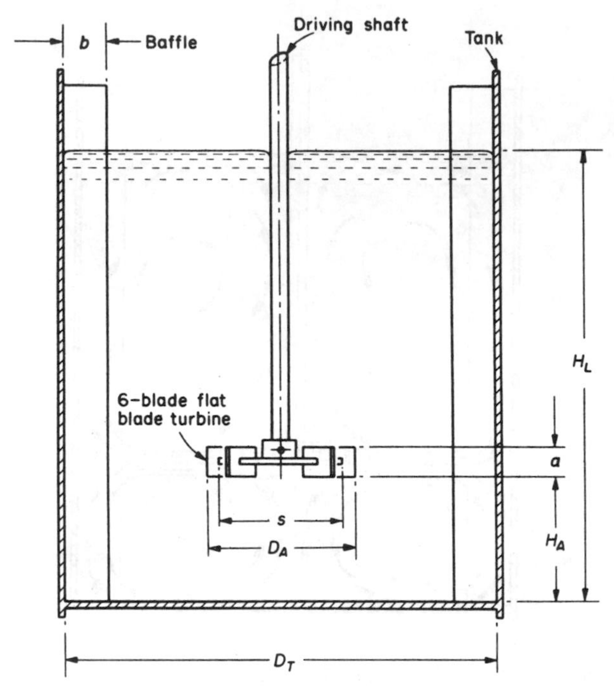

Figure 5.5
Standard tank configuration

immediately adjacent to the wall have a width b. The agitator has a blade width a and blade length r and the blades are mounted on a central disc of diameter s. A typical turbine mixing system is the standard configuration defined by the following geometrical relationships:

1 a six-blade flat blade turbine agitator
2 $D_A = D_T/3$
3 $H_A = D_T/3$
4 $a = D_T/5$
5 $r = D_T/4$
6 $H_L = D_T$
7 4 symmetrical baffles
8 $b = D_T/10$

Processing considerations sometimes necessitate deviations from the standard configuration.

Agitator tip speeds u_T given by equation 5.3 are commonly used as a measure of the degree of agitation in a liquid mixing system.

$$u_T = \pi D_A N \tag{5.3}$$

Tip speed ranges for turbine agitators are recommended as follows:

2.5 to 3.3 m/s for low agitation

3.3 to 4.1 m/s for medium agitation

and

4.1 to 5.6 m/s for high agitation

If turbine or marine propeller agitators are used to mix relatively low viscosity liquids in unbaffled tanks, vortexing develops. In this case the liquid level falls in the immediate vicinity of the agitator shaft. Vortexing increases with rotational speed N until eventually the vortex passes through the agitator. As the liquid viscosity increases, the need for baffles to reduce vortexing decreases.

A marine propeller can be considered as a caseless pump. In this case its volumetric circulating capacity Q_A is related to volumetric displacement per revolution V_D by the equation

$$Q_A = \eta V_D N \tag{5.4}$$

where η is a dimensionless efficiency factor which is approximately 0.6 [Weber (1963)]. V_D is related to the propeller pitch p and the propeller diameter D_A by equation 5.5

$$V_D = \frac{\pi D_A^2 p}{4} \tag{5.5}$$

Most propellers are square pitch propellers where $p = D_A$ so that equation 5.5 becomes

$$V_D = \frac{\pi D_A^3}{4} \tag{5.6}$$

Combining equations 5.4 and 5.6 gives

$$Q_A = \frac{\eta \pi N D_A^3}{4} \tag{5.7}$$

which is analogous to equation 4.19 for centrifugal pumps.

Weber (1963) defined a tank turnover rate I_T by the equation

$$I_T = \frac{Q_A}{V} \tag{5.8}$$

where V is the tank volume and I_T is the number of turnovers per unit time. To get the best mixing, I_T should be at a maximum. For a given tank volume V, this means that the circulating capacity Q_A should have the highest possible value for the minimum consumption of power.

The head developed by the rotating agitator h_A can be written as

$$h_A = C_1 N^2 D_A^2 \tag{5.9}$$

where C_1 is a constant. Equation 5.9 is analogous to equation 4.20 for centrifugal pumps.

Combining equations 5.7 and 5.9 gives the ratio

$$\frac{Q_A}{h_A} = \frac{C D_A}{N} \tag{5.10}$$

where C is a constant.

Since the mean shear rate in a mixing tank $\dot{\gamma}_m$ is given by equation 5.1

$$\dot{\gamma}_m = kN \tag{5.1}$$

equation 5.10 can also be written in the form

$$\frac{Q_A}{h_A} = \frac{C' D_A}{\dot{\gamma}_m} \tag{5.11}$$

where C' is also a constant.

It should be noted that the constants C_1 and C' in equations 5.9 and 5.11 respectively are dimensional: dimensionless forms can be defined as was done for the analogous case with pumps, see equation 4.22.

The ratio of circulating capacity to head Q_A/h_A is low for high shear agitators. For mixing shear thinning liquids a high circulating capacity Q_A and a high shear rate $\dot{\gamma}_m$ or head h_A are both desirable. In this case a compromise has to be made.

5.3 Large blade low speed agitators

Large blade low speed agitators include anchors, gates, paddles, helical ribbons and helical screws. They are used to mix relatively high viscosity liquids and depend on a large blade area to produce liquid movement throughout a tank. Since they are low shear agitators they are useful for mixing shear thickening liquids.

A gate type anchor agitator is shown in Figure 5.6. Anchor agitators

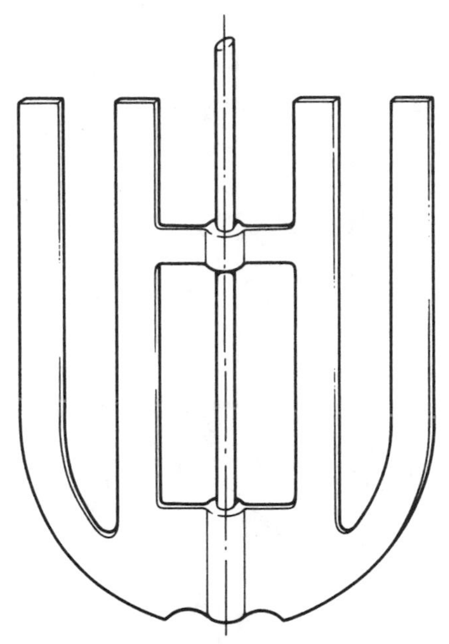

Figure 5.6
Gate type anchor agitator

operate within close proximity to the tank wall. The shearing action of the anchor blades past the tank wall produces a continual interchange of liquid between the bulk liquid and the liquid film between the blades and the wall [Holland and Chapman (1966)]. Anchors have successfully been used to mix liquids with dynamic viscosities up to 100 Pa s, [Brown *et al.* (1947), Uhl and Voznick (1960)]. For heat transfer applications, anchors may be fitted with wall scrapers to prevent the build up of a stagnant film between the anchor and the tank wall.

Uhl and Voznick showed that the mixing effectiveness of a particular anchor agitator in a Newtonian liquid of dynamic viscosity 40 Pa s was the same as for a particular turbine agitator in a Newtonian liquid of dynamic viscosity 15 Pa s.

Helical screws normally function by pumping liquid from the bottom of a tank to the liquid surface. The liquid then returns to the bottom of the tank to fill the void created when fresh liquid is pumped to the surface. A rotating helical screw positioned vertically in the centre of an unbaffled cylindrical tank produces a mild swirling motion in the liquid. Since the liquid velocity decreases towards the tank wall, the liquid at the wall of an

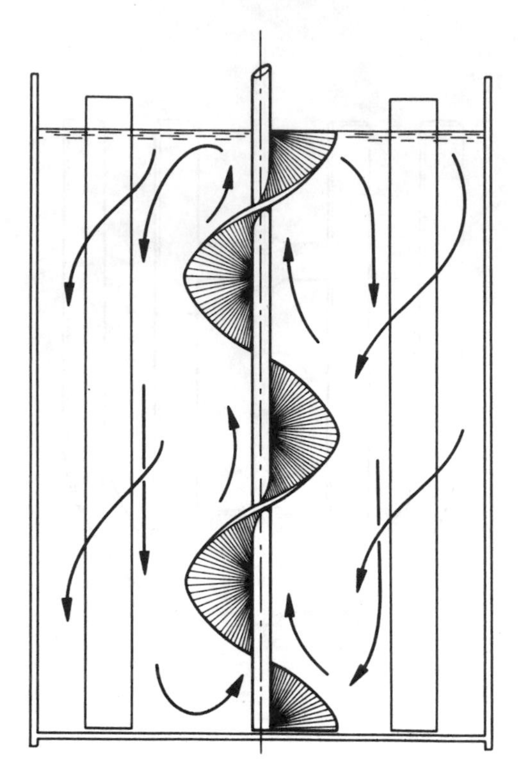

Figure 5.7
Flow pattern in a baffled helical screw system

unbaffled tank is nearly motionless. Baffles set away from the tank wall create turbulence and facilitate the entrainment of liquid in contact with the tank wall. The flow pattern in a baffled helical screw system is shown in Figure 5.7. Baffles are not required if the helical screw is placed in an off-centred position since in this case the system becomes self-baffling. However, off-centred helical screws require more power to produce a comparable mixing result.

Gray (1963) investigated the mixing times of helical ribbon agitators and found the following equation to hold:

$$Nt = 30 \tag{5.12}$$

where N is the rotational speed of the helical ribbon agitator and t is the batch mixing time.

5.4 Dimensionless groups for mixing

In the design of liquid mixing systems the following dimensionless groups are of importance.

The power number

$$Po = \frac{P_A}{\rho N^3 D_A^5} \tag{5.13}$$

The Reynolds number for mixing Re_M represents the ratio of the applied to the opposing viscous drag forces.

$$Re_M = \frac{\rho N D_A^2}{\mu} \tag{5.14}$$

The Froude number for mixing Fr_M represents the ratio of the applied to the opposing gravitational forces.

$$Fr_M = \frac{N^2 D_A}{g} \tag{5.15}$$

The Weber number for mixing We_M represents the ratio of the applied to the opposing surface tension forces.

$$We_M = \frac{\rho N^2 D_A^3}{\sigma} \tag{5.16}$$

In the above equations, ρ, μ and σ are the density, dynamic viscosity and surface tension respectively of the liquid; P_A, N and D_A are the power consumption, rotational speed and diameter respectively of the agitator.

The terms in equations 5.13 to 5.16 must be in consistent units. In the SI system ρ is in kg/m^3, μ in Pa s and σ in N/m; P_A is in W, N in rev/s and D_A in m.

It can be shown by dimensional analysis [Holland and Chapman] that the power number Po can be related to the Reynolds number for mixing Re_M, and the Froude number for mixing Fr_M, by the equation

$$Po = C Re_M^x Fr_M^y \tag{5.17}$$

where C is an overall dimensionless shape factor which represents the geometry of the system.

Equation 5.17 can also be written in the form

$$\phi = \frac{Po}{Fr_M^y} = C Re_M^x \tag{5.18}$$

where ϕ is defined as the dimensionless power function.

In liquid mixing systems, baffles are used to suppress vortexing. Since vortexing is a gravitational effect, the Froude number is not required to describe baffled liquid mixing systems. In this case the exponent y in equations 5.17 and 5.18 is zero and $Fr_M^y = 1$.

Thus for non-vortexing systems equation 5.18 can be written either as

$$\phi = Po = CRe_M^x \tag{5.19}$$

or as

$$\log Po = \log C + x \log Re_M \tag{5.20}$$

The Weber number for mixing We_M is only of importance when separate physical phases are present in the liquid mixing system as in liquid–liquid extraction.

5.5 Power curves

A power curve is a plot of the power function ϕ or the power number Po against the Reynolds number for mixing Re_M on log–log coordinates. Each geometrical configuration has its own power curve and since the plot involves dimensionless groups it is independent of tank size. Thus a power curve used to correlate power data in a 1 m^3 tank system is also valid for a 1000 m^3 tank system provided that both tank systems have the same geometrical configuration.

Figure 5.8 shows the power curve for the standard tank configuration geometrically illustrated in Figure 5.5. Since this is a baffled non-vortexing system, equation 5.20 applies.

$$\log Po = \log C + x \log Re_M \tag{5.20}$$

The power curve for the standard tank configuration is linear in the laminar flow region AB with a slope of -1.0. Thus in this region for $Re_M < 10$, equation 5.20 can be written as

$$\log Po = \log C - \log Re_M \tag{5.21}$$

which can be rearranged to

$$P_A = \mu C N^2 D_A^3 \tag{5.22}$$

where $C = 71.0$ for the standard tank configuration. Thus for laminar flow, power is directly proportional to dynamic viscosity for a fixed agitator speed.

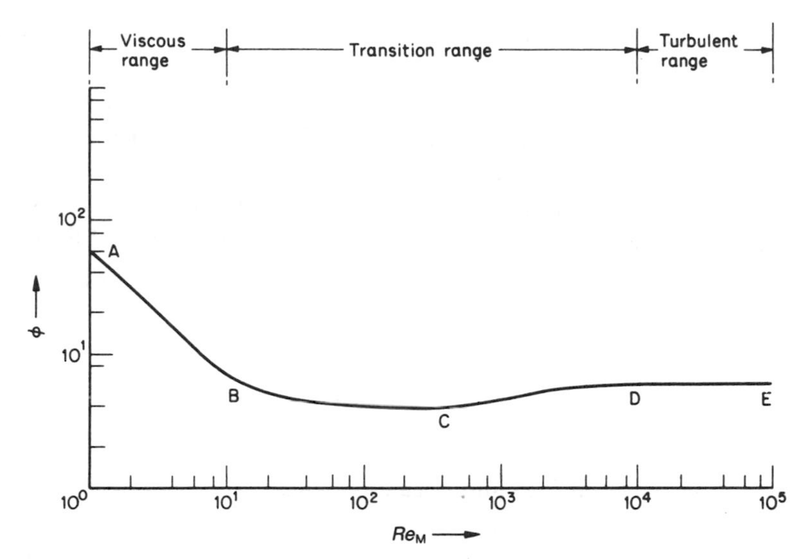

Figure 5.8
Power curve for the standard tank configuration

For the transition flow region BCD which extends up to $Re_M = 10\,000$, the parameters C and x in equation 5.20 vary continuously.

In the fully turbulent flow region DE, the curve becomes horizontal and the power function ϕ is independent of the Reynolds number for mixing Re_M. For the region $Re_M > 10\,000$

$$\phi = Po = 6.3 \tag{5.23}$$

At point C on the power curve for the standard tank configuration given in Figure 5.8, enough energy is being transferred to the liquid for vortexing to start. However the baffles in the tank prevent this. If the baffles were not present vortexing would develop and the power curve would be as shown in Figure 5.9.

The power curve in Figure 5.8 for the baffled system is identical with the power curve in Figure 5.9 for the unbaffled system up to point C where $Re_M \cong 300$. As the Reynolds number for mixing Re_M increases beyond point C in the unbaffled system, vortexing increases and the power falls sharply.

Equation 5.17 can be written in the form

$$\log Po = \log C + x \log Re_M + y \log Fr_M \tag{5.24}$$

For the unbaffled system, $\phi = Po$ at $Re_M < 300$ and $\phi = Po/Fr_M^y$ at $Re_M > 300$.

A plot of Po against Re_M on log–log coordinates for the unbaffled system gives a family of curves at $Re_M > 300$. Each curve has a constant Froude number for mixing Fr_M.

A plot of Po against Fr_M on log–log coordinates is a straight line of slope y at a constant Reynolds number for mixing Re_M. A number of lines can be plotted for different values of Re_M. A plot of y against log Re_M is also a straight line. If the slope of the line is $-1/\beta$ and the intercept at $Re_M = 1$ is α/β then

$$y = \frac{\alpha - \log Re_M}{\beta} \tag{5.25}$$

Substituting equation 5.25 into equation 5.18 gives

$$\phi = \frac{Po}{Fr_M^{[(\alpha - \log Re_M)/\beta]}} \tag{5.26}$$

Rushton, Costich, and Everett (1950) have listed values of α and β for various vortexing systems. For a six-blade flat blade turbine agitator 0.1 m in diameter $\alpha = 1.0$ and $\beta = 40.0$.

If a power curve is available for a particular system geometry, it can be used to calculate the power consumed by an agitator at various rotational speeds, liquid viscosities and densities. The procedure is as follows: calculate the Reynolds number for mixing Re_M; read the power number Po or the power function ϕ from the appropriate power curve and calculate the power P_A from either equation 5.13 rewritten in the form

$$P_A = Po\, \rho N^3 D_A^5 \tag{5.27}$$

or equation 5.18 rewritten in the form

$$P_A = \phi \rho N^3 D_A^5 \left(\frac{N^2 D_A}{g}\right)^y \tag{5.28}$$

Equations 5.27 and 5.28 can be used to calculate only the power consumed by the agitator. Additional power is required to overcome electrical and mechanical losses which occur in all mixing systems.

The power curves given in Figures 5.8 and 5.9 were obtained for experiments using Newtonian liquids.

It is possible to calculate the apparent viscosities of non-Newtonian liquids in agitated tanks from the appropriate power curves for Newtonian

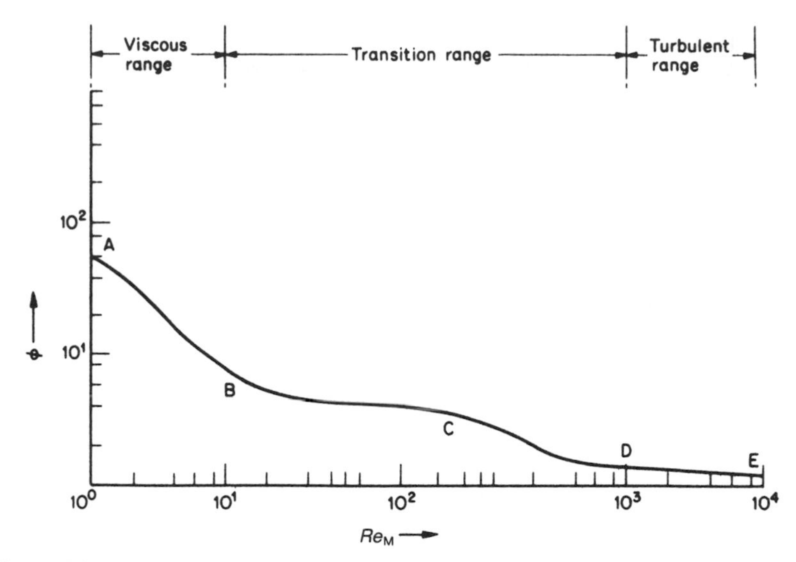

Figure 5.9
Power curve for the standard tank configuration without baffles

liquids. Metzner and Otto (1957) used this procedure to obtain the dimensionless proportionality constant k in equation 5.1 and a non-Newtonian power curve for a particular system geometry.

$$\dot{\gamma}_m = kN \tag{5.1}$$

The procedure is as follows.

1 Obtain power data using a non-Newtonian liquid and calculate the power number Po from equation 5.13 for various agitator speeds N.
2 Read the Reynolds number for mixing Re_M from the appropriate Newtonian power curve for each value of Po and N.
3 For each value of Re_M and N in the laminar flow region calculate the apparent viscosity μ_a from equation 5.14 rewritten in the form

$$\mu_a = \frac{\rho N D_A^2}{Re_M} \tag{5.29}$$

4 Compare a log–log plot of μ_a against N with a log–log plot of μ_a against $\dot{\gamma}$ experimentally determined using a viscometer. Plot $\dot{\gamma}$ against N on ordinary Cartesian coordinates for corresponding values of μ_a. The plot is a straight line of slope k which is the dimensionless proportionality constant in equation 5.1.
5 For various values of power number Po and corresponding agitator

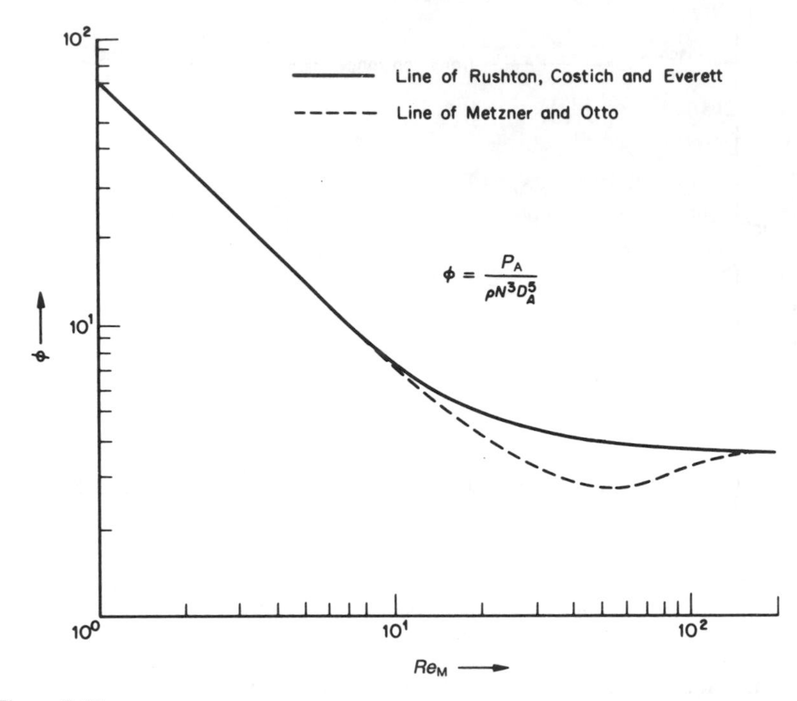

Figure 5.10
Deviation from Newtonian power curve for shear thinning liquids

speeds N beyond the laminar flow region calculate values of shear rate $\dot{\gamma}$ from equation 5.1. Read the corresponding values of apparent viscosity μ_a from the log–log plot of μ_a against $\dot{\gamma}$ and calculate the Reynolds number for mixing Re_M for each value of N and Po. Extend the power curve beyond the laminar flow region by plotting these values of Po and Re_M.

Figure 5.10 illustrates the use of this method to extend a Newtonian power curve in the laminar region into a non-Newtonian power curve. The full line is the Newtonian power curve obtained by Rushton, Costich and Everett for a flat blade turbine system. The dashed line is a plot of the data of Metzner and Otto for shear thinning liquids.

Figure 5.10 shows that at no point is the shear thinning power curve higher than the Newtonian power curve. Thus the use of the Newtonian power curve to calculate powers will give conservative values when used for shear thinning liquids. Figure 5.10 also shows that the laminar flow region for shear thinning liquids extends to higher Reynolds numbers than that for Newtonian liquids.

Nienow and Elson (1988) have reviewed work done mainly by them and their co-workers on the mixing of non-Newtonian liquids in tanks. The above approach for inelastic, shearing thinning liquids has been largely substantiated but considerable doubt has been cast over using this method for dilatant, shear thickening materials.

In the case of highly elastic liquids mixed by a Rushton turbine, flow reversal may occur in the low Reynolds number region, $Re_M < 30$, leading to values of the power number as much as 60 per cent higher than for inelastic liquids. In the intermediate region, $50 \leqslant Re_M \leqslant 1000$, the power curve lies below the Newtonian one, rather similar to that shown in Figure 5.10 for inelastic, shear thinning liquids, except that the fairly sharp minimum shown there is much broader for elastic liquids. At higher values of Re_M the power curve lies slightly below that for a Newtonian liquid.

Mixing of elastic liquids is very strongly dependent on the type of mixer and the tank geometry, as well as the rheological properties of the liquid so that at a particular value of Re_M the power drawn may be lower or higher than that for a Newtonian liquid.

When mixing a liquid exhibiting a yield stress, it is clear that material near the impeller will be fluid while that further away, where the shear stress has fallen below the yield stress τ_y, will be stagnant. Mixing therefore occurs only in a 'cavern' around the impeller. The cavern diameter D_c for a flat blade single impeller can be calculated from the equation

$$\frac{D_c}{D_A} = \left[\frac{1.36}{\pi^2}\left(\frac{Po\rho N^2 D_A^2}{\tau_y}\right)\right]^{1/3} \tag{5.30}$$

for $Re_M > 10$ [Nienow and Elson (1988)]. Thus increasing the value of $Po\rho N^2 D_A^2/\tau_y$ causes the size of the mixing cavern to increase until the cavern fills the tank.

Example 5.1

Calculate the theoretical power for a six-blade flat blade turbine agitator with diameter $D_A = 3$ m running at a speed of $N = 0.2$ rev/s in a tank system conforming to the standard tank configuration illustrated in Figure 5.5. The liquid in the tank has a dynamic viscosity $\mu = 1.0$ Pa s and a density of $\rho = 1000$ kg/m^3.

Calculations

The Reynolds number for mixing is

$$Re_M = \frac{\rho N D_A^2}{\mu} \tag{5.14}$$

Substituting the given values

$$Re_M = \frac{(1000\ \text{kg/m}^3)(0.2\ \text{rev/s})(9.0\ \text{m}^2)}{1.0\ \text{Pa s}}$$

$$Re_M = 1800$$

From the graph of ϕ against Re_M in Figure 5.8

$$\phi = Po = 4.5$$

The theoretical power for mixing is

$$\begin{aligned} P_A &= Po\,\rho N^3 D_A^5 \\ &= (4.5)(1000\ \text{kg/m}^3)(0.008\ \text{rev}^3/\text{s}^3)(243\ \text{m}^5) \\ &= \underline{8748\ \text{W}} \end{aligned}$$

Example 5.2
Calculate the theoretical power for a six-blade flat blade turbine agitator with diameter $D_A = 0.1$ m running at $N = 16$ rev/s in a tank system without baffles but otherwise conforming to the standard tank configuration illustrated in Figure 5.5. The liquid in the tank has a dynamic viscosity of $\mu = 0.08$ Pa s and a density of $\rho = 900$ kg/m^3. For this configuration $\alpha = 1.0$ and $\beta = 40.0$.

Calculations
The Reynolds number for mixing is given by

$$Re_M = \frac{\rho N D_A^2}{\mu} \tag{5.14}$$

$$= \frac{(900\ \text{kg/m}^3)(16\ \text{rev/s})(0.01\ \text{m}^2)}{0.08\ \text{Pa s}}$$

$$= 1800$$

From the graph of ϕ against Re_M in Figure 5.9, $\phi = 2.2$. The theoretical power for mixing is given by

$$P_A = \phi \rho N^3 D_A^5 \left(\frac{N^2 D_A}{g}\right)^y \tag{5.28}$$

Now

$$y = \frac{\alpha - \log Re_M}{\beta} \tag{5.25}$$

with

$$\alpha = 1.0 \text{ and } \beta = 40.0$$

and

$$\log 1800 = 3.2553$$

Therefore

$$y = \frac{-2.2553}{40} = -0.05638$$

Substituting known values

$$\frac{N^2 D_A}{g} = \frac{(256 \text{ rev}^2/\text{s}^2)(0.1 \text{ m})}{9.81 \text{ m/s}^2}$$

$$= 2.610$$

So

$$\left(\frac{N^2 D_A}{g}\right)^y = 2.610^{(-0.05638)} = 0.9479$$

Therefore

$$P_A = (2.2)(900 \text{ kg/m}^3)(4096 \text{ rev}^3/\text{s}^3)(0.00001 \text{ m}^5)\ (0.9479)$$

$$= \underline{76.88 \text{ W}}$$

5.6 Scale-up of liquid mixing systems

The principle of similarity [Holland (1964), Johnstone and Thring (1957)] together with the use of dimensionless groups is the essential basis of scale-up. The types of similarity relevant to liquid mixing systems together with their definitions are listed as follows.

Geometrical similarity exists between two systems of different sizes when the ratios of corresponding dimensions in one system are equal to those in the other. Hence, geometrical similarity exists between two pieces of equipment of different sizes when both have the same shape.

Kinematic similarity exists between two systems of different sizes when they are not only geometrically similar but when the ratios of velocities between corresponding points in one system are equal to those in the other.

Dynamic similarity exists between two systems when, in addition to being geometrically and kinematically similar, the ratios of forces between corresponding points in one system are equal to those in the other.

Dimensionless groups provide a convenient way of correlating scientific and engineering data.

The classical principle of similarity can be expressed by equations of the form

$$N_1 = f(N_2, N_3 \ldots) \tag{5.31}$$

where a dimensionless group N_1 is a function of other dimensionless groups N_2, N_3, etc. Equation 5.31 is derived for a particular case by dimensional analysis, which is a technique for expressing the behaviour of a physical system in terms of the minimum number of independent variables.

Each dimensionless group represents a rule for scale-up. Frequently these individual scale-up rules conflict. For example, scale-up on dynamic similarity should depend chiefly upon a single dimensionless group that represents the ratio of the applied to the opposing forces. The Reynolds, Froude and Weber numbers are the ratios of the applied to the resisting viscous, gravitational and surfaces forces, respectively.

For scale-up from system 1 to system 2 for the same liquid properties and system geometry, equation 5.14 which defines the Reynolds number for mixing Re_M, equation 5.15 which defines the Froude number for mixing Fr_M, and equation 5.16 which defines the Weber number for mixing We_M, can be written respectively in the following forms:

$$N_1 D_{A1}^2 = N_2 D_{A2}^2 \tag{5.32}$$

$$N_1^2 D_{A1} = N_2^2 D_{A2} \tag{5.33}$$

$$N_1^2 D_{A1}^3 = N_2^2 D_{A2}^3 \tag{5.34}$$

Clearly the scale-up rules represented by equations 5.32, 5.33 and 5.34

conflict. In order to scale up with accuracy, it is often necessary to design pilot equipment so that the effects of certain dimensionless groups are deliberately suppressed in favour of a particular dimensionless group [Holland and Chapman (1966)]. For example, baffles can be used to eliminate vortexing so that the Froude number need not be considered.

Frequently it is not possible to achieve the desired similarity when scaling up from small to large scale units. In this case, results on the small scale unit must be extrapolated to dissimilar conditions on the large scale.

In order to extrapolate, use is made of what is known as the extended principle of similarity, where equations of the form

$$N_1 = CN_2^x N_3^y \dots \tag{5.35}$$

are used. Here the dimensionless group N_1 is proportional to the dimensionless group N_2 to the xth power and the dimensionless group N_3 to the yth power etc. C is a constant that depends on the geometry of the system and is consequently a shape factor, which usually must be determined by experiment.

Equation 5.17 for liquid mixing systems is in the same form as equation 5.35

$$Po = CRe_M^x Fr_M^y \tag{5.17}$$

The scale-up of liquid mixing systems can be divided into two categories: the scale-up of process result and the scale-up of power data.

The type of agitator and tank geometry required to achieve a particular process result, is determined from pilot plant experiments. The desired process result may be the dispersion or emulsification of immiscible liquids, the completion of a chemical reaction, the suspension of solids in a liquid or any one of a number of other processes [Holland and Chapman (1966)].

Once the process result has been satisfactorily obtained in the pilot size unit, it is necessary to predict the agitator speed in a geometrically similar production size unit using a suitable rule for scale-up.

Mutually conflicting scale-up rules are given by equations 5.32, 5.33 and 5.34. Other possible ways of scaling up are a constant tip speed u_T, and a constant ratio of circulating capacity to head Q_A/h_A, and a constant power per unit volume P_A/V. Since P_A is proportional to $N^3D_A^5$ and V is proportional to D_A^3, the ratio P_A/V is proportional to $N^3D_A^2$.

For scale-up from system 1 to system 2 for the same liquid properties and geometrically similar tanks the following equations can be written:

$$N_1 D_{A1} = N_2 D_{A2} \tag{5.36}$$

for a constant tip speed u_T

$$\frac{D_{A1}}{N_1} = \frac{D_{A2}}{N_2} \tag{5.37}$$

for a constant ratio of circulating capacity to head Q_A/h_A and

$$N_1^3 D_{A1}^2 = N_2^3 D_{A2}^2 \tag{5.38}$$

for a constant power per unit volume P_A/V. The scale-up rules given by equations 5.32 to 5.34 and 5.36 to 5.38 are all mutually conflicting.

In practice, the process result and corresponding agitator speeds can be obtained in three small geometrically similar tank systems of different sizes. These data can then be extrapolated to give the agitator speed in a geometrically similar production size tank system which will give the desired process result.

The power curve obtained on a pilot size unit can be used directly to obtain the power requirements for a geometrically similar production size unit once the agitator speed is known.

Consider the scale-up of the rotational speed of marine propellers for the same power consumption and Reynolds number for mixing.

The power consumption P_A is given by equation 5.27.

$$P_A = Po\,\rho N^3 D_A^5 \tag{5.27}$$

For scale-up from system 1 to system 2 for the same liquid properties and system geometry equation 5.27 can be written in the form

$$\frac{Po_1 N_1^3 D_{A1}^5}{P_{A1}} = \frac{Po_2 N_2^3 D_{A2}^5}{P_{A2}} \tag{5.39}$$

which for the same power consumption and Reynolds number and hence power number becomes

$$N_1^3 D_{A1}^5 = N_2^3 D_{A2}^5 \tag{5.40}$$

The corresponding equation for equality of Reynolds numbers for mixing has already been shown to be

$$N_1 D_{A1}^2 = N_2 D_{A2}^2 \tag{5.32}$$

Dividing equation 5.40 by equation 5.32 gives

$$N_1^2 D_{A1}^3 = N_2^2 D_{A2}^3 \tag{5.41}$$

The circulating capacity Q_A of a square pitch propeller has already been shown to be

$$Q_A = \frac{\eta\pi N D_A^3}{4} \tag{5.7}$$

For scale-up from system 1 to system 2, equation 5.7 can be written in the form

$$\frac{Q_{A1}}{\eta_1 N_1 D_{A1}^3} = \frac{Q_{A2}}{\eta_2 N_2 D_{A2}^3} \tag{5.42}$$

Combining equations 5.41 and 5.42 gives

$$Q_{A2} = \left(\frac{\eta_2}{\eta_1}\right)\left(\frac{N_1}{N_2}\right) Q_{A1} \tag{5.43}$$

Equation 5.43 shows that the circulation capacity of low speed square pitch propellers greatly exceeds that of high speed propellers for the same power consumption and Reynolds number [Holland and Chapman (1966)].

5.7 The purging of stirred tank systems

In industry it is common practice to use a number of tanks equipped with agitators in series. Frequently it is necessary to know the time required to reduce the concentration of off-quality material in the system below a certain acceptable limit.

Let a mass m of solute be dissolved in a liquid volume V in the first stirred tank, there being no solute in the other tanks. Let solute free liquid flow into the tank at a volumetric flow rate Q. Let liquid flow from the tank at a volumetric flow rate Q. If the liquid in the tank is uniformly mixed the discharge liquid contains a concentration of solute m/V which is the same as the solute concentration in the tank.

The rate of change of the mass of solute in the tank is given by the equation

$$\frac{\mathrm{d}m}{\mathrm{d}t} = -\frac{m}{V} Q \tag{5.44}$$

which can be integrated to give

$$\frac{m}{m_0} = \frac{C_{1t}}{C_{10}} = \mathrm{e}^{-Qt/V} \tag{5.45}$$

where C_{1t} is the solute concentration after time t and C_{10} is the initial solute concentration at time zero.

Equation 5.45 can also be written in the form

$$C_{1t} = C_{10}\,\mathrm{e}^{-\alpha t} \tag{5.46}$$

where $\alpha = Q/V$ is the reciprocal of the nominal holding time for the liquid in the tank.

The fraction x of the original solute which has been purged from the tank after a time t is given by the equation

$$x = \frac{m_0 - m}{m_0} = \frac{C_{10} - C_{1t}}{C_{10}} = 1 - \mathrm{e}^{-\alpha t} \tag{5.47}$$

For a second tank of the same size in series, the rate of change of solute concentration is given by the equation

$$V\frac{\mathrm{d}C_{2t}}{\mathrm{d}t} = Q(C_{1t} - C_{2t}) \tag{5.48}$$

which can be rewritten as

$$\frac{\mathrm{d}C_{2t}}{\mathrm{d}t} = \alpha(C_{1t} - C_{2t}) \tag{5.49}$$

At time $t = 0$, $C_{2t} = 0$. For this case substitute equation 5.46 into equation 5.49 and integrate to give

$$C_{2t} = \alpha C_{10} t\,\mathrm{e}^{-\alpha t} \tag{5.50}$$

Similarly for a third tank of the same size in series, the rate of change of solute concentration is

$$\frac{\mathrm{d}C_{3t}}{\mathrm{d}t} = \alpha(C_{2t} - C_{3t}) \tag{5.51}$$

At time $t = 0$, $C_{3t} = 0$. For this case substituting equation 5.50 into equation 5.51 and integrating gives

$$C_{3t} = \alpha^2 C_{10}\left(\frac{t^2}{2!}\right)\mathrm{e}^{-\alpha t} \tag{5.52}$$

Similarly, the concentration of solute in an nth tank of the same size in series can be written as

$$C_{nt} = \alpha^{n-1} C_{10}\left[\frac{t^{n-1}}{(n-1)!}\right]\mathrm{e}^{-\alpha t} \tag{5.53}$$

where at time $t = 0$, $C_{nt} = 0$. Adding equations 5.46, 5.50, 5.52 and 5.53 gives the equation

$$V(C_{1t}+C_{2t}+C_{3t}+\ldots+C_{nt}) = VC_{10}\,\mathrm{e}^{-\alpha t}\left[1+\alpha t+\frac{\alpha^2 t^2}{2!}+\ldots+\frac{\alpha^{n-1}t^{n-1}}{(n-1)!}\right] \quad (5.54)$$

Equation 5.54 gives the total amount of solute remaining in a system of n equal size tanks after a time t where at time $t = 0$ the concentration in the first tank was C_{10} and the concentration in all other tanks was zero.

The amount of solute purged from the system after time t is given by the equation

$$m_t = V\left\{C_{10}-C_{10}\,\mathrm{e}^{-\alpha t}\left[1+\alpha t+\frac{\alpha^2 t^2}{2!}+\ldots+\frac{\alpha^{n-1}t^{n-1}}{(n-1)!}\right]\right\} \quad (5.55)$$

The fraction x of the original solute which has been purged from the system after time t is

$$x = 1-\mathrm{e}^{-\alpha t}\left[1+\alpha t+\frac{\alpha^2 t^2}{2!}+\ldots+\frac{\alpha^{n-1}t^{n-1}}{(n-1)!}\right] \quad (5.56)$$

Equation 5.56 is known as the purging time equation for a system of continuous mixing vessels of equal size in series.

References

Brown, R.W., Scott, R. and Toyne, C., An investigation of heat transfer in agitated jacketed cast iron vessels, *Transactions of the Institution of Chemical Engineers*, **25**, pp. 181–8 (1947).

Gray, J.B., Batch mixing of viscous liquids, *Chemical Engineering Progress*, **59**, No. 3, pp. 55–9 (1963).

Holland, F.A., Scale-up in chemical engineering, *Chemical and Process Engineering*, **45**, pp. 121–4 (1964).

Holland, F.A. and Chapman, F.S., *Liquid Mixing and Processing in Stirred Tanks*, New York , Reinhold Publishing Corporation (1966).

Johnstone, R.E. and Thring, M.W., *Pilot Plants, Models and Scale-up Methods in Chemical Engineering*, New York, McGraw-Hill Book Company, Inc. (1957).

Metzner, A.B. and Otto, R.E., Agitation of non-Newtonian fluids, *AIChE Journal*, **3**, pp. 3–10 (1957).

Nienow, A.W. and Elson, T.P., Aspects of mixing in rheologically complex fluids, *Chemical Engineering Research and Design*, **66**, pp. 5–15 (1988).

Norwood, K.W. and Metzner, A.B., Flow patterns and mixing rates in agitated vessels, *AIChE Journal*, **6**, pp. 432–7 (1960).

Parker, N.H., Mixing, *Chemical Engineering*, **71**, pp. 165–220 (8 June 1964).

Quillen, C.S., Mixing: liquids, pastes, plastics, solids, continuous mixing, *Chemical Engineering*, **61**, pp. 178–224 (June 1954).

Rushton, J.H., Costich, E.W. and Everett, H.J., Power characteristics of mixing impellers, *Chemical Engineering Progress*, **46**, pp. 395–404 (1950).
Uhl, V.W. and Voznick, H.P., Anchor agitator, *Chemical Engineering Progress*, **56**, pp. 72–7 (1960).
Weber, A.P., Selecting propeller mixers, *Chemical Engineering*, **70**, pp. 91–8 (2 Sept 1963).

6 Flow of compressible fluids in conduits

When a compressible fluid, ie a gas, flows from a region of high pressure to one of low pressure it expands and its density decreases. It is necessary to take this variation of density into account in compressible flow calculations. In a pipe of constant cross-sectional area, the falling density requires that the fluid accelerate to maintain the same mass flow rate. Consequently, the fluid's kinetic energy increases.

It is found convenient to base compressible flow calculations on an energy balance per unit mass of fluid and to work in terms of the fluid's specific volume V rather than the density ρ. The specific volume is the volume per unit mass of fluid and is simply the reciprocal of the density:

$$V = \frac{1}{\rho} \tag{6.1}$$

6.1 Energy relationships

The total energy E per unit mass of fluid is given by either of the following equations:

$$E = U + zg + \frac{P}{\rho} + \frac{v^2}{2} \tag{1.8}$$

or

$$E = U + zg + PV + \frac{v^2}{2} \tag{6.2}$$

where U, zg, P/ρ and $v^2/2$ are the internal, potential, pressure and kinetic energies per unit mass respectively.

Consider unit mass of fluid flowing in steady state from a point 1 to a point 2. Between these two points, let a net amount of heat energy q be added to the fluid and let a net amount of work W be done on the fluid.

An energy balance for unit mass of fluid can be written either as

$$E_1 + q + W = E_2 \tag{6.3}$$

or as

$$(U_2 - U_1) + (z_2 - z_1)g + (P_2V_2 - P_1V_1) + \frac{(v_2^2 - v_1^2)}{2} = q + W \tag{6.4}$$

For steady flow in a pipe or tube the kinetic energy term can be written as $u^2/(2\alpha)$ where u is the volumetric average velocity in the pipe or tube and α is a dimensionless correction factor which accounts for the velocity distribution across the pipe or tube. Fluids that are treated as compressible are almost always in turbulent flow and α is approximately 1 for turbulent flow. Thus for a compressible fluid flowing in a pipe or tube, equation 6.4 can be written as

$$(U_2 - U_1) + (z_2 - z_1)g + (P_2V_2 - P_1V_1) + \frac{(u_2^2 - u_1^2)}{2} = q + W \tag{6.5}$$

where in SI units each term is in J/kg.

Since the enthalpy per unit mass of a fluid H is defined by the equation

$$H = U + PV \tag{6.6}$$

equation 6.5 can be written in the alternative form

$$(H_2 - H_1) + (z_2 - z_1)g + \frac{(u_2^2 - u_1^2)}{2} = q + W \tag{6.7}$$

The work term W in equations 6.5 and 6.7 is positive if work is done on the fluid by a pump or compressor. W is negative if the fluid does work in a turbine. W is often referred to as shaft work since it is transmitted into or out of a system by means of a shaft.

The differential form of equation 6.5 is

$$\mathrm{d}U + g\mathrm{d}z + P\mathrm{d}V + V\mathrm{d}P + \mathrm{d}\left(\frac{u^2}{2}\right) = \mathrm{d}q + \mathrm{d}W \tag{6.8}$$

For a reversible change, the first law of thermodynamics can be expressed by the equation

$$\mathrm{d}q = \mathrm{d}U + P\mathrm{d}V \tag{6.9}$$

where $\mathrm{d}U$ is the increase in internal energy per unit mass of fluid and $P\mathrm{d}V$ is the work of expansion on the fluid layers ahead for a net addition of heat $\mathrm{d}q$ to the system.

In flow, energy is required to overcome friction. The effect of friction is to generate heat in a system by converting mechanical to thermal energy. Thus where friction is involved, equation 6.9 can be written as

$$dq = dU + PdV - dF \tag{6.10}$$

where dF is the energy per unit mass required to overcome friction. Substituting equation 6.10 into equation 6.8 gives

$$g\,dz + VdP + d\left(\frac{u^2}{2}\right) + dF = dW \tag{6.11}$$

Equation 6.11 can be integrated between states 1 and 2 to give

$$(z_2 - z_1)g + \int_1^2 VdP + \frac{u_2^2 - u_1^2}{2} + F = W \tag{6.12}$$

where in SI units each term is in J/kg.

Equations 6.5, 6.7 and 6.12 all relate to the energy changes involved for a fluid in steady turbulent flow. The most appropriate equation is selected for each particular application: equation 6.12 is a convenient form from which a basic flow rate–pressure drop equation will be derived.

Due to the change in the average velocity u, it is more convenient in calculations for compressible flow in pipes of constant cross-sectional area to work in terms of the mass flux G. This is the mass flow rate per unit flow area and is sometimes called the mass velocity. If the mass flow rate is constant, as will usually be the case, then G is constant when the area is constant. The relationship between G and u is given by

$$G = \frac{M}{S} = \rho u = \frac{u}{V} \tag{6.13}$$

In SI units G is in kg/(m^{2}s).

Writing equation 6.11 in terms of G, noting that $d(V^2/2) = VdV$, gives

$$g\,dz + VdP + G^2VdV + dF = dW \tag{6.14}$$

The pressure drop ΔP_f due to friction in a pipe of length L and inside diameter d_i is given by equation 2.13

$$\Delta P_f = 4f\left(\frac{L}{d_i}\right)\frac{\rho u^2}{2} \tag{2.13}$$

where f is the Fanning friction factor.

For an element of length dx of the pipe, equation 2.13 can be written as

$$\mathrm{d}P_f = 2f\left(\frac{\mathrm{d}x}{d_i}\right)\rho u^2 \tag{6.15}$$

The corresponding energy required to overcome friction is $\mathrm{d}f = V\mathrm{d}P_f$. Thus equation 6.15 gives $\mathrm{d}F$ as

$$\mathrm{d}F = 2f\left(\frac{\mathrm{d}x}{d_i}\right)u^2 = 2f\left(\frac{\mathrm{d}x}{d_i}\right)G^2V^2 \tag{6.16}$$

where advantage has been taken of equations 6.1 and 6.13.

Substituting for $\mathrm{d}F$ in equation 6.14 gives

$$g\mathrm{d}z + V\mathrm{d}P + G^2V\mathrm{d}V + 2f\left(\frac{\mathrm{d}x}{d_i}\right)G^2V^2 = \mathrm{d}W \tag{6.17}$$

Dividing equation 6.17 throughout by V^2 and integrating between states 1 and 2 over a length L of pipe gives

$$\int_1^2 \frac{g}{V^2}\mathrm{d}z + \int_1^2 \frac{\mathrm{d}P}{V} + G^2 \ln\left(\frac{V_2}{V_1}\right) + \frac{2fG^2L}{d_i} = \int_1^2 \frac{\mathrm{d}W}{V_2} \tag{6.18}$$

In integrating the frictional term it has been assumed that the value of the friction factor is constant: this is a good approximation because the Reynolds number will usually be very high, a condition for which f is independent of Re.

In almost all cases, the change in potential energy will be negligible for gas flow. Also, it is convenient to treat a compressor separately from flow in the pipe, ie equation 6.18 will be applied to a section of pipe in which no shaft work is done. Consequently, 6.18 can be written in a reduced form:

$$\int_1^2 \frac{\mathrm{d}P}{V} + G^2\left[\ln\left(\frac{V_2}{V_1}\right) + \frac{2fL}{d_i}\right] = 0 \tag{6.19}$$

Equation 6.19 is that required for most calculations involving compressible flow in a pipe. The three terms represent, respectively, changes in pressure energy, kinetic energy and the conversion of mechanical energy to thermal energy by frictional dissipation. The terms in the square brackets are necessarily positive so the pressure energy term must be negative: this reflects the fact that the pressure falls in the direction of flow.

In most cases the kinetic energy term will be negligible compared with the frictional term. This is useful when calculating the pressure drop for a given flow rate because in this case one of the pressures and therefore the corresponding specific volume will be unknown. An approximate calculation can be made neglecting the kinetic energy term then, when the pressures are known, the value of that term can be calculated to check whether it was in fact negligible.

In equation 6.19, and all other equations in this chapter, P denotes the absolute pressure.

In order to make use of equation 6.19 or equation 6.18 it is necessary to know the relationship between the pressure P and the specific volume V so that terms such as $\int_1^2 \mathrm{d}P/V$ can be evaluated. The relationship between P and V is known as the equation of state.

6.2 Equations of state

An ideal or perfect gas obeys the equation

$$PV = \frac{RT}{\mathrm{RMM}} \tag{6.20}$$

where R is the universal gas constant, T the absolute temperature and RMM the relative molecular mass converson factor for the gas. In SI units R = 8314.3 J/(kmol K) and T is in K. The conversion factor RMM has the numerical value of the relative molecular mass and the units kg/kmol in the SI system. Equation 6.20, which is a combination of Boyle's and Charles's laws, will be more familiar in the form

$$P\hat{V} = RT \tag{6.21}$$

where $\hat{V}$ is the molar volume of the gas. The relative molecular mass has to be introduced in equation 6.20 because V is the specific volume, ie the volume per unit mass. It is convenient to define a specific gas constant R' by

$$R' = \frac{R}{\mathrm{RMM}} \tag{6.22}$$

so that equation 6.20 can be written as

$$PV = R'T \tag{6.23}$$

It is essential to remember that in equation 6.23 both V and R' are values per unit mass of gas and they must not be confused with the molar equivalents. The value of R' is different for gases of different relative molecular masses.

Many gases obey equation 6.23 up to a few atmospheres pressure. At high pressures it is necessary to modify equation 6.23 by introducing the compressibility factor Z:

$$PV = ZR'T \tag{6.24}$$

The compressibility factor is a function of the reduced pressure P_r and the reduced temperature T_r of the gas. P_r is the ratio of the actual pressure P to the critical pressure P_c of the gas:

$$P_r = \frac{P}{P_c} \tag{6.25}$$

and T_r is the ratio of the actual temperature T to the critical temperature T_c of the gas:

$$T_r = \frac{T}{T_c} \tag{6.26}$$

Plots of Z against P_r at constant T_r are available [Perry (1984), Smith and Van Ness (1987)].

When ideal gases are compressed or expanded they obey the following general equation:

$$PV^k = \text{constant} \tag{6.27}$$

Thus, for two states 1 and 2, equation 6.27 gives

$$P_1 V_1^k = P_2 V_2^k \tag{6.28}$$

and

$$\frac{P_2}{P_1} = \left(\frac{V_1}{V_2}\right)^k \tag{6.29}$$

Combining equation 6.29 with equation 6.23 gives the relationship between pressure and temperature:

$$\frac{P_2}{P_1} = \left(\frac{T_2}{T_1}\right)^{k/(k-1)} \tag{6.30}$$

Equation 6.30 shows that, in general, expansion or compression of a gas is accompanied by a change of temperature.

A change of state according to equation 6.27 is called a polytropic change. Two special cases are the isothermal change and the adiabatic change.

As the name implies, an isothermal change takes place at constant temperature. This requires that the process be relatively slow and heat transfer between the gas and the surroundings be rapid. An isothermal change corresponds to $k = 1$ and equation 6.27 becomes

$$PV = \text{constant} \tag{6.31}$$

for an ideal gas.

The other extreme case is the adiabatic change, which occurs with no heat transfer between the gas and the surroundings. For a reversible adiabatic change, $k = \gamma$ where $\gamma = C_p/C_v$, the ratio of the specific heat capacities at constant pressure (C_p) and at constant volume (C_v). For a reversible adiabatic change of an ideal gas, equation 6.27 becomes

$$PV^{\gamma} = \text{constant} \tag{6.32}$$

and equation 6.29 becomes

$$\frac{P_2}{P_1} = \left(\frac{V_1}{V_2}\right)^{\gamma} \tag{6.33}$$

From equation 6.30, the temperature change is given by

$$\frac{P_2}{P_1} = \left(\frac{T_2}{T_1}\right)^{\gamma/(\gamma-1)} \tag{6.34}$$

In a reversible adiabatic change the entropy remains constant and therefore this type of change is called an isentropic change. Although not rigorously valid for irreversible changes, equations 6.32 to 6.34 are good approximations for these conditions.

Approximate values of γ at ordinary temperatures and pressures are 1.67 for monatomic gases such as helium and argon, 1.40 for diatomic gases such as hydrogen, carbon monoxide and nitrogen, and 1.30 for triatomic gases such as carbon dioxide. Gases and vapours of complex molecules can have significantly lower values of γ, for example 1.05 for *n*-heptane and 1.03 for *n*-decane.

6.3 Isothermal flow of an ideal gas in a horizontal pipe

For steady flow of a gas between points 1 and 2, distance L apart, in a horizontal pipe of constant cross-sectional area in which no shaft work is done, the energy relationships are given by equation 6.19:

$$\int_1^2 \frac{dP}{V} + G^2\left[\ln\left(\frac{V_2}{V_1}\right) + \frac{2fL}{d_i}\right] = 0 \tag{6.19}$$

In the case of isothermal flow of an ideal gas, the equation of state can be written as

$$PV = P_1V_1 \tag{6.35}$$

and evaluation of the integral in equation 6.19 gives

$$\int_1^2 \frac{dP}{V} = \frac{1}{P_1V_1}\int_1^2 P\,dP = \frac{P_2^2 - P_1^2}{2P_1V_1} \tag{6.36}$$

Often the upstream pressure P_1 will be unknown but for an isothermal change P_1V_1 can be replaced by any known value of PV at the same temperature, for example the downstream conditions P_2V_2 if P_2 is specified.

In the kinetic energy term, from equation 6.35

$$\ln\left(\frac{V_2}{V_1}\right) = \ln\left(\frac{P_1}{P_2}\right) \tag{6.37}$$

Substituting equations 6.36 and 6.37 into equation 6.19 gives the following working equation for isothermal flow of an ideal gas:

$$\frac{P_2^2 - P_1^2}{2P_1V_1} + G^2\left[\ln\left(\frac{P_1}{P_2}\right) + \frac{2fL}{d_i}\right] = 0 \tag{6.38}$$

As noted previously, the first term is negative, as must be the case because $P_1 > P_2$.

Equation 6.38 is the basic form of the energy equation to be used for isothermal conditions, however it is instructive to write the equation in a slightly different form that allows easy comparison with incompressible flow.

The pressure energy term can be written in the following form:

$$\frac{P_2^2 - P_1^2}{2P_1V_1} = \frac{(P_2 + P_1)(P_2 - P_1)}{2P_1V_1} = \frac{P_m(P_2 - P_1)}{P_1V_1} \tag{6.39}$$

where $P_m = (P_2 + P_1)/2$ is the arithmetic mean pressure in the pipe. By equation 6.35

$$P_mV_m = P_1V_1 \tag{6.40}$$

so that

$$\frac{P_m(P_2-P_1)}{P_1V_1}=\frac{P_2-P_1}{V_m} \tag{6.41}$$

where V_m is the specific volume at the mean pressure P_m. Thus, equation 6.38 can be written as

$$\frac{P_2-P_1}{V_m}+G^2\left[\ln\left(\frac{P_1}{P_2}\right)+\frac{2fL}{d_i}\right]=0 \tag{6.42}$$

As noted previously, the kinetic energy term is usually negligible compared with the frictional term and this is certainly true when the pressure drop $\Delta P = P_1-P_2$ is small compared with P_1. In this case, equation 6.42 can be approximated by

$$\frac{P_2-P_1}{V_m}+\frac{2fLG^2}{d_i}=0 \tag{6.43}$$

or

$$\Delta P_f=P_1-P_2=\frac{2fLG^2V_m}{d_i}=2f\left(\frac{L}{d_i}\right)\rho_m u_m^2 \tag{6.44}$$

where ρ_m and u_m are the density and average velocity at the mean pressure P_m. Equation 6.44 will be recognized as being of the same form as equation 2.13 for incompressible flow, except that it is written in terms of average properties.

Thus, when the pressure drop is small compared with the mean pressure in the pipe, the gas flow may be treated as incompressible flow. For large values of the pressure drop it is necessary to use equation 6.38.

In order to maintain isothermal flow it is necessary for heat to be transferred across the pipe wall. From equation 6.7, for flow in a section with no shaft work and negligible change in elevation, the energy equation takes the form

$$(H_2-H_1)+\frac{u_2^2-u_1^2}{2}=q \tag{6.45}$$

For an ideal gas under isothermal conditions, the enthalpy remains constant and hence it follows from equation 6.45 that the required heat leak into the pipe is equal to the increase in kinetic energy. This is usually a small quantity and therefore flow in long, uninsulated pipes will be virtually isothermal.

Example 6.1

Hydrogen is to be pumped from one vessel through a pipe of length 400 m to a second vessel, which is at a pressure of 20 bar absolute. The required flow rate is 0.2 kg/s and the allowable pressure at the pipe inlet is 25 bar. The flow conditions are isothermal and the gas temperature is 25°C. If the friction factor may be assumed to have a value of 0.005, what diameter of pipe is required?

Calculations

$$\int_1^2 \frac{dP}{V} + G^2\left[\ln\left(\frac{V_2}{V_1}\right) + \frac{2fL}{d_i}\right] = 0 \tag{6.19}$$

For isothermal conditions

$$\int_1^2 \frac{dP}{V} = \frac{P_2^2 - P_1^2}{2P_1V_1} \qquad \text{and} \qquad \ln\left(\frac{V_2}{V_1}\right) = \ln\left(\frac{P_1}{P_2}\right)$$

Equation of state

$$PV = R'T = \frac{RT}{\text{RMM}}$$

$T = 298$ K and for hydrogen RMM $= 2$ kg/kmol. Therefore

$$PV = \frac{8314.3 \text{ J/(kmol K)} \times 298 \text{ K}}{2 \text{ kg/kmol}} = 2.239 \times 10^6 \text{ J/kg}$$

The product PV is constant for isothermal conditions. Therefore

$$\frac{P_2^2 - P_1^2}{2P_1V_1} = \frac{(20 \times 10^5 \text{ Pa})^2 - (25 \times 10^5 \text{ Pa})^2}{2(2.239 \times 10^6 \text{ J/kg})} = -5.025 \times 10^5 \text{ kg}^2/(\text{m}^4\text{s}^2)$$

Mass flux is given by

$$G = \frac{M}{S} = \frac{0.2 \text{ kg/s}}{(\pi d_i^2/4) \text{ m}^2} = \frac{0.255}{d_i^2} \text{ kg/m}^2\text{s} \tag{6.13}$$

From the given values

$$\ln\frac{P_1}{P_2} = 0.223 \qquad \text{and} \qquad \frac{2fL}{d_i} = \frac{4}{d_i}$$

Substituting these values into equation 6.19 gives

$$\left(\frac{6.485 \times 10^{-2}}{d_i^4}\right)\left(0.223 + \frac{4}{d_i}\right) = 5.025 \times 10^5$$

It may be anticipated that $0.223 \ll 4/d_i$, ie $d_i \ll 17.9$ m. Thus the calculation may be simplified by neglecting the kinetic energy term, so that

$$d_i^5 = \frac{4 \times 6.485 \times 10^{-2}}{5.025 \times 10^5} = 5.162 \times 10^{-7} \text{ m}^5$$

and

$$d_i = \underline{0.0553 \text{ m}}$$

6.4 Non-isothermal flow of an ideal gas in a horizontal pipe

For steady flow of an ideal gas between points 1 and 2, distance L apart, in a pipe of constant cross-sectional area in which no shaft work is done, the energy equation is given by equation 6.19. For the general case of polytropic flow, from equation 6.27, the equation of state can be written as

$$PV^k = P_1 V_1^k \tag{6.46}$$

The pressure energy term in equation 6.19 can be written as

$$\int_1^2 \frac{dP}{V} = \left(\frac{k}{k+1}\right)\left(\frac{P_1}{V_1}\right)\left[\left(\frac{P_2}{P_1}\right)^{(k+1)/k} - 1\right] \tag{6.47}$$

Also

$$\ln\left(\frac{V_2}{V_1}\right) = \frac{1}{k} \ln\left(\frac{P_1}{P_2}\right) \tag{6.48}$$

Thus, equation 6.19 becomes

$$\left(\frac{k}{k+1}\right)\left(\frac{P_1}{V_1}\right)\left[\left(\frac{P_2}{P_1}\right)^{(k+1)/k} - 1\right] + G^2\left[\frac{1}{k}\ln\left(\frac{P_1}{P_2}\right) + \frac{2fL}{d_i}\right] = 0 \tag{6.49}$$

Equation 6.49 is the general equation for polytropic flow of an ideal gas in a horizontal pipe with no shaft work. On putting $k = 1$, equation 6.38 for isothermal flow is obtained.

Putting $k = \gamma$ gives an approximate equation for adiabatic flow. The result is only approximate because it implies an isentropic change, ie a reversible adiabatic change, but this is not the case owing to friction. A rigorous solution for adiabatic flow is given in Section 6.5.

6.5 Adiabatic flow of an ideal gas in a horizontal pipe

Equation 6.19 is the basic equation relating the pressure drop to the flow rate. The difficulty that arises in the case of adiabatic flow is that the equation of state is unknown. The relationship, $PV^{\gamma} =$ constant, is valid for a *reversible* adiabatic change but flow with friction is irreversible. Thus a difficulty arises in determining the integral in equation 6.19: an alternative method of finding an expression for $\mathrm{d}P/V$ is sought.

For adiabatic flow in a horizontal pipe with no shaft work, equation 6.7 reduces to

$$H_2 - H_1 + \frac{u_2^2 - u_1^2}{2} = 0 \tag{6.50}$$

The differential form of equation 6.50 with u expressed in terms of G and V is

$$\mathrm{d}H + G^2 V \mathrm{d}V = 0 \tag{6.51}$$

The enthalpy change can be found from the following two fundamental thermodynamic relationships which, in the case of ideal gases, are valid for irreversible processes as well as reversible ones:

$$\mathrm{d}U = C_v \mathrm{d}T \tag{6.52}$$

$$\mathrm{d}H = C_p \mathrm{d}T \tag{6.53}$$

From equation 6.6

$$\mathrm{d}H = \mathrm{d}U + \mathrm{d}(PV) \tag{6.54}$$

or

$$C_p \mathrm{d}T = C_v \mathrm{d}T + \mathrm{d}(PV) \tag{6.55}$$

Thus

$$\mathrm{d}T = \mathrm{d}(PV)/(C_p - C_v) \tag{6.56}$$

and

$$\mathrm{d}H = C_v \mathrm{d}T + \mathrm{d}(PV) = \left(\frac{C_v}{C_p - C_v} + 1\right)\mathrm{d}(PV) \tag{6.57}$$

ie

$$dH = \left(\frac{\gamma}{\gamma - 1}\right) d(PV) \tag{6.58}$$

Thus, equation 6.51 can be written as

$$\left(\frac{\gamma}{\gamma - 1}\right) d(PV) + G^2 V dV = 0 \tag{6.59}$$

Integrating equation 6.59 gives the desired relationship between P and V:

$$PV + \left(\frac{\gamma - 1}{2\gamma}\right) G^2 V^2 = \text{constant} = C \tag{6.60}$$

Thus

$$\frac{1}{V}\frac{dP}{dV} = -\frac{C}{V^3} - \left(\frac{\gamma - 1}{\gamma}\right)\frac{G^2}{2}\frac{1}{V} \tag{6.61}$$

and the integral in equation 6.19 is readily found by integrating equation 6.61:

$$\int_1^2 \frac{dP}{V} = \frac{C}{2}\left(\frac{1}{V_2^2} - \frac{1}{V_1^2}\right) - \left(\frac{\gamma - 1}{\gamma}\right)\frac{G^2}{2} \ln\left(\frac{V_2}{V_1}\right) \tag{6.62}$$

Substituting this result in equation 6.19 gives

$$C\left(\frac{1}{V_2^2} - \frac{1}{V_1^2}\right) + G^2\left[\left(\frac{\gamma + 1}{\gamma}\right)\ln\left(\frac{V_2}{V_1}\right) + \frac{4fL}{d_i}\right] = 0 \tag{6.63}$$

where, from equation 6.60

$$C = P_1 V_1 + \left(\frac{\gamma - 1}{2\gamma}\right) G^2 V_1^2 = P_2 V_2 + \left(\frac{\gamma - 1}{2\gamma}\right) G^2 V_2^2 \tag{6.64}$$

Calculations for adiabatic flow require the use of equations 6.63 and 6.64. For example, if the upstream conditions P_1 and V_1 are known, and G and d_i are specified, C can be calculated from equation 6.64 then V_2 from equation 6.63. Substituting this value of V_2 in equation 6.64 gives P_2. If the logarithmic term is not negligible, an iterative calculation will be needed to determine V_2 from equation 6.63.

Rapid flow in relatively short sections of pipe-work may approach adiabatic conditions. The flow rate for an ideal gas for a given pressure drop is greater for adiabatic conditions than for isothermal conditions. Although the maximum possible difference is 20 per cent, for ratios $L/d_i > 1000$ the difference is seldom more than 5 per cent. Consequently, it is common practice to assume isothermal conditions, any departure providing a small bonus.

6.6 Speed of sound in a fluid

The speed u_w with which a small pressure wave propagates through a fluid can be shown [Shapiro (1953)] to be related to the compressibility of the fluid $\partial\rho/\partial P$ by equation 6.65:

$$u_w = \sqrt{\frac{\partial P}{\partial \rho}} \tag{6.65}$$

Assuming that the pressure wave propagates through the fluid polytropically, then the equation of state is

$$PV^k = \text{constant} = K \tag{6.66}$$

from which

$$P = \rho^k K \tag{6.67}$$

Thus

$$\frac{\partial P}{\partial \rho} = k\rho^{k-1}K = \frac{kP}{\rho} = kPV \tag{6.68}$$

The propagation speed u_w of the pressure wave is therefore given by

$$u_w = \sqrt{kPV} \tag{6.69}$$

where P, V, are the local pressure and specific volume of the fluid through which the wave is propagating. Note that u_w is relative to the gas. If the wave were to propagate isothermally, its speed would be $\sqrt{PV}$.

In practice, small pressure waves (such as sound waves) propagate virtually isentropically. The reasons for this are that, being a very small disturbance, the change is almost reversible and, by virtue of the high speed, there is very little heat transfer. Thus the speed of sound c is equal to the speed at which a small pressure wave propagates isentropically, so from equation 6.69

$$c = \sqrt{\gamma PV} \tag{6.70}$$

As will be seen in Section 6.10, there is a fundamental difference between flow in which the fluid's speed u is less than c, ie subsonic flow, and that when u is greater than c, ie supersonic flow. It is therefore useful to define the Mach number Ma:

$$Ma = \frac{u}{c} \tag{6.71}$$

Subsonic flow corresponds to $Ma < 1$ and supersonic flow to $Ma > 1$. The conditions of incompressible flow are approached as $Ma \to 0$.

6.7 Maximum flow rate in a pipe of constant cross-sectional area

Consider the case of steady polytropic flow in a horizontal pipe described by equation 6.49:

$$\left(\frac{k}{k+1}\right)\left(\frac{P_1}{V_1}\right)\left[\left(\frac{P_2}{P_1}\right)^{(k+1)/k} - 1\right] + G^2\left[\frac{1}{k}\ln\left(\frac{P_1}{P_2}\right) + \frac{2fL}{d_i}\right] = 0 \tag{6.49}$$

If the upstream pressure P_1 is kept constant and the downstream pressure P_2 is gradually reduced, the flow rate will gradually increase. However, as $P_2 \to 0$ the density $\rho \to 0$ and consequently the mass flow rate must approach zero if the gas speed u remains finite. Thus, at some value of P_2 satisfying the condition $0 < P_2 < P_1$, the mass flow rate must reach a maximum.

Differentiating equation 6.49 with respect to P_2 gives

$$\frac{1}{P_2}\frac{P_1}{V_1}\left(\frac{P_2}{P_1}\right)^{(k+1)/k} + 2G\frac{\partial G}{\partial P_2}\left[\frac{1}{k}\ln\left(\frac{P_1}{P_2}\right) + \frac{2fL}{d_i}\right] - \frac{G^2}{kP_2} = 0 \tag{6.72}$$

Putting $\partial G/\partial P_2 = 0$, the conditions giving the maximum flow rate G_w satisfy the equation

$$\frac{P_1}{V_1}\left(\frac{P_w}{P_1}\right)^{(k+1)/k} = \frac{G_w^2}{k} \tag{6.73}$$

where P_w is the value of P_2 at which the maximum flow rate occurs. Combining the equation of state (equation 6.27) with equation 6.73 gives the maximum mass flux as

$$G_w = \sqrt{kP_w/V_w} \tag{6.74}$$

The corresponding gas speed u_w is therefore given by

$$u_w = \sqrt{kP_w V_w} \tag{6.75}$$

Recalling equation 6.69, it will be seen that the maximum flow rate is achieved when the gas speed reaches the speed at which a small pressure wave propagates through the gas. (This is why the subscript w has been used to denote this condition.) If the downstream pressure P_2 is reduced further, there can be no increase in the flow rate and the flow is said to be choked.

A simple interpretation of this choking condition is as follows. The gas flows as a result of the pressure difference $P_1 - P_2$. When the gas speed reaches the speed at which a pressure wave propagates relative to the gas, any pressure wave generated will be unable to travel upstream but will remain stationary relative to the pipe. Thus, if the pressure in the reservoir into which the gas discharges is reduced below P_w, the fact cannot be transmitted upstream and so the flow rate will not change.

Putting $k = 1$ gives the value of u_w for isothermal conditions but the result is of doubtful value because it would be extremely difficult, if not impossible, to maintain isothermal conditions at the very high speeds involved.

Putting $k = \gamma$ in equation 6.75 gives the gas speed at the maximum flow rate as $\sqrt{\gamma P_w V_w}$. That this is the correct result for adiabatic flow can be confirmed from equations 6.63 and 6.64. Differentiating equation 6.63 with respect to P_2 gives

$$-\frac{2C}{V_2^3}\frac{\partial V_2}{\partial P_2} + 2G\frac{\partial G}{\partial P_2}\left[\left(\frac{\gamma+1}{\gamma}\right)\ln\left(\frac{V_2}{V_1}\right) + \frac{4fL}{d_i}\right] + G^2\left(\frac{\gamma+1}{\gamma}\right)\frac{1}{V_2}\frac{\partial V_2}{\partial P_2} = 0 \tag{6.76}$$

Putting $\partial G/\partial P_2 = 0$ to determine the maximum mass flux G_w gives

$$\left[G_w^2\left(\frac{\gamma+1}{\gamma}\right)\frac{1}{V_w} - \frac{2C}{V_w^3}\right]\frac{\partial V_2}{\partial P_2} = 0 \tag{6.77}$$

Noting that $\partial V_2/\partial P_2 \neq 0$, the quantity in square brackets must be equal to zero. From equation 6.60, the constant C is given by

$$C = P_w V_w + \left(\frac{\gamma-1}{2\gamma}\right)G_w^2 V_w^2 \tag{6.78}$$

When this value is substituted into equation 6.77, it is found that

$$G_w = \sqrt{\gamma P_w / V_w} \tag{6.79}$$

and therefore

$$u_w = \sqrt{\gamma P_w V_w} \tag{6.80}$$

6.8 Adiabatic stagnation temperature for an ideal gas

For adiabatic flow with negligible change of elevation and no shaft work, the energy equation reduces to

$$H_2 - H_1 + \frac{u_2^2 - u_1^2}{2} = 0 \tag{6.50}$$

which may be written in differential form as

$$\mathrm{d}H + \mathrm{d}\left(\frac{u^2}{2}\right) = 0 \tag{6.81}$$

Substituting for dH using equation 6.53 gives

$$C_p \mathrm{d}T + \mathrm{d}\left(\frac{u^2}{2}\right) = 0 \tag{6.82}$$

Equations 6.81 and 6.82 show that as the velocity rises the kinetic energy increases at the expense of the enthalpy and consequently the temperature falls. Over the relatively small temperature changes involved, C_p may be taken as constant. Integrating equation 6.83 with the condition $T = T_0$ when $u = 0$ gives

$$C_p(T - T_0) + \frac{u^2}{2} = 0 \tag{6.83}$$

or

$$T = T_0 - \frac{u^2}{2C_p} \tag{6.84}$$

The temperature T_0 corresponding to zero velocity is known as the adiabatic stagnation temperature: it is the temperature that the flowing gas would attain if it were brought to rest adiabatically without doing any shaft work. It is sometimes called the total temperature. The temperature difference is small: for air flowing at a speed of 100 m/s, $T_0 - T = 5$ K.

When a thermometer is placed in a flowing gas stream, most of the thermometer's surface has gas flowing past it but a stagnation point occurs at its upstream side. Thus instead of measuring the temperature T, it measures a value that is slightly higher. This can be accommodated by introducing a correction factor known as the recovery factor r_f:

$$T_m = T_0 - \frac{r_f u^2}{2C_p} \tag{6.85}$$

where T_m is the measured temperature. For thermometers of conventional design, $r_f = 0.88$, Barna (1969).

6.9 Gas compression and compressors

Compressors are devices for supplying energy or pressure head to a gas. For the most part, compressors like pumps can be classified into centrifugal and positive displacement types. Centrifugal compressors impart a high velocity to the gas and the resultant kinetic energy provides the work for compression. Positive displacement compressors include rotary and reciprocating compressors although the latter are the most important for high pressure applications.

From equation 6.12, the shaft work of compression W required to compress unit mass of gas from pressure P_1 to pressure P_2 in a reversible frictionless process, in which changes in potential and kinetic energy are negligible, is

$$W = \int_1^2 V\mathrm{d}P \tag{6.86}$$

Although isothermal compression is desirable, in practice the heat of compression is never removed fast enough to make this possible. In actual compressors only a small fraction of the heat of compression is removed and the process is almost adiabatic.

When ideal gases are compressed under reversible adiabatic conditions they obey equation 6.32, which can be written as

$$PV^{\gamma} = P_1 V_1^{\gamma} \tag{6.87}$$

so that the specific volume is given by

$$V = \frac{P_1^{1/\gamma} V_1}{P^{1/\gamma}} \tag{6.88}$$

Substituting for V in equation 6.86 and integrating gives

$$W = \left(\frac{\gamma}{\gamma - 1}\right) P_1 V_1 \left[\left(\frac{P_2}{P_1}\right)^{(\gamma-1)/\gamma} - 1\right] \tag{6.89}$$

Equation 6.89 gives the theoretical adiabatic work of compression from pressure P_1 to pressure P_2.

Compression is often done in several stages with the gas being cooled between stages. For two-stage compression from P_1 to P_2 to P_3, with the gas cooled to the initial temperature T_1 at constant pressure, equation 6.89 becomes

$$W = \left(\frac{\gamma}{\gamma-1}\right)P_1V_1\left\{\left[\left(\frac{P_2}{P_1}\right)^{(\gamma-1)/\gamma}-1\right]+\left[\left(\frac{P_3}{P_2}\right)^{(\gamma-1)/\gamma}-1\right]\right\} \quad (6.90)$$

In the case of compression from pressure P_1 to pressure P_2 through n stages each having the same pressure ratio $(P_2/P_1)^{1/n}$, the compression work is given by

$$W = \left(\frac{n\gamma}{\gamma-1}\right)P_1V_1\left[\left(\frac{P_2}{P_1}\right)^{(\gamma-1)/n\gamma}-1\right] \quad (6.91)$$

Equations 6.89 to 6.91 give the work required to compress *unit* mass of the gas. It should be noted that the work required depends on the pressure ratio so that compression from 1 bar to 10 bar requires as much power as compressing the same mass of gas, with the same initial temperature, from 10 bar to 100 bar.

In practice it is possible to approach more nearly isothermal compression by carrying out the compression in a number of stages with cooling of the gas between stages.

When ideal gases are compressed under reversible adiabatic conditions the temperature rise from T_1 to T_2 is given by equation 6.34:

$$\frac{P_2}{P_1} = \left(\frac{T_2}{T_1}\right)^{\gamma/(\gamma-1)} \quad (6.34)$$

So far only reversible adiabatic compression of an ideal gas has been considered. For the irreversible adiabatic compression of an actual gas, the shaft work W required to compress the gas from state 1 to state 2 can be obtained from equation 6.7, which in this case becomes

$$W = H_2 - H_1 \quad (6.92)$$

where H is the enthalpy per unit mass of gas.

The actual work of compression is greater than the theoretical work because of clearance gases, back leakage and friction.

Example 6.2

Calculate the theoretical work required to compress 1 kg of a diatomic ideal gas initially at a temperature of 200 K adiabatically from a pressure of 10000 Pa to a pressure of 100000 Pa in (i) a single stage, (ii) a compressor with two equal stages and (iii) a compressor with three equal stages. The relative molecular mass of the gas is 28.0 and the ratio of specific heat capacities γ is 1.40.

Calculations

(i) For a single stage compression,

$$W = \left(\frac{\gamma}{\gamma - 1}\right) P_1 V_1 \left[\left(\frac{P_2}{P_1}\right)^{(\gamma-1)/\gamma} - 1\right] \tag{6.89}$$

From given values

$$\frac{P_2}{P_1} = 10$$

Therefore

$$\left(\frac{P_2}{P_1}\right)^{(\gamma-1)/\gamma} = 10^{0.2857} = 1.931$$

Equation of state

$$PV = R'T = \frac{RT}{\text{RMM}}$$

Therefore

$$P_1 V_1 = \frac{RT_1}{\text{RMM}}$$

$$= \frac{8314.3 \text{ J/(kmol K)} \times 200 \text{ K}}{28.0 \text{ kg/kmol}}$$

$$= 5.939 \times 10^4 \text{ J/kg}$$

Also

$$\frac{\gamma}{\gamma - 1} = 3.5$$

Substituting these values into equation 6.89

$$W\,(1 \text{ stage}) = (3.5)(5.939 \times 10^4 \text{ J/kg})(1.931 - 1)$$

$$W\,(1 \text{ stage}) = 1.935 \times 10^5 \text{ J/kg} = \underline{193.5 \text{ kJ/kg}}$$

(ii) For adiabatic compression of an ideal gas in n equal stages

$$W=\left(\frac{n\gamma}{\gamma-1}\right)P_1V_1\left[\left(\frac{P_2}{P_1}\right)^{(\gamma-1)/n\gamma}-1\right] \tag{6.91}$$

For $n = 2$

$$\left(\frac{P_2}{P_1}\right)^{(\gamma-1)/n\gamma}=10^{0.1429}=1.389$$

Since

$$\frac{n\gamma}{\gamma-1}=7.0$$

and as before

$$P_1V_1 = 5.939\times10^4 \text{ J/kg}$$

it follows that

$$W\,(2\text{ stages}) = (7.0)(5.939\times10^4\text{ J/kg})(1.389-1)$$

$$W\,(2\text{ stages}) = 1.617\times10^5\text{ J/kg} = \underline{161.7\text{ kJ/kg}}$$

(iii) Repeating the above calculation for $n = 3$ gives

$$W\,(3\text{ stages}) = \underline{152.8\text{ kJ/kg}}$$

6.10 Compressible flow through nozzles and constrictions

High speed gas flow through nozzles and other constrictions is essentially adiabatic because there is insufficient time for heat transfer between the surroundings and the gas to occur to any significant extent. In a well-designed nozzle, frictional effects will be negligible and the flow is therefore reversible and adiabatic, ie an isentropic process. Flow through an orifice in a pipe may be treated in the same way, the flow being nearly frictionless up to the location where the flow reattaches to the pipe walls. As will be seen, conditions at the throat of a convergent or convergent–divergent nozzle are of great importance. In the case of flow through an orifice, a *vena contracta* forms downstream of the orifice: this point of minimum flow area must be treated as the equivalent of the throat of a nozzle.

Consider steady flow from a large reservoir, where the gas speed is negligible, and through a convergent nozzle that discharges into another large reservoir. The arrangement is shown in Figure 6.1. If the upstream pressure P_0 is held constant and the back pressure P_B, ie the pressure in

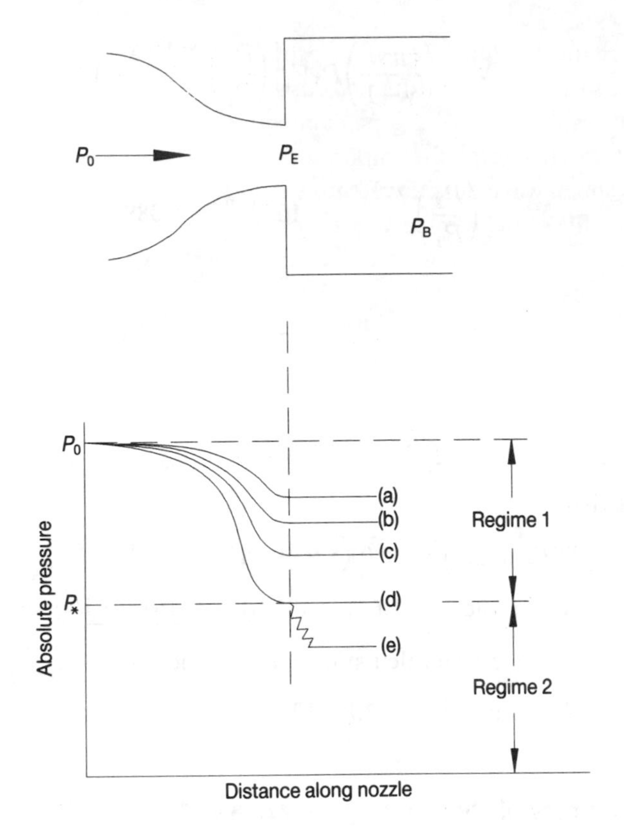

Figure 6.1
Pressure profiles for compressible flow through a convergent nozzle

the discharge reservoir, is gradually reduced below P_0, the gas flow through the nozzle will gradually increase.

This is illustrated by conditions (a) to (c). In each case the pressure P_E at the exit plane is equal to the back pressure P_B. Flow is subsonic throughout the nozzle. This type of behaviour in which the flow rate increases as the back pressure is reduced (P_0 held constant) continues until a critical value of the pressure ratio P/P_0 is reached at the throat of the nozzle, condition (d). At the critical pressure ratio P_*/P_0, the gas reaches the speed of sound at the throat. It will be shown that P_*/P_0 is a function of γ only.

If the back pressure is reduced below P_*, there is no increase in the flow rate through the nozzle, ie the flow is choked. In a convergent nozzle it is impossible for the gas speed to exceed the speed of sound. This case is illustrated in condition (e). The profiles of pressure and gas speed are identical to those of condition (d) up to the exit plane. For condition (e) the exit plane pressure P_E is equal to P_* and it is necessary for the pressure to change to the imposed back pressure P_B: this occurs in an oblique shock wave outside the nozzle, indicated by the jagged line.

The profile of the gas speed is a mirror image of the pressure profile: the speed increases where the pressure falls and vice versa. Thus, there are two regimes of flow through a convergent nozzle: regime 1 where $P_* \leq P_B \leq P_0$ and the flow rate depends on the imposed back pressure for a fixed supply pressure, and regime 2 where $P_B < P_*$ and the flow rate is independent of the back pressure for a fixed supply pressure. This latter regime can be used to advantage when it is necessary to maintain a constant flow rate into a vessel in which the pressure varies. Provided the supply pressure upstream of a convergent nozzle can be kept sufficiently high so that the back pressure never exceeds the critical pressure P_*, the flow will be choked. The flow will remain constant if the supply pressure is held constant.

The case of flow through a convergent–divergent nozzle is shown in Figure 6.2. On reducing the back pressure P_B, while keeping the supply

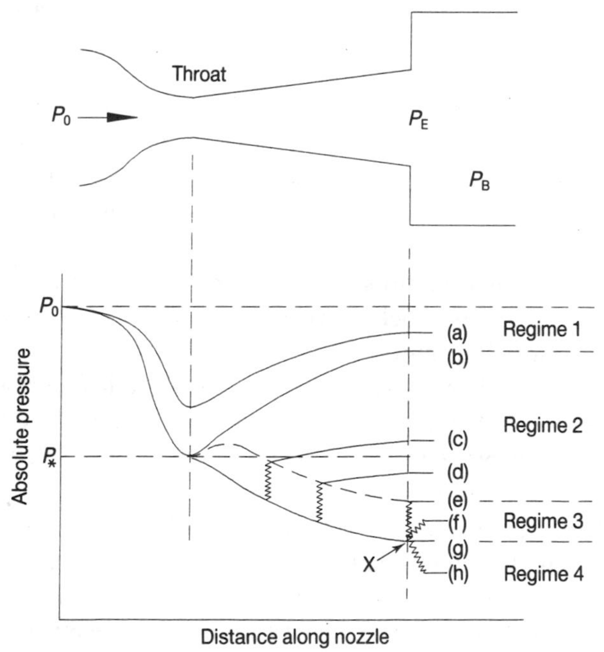

Figure 6.2
Pressure profiles for compressible flow through a convergent–divergent nozzle

pressure P_0 constant, flow in the converging section is as discussed above for a convergent nozzle.

When the back pressure is only slightly lower than the supply pressure, as in condition (a) in Figure 6.2, the pressure passes through a minimum value at the throat, where the gas speed is a maximum, but pressure recovery occurs in the diverging section as the gas decelerates. This type of behaviour, denoted as regime 1, is observed until the back pressure is reduced sufficiently to cause the critical pressure P_* to occur at the throat of the nozzle [condition (b)]. Reducing the back pressure further causes regime 2 to be entered. Here, the pressure continues to fall in the diverging section and the flow is supersonic. However, if isentropic flow were to continue throughout the nozzle, the pressure curve would have to be followed to the point X at the exit plane. In region 2 the back pressure is not low enough for this to happen and a stationary shock wave occurs inside the nozzle indicated by the vertical jagged lines for conditions (c) and (d). Downstream of the shock wave, the flow is subsonic and pressure recovery occurs as the gas decelerates in the remaining part of the nozzle.

The shock wave occurs in such a position as to allow the exit pressure to equal the imposed back pressure. The back pressure being lower for condition (d) than for condition (c), the shock wave for condition (d) is closer to the exit plane.

The end of regime 2 is when the shock wave occurs at the exit plane, this is shown as condition (e). In regime 2 the shock waves are perpendicular to the flow and are therefore called normal shock waves. In regime 3, for example condition (f), the adjustment from the exit-plane pressure to the back pressure occurs outside the nozzle as an oblique compression shock wave.

Condition (g) is that condition for which the exit pressure is equal to the back pressure and no shock wave occurs. This is called the design condition for supersonic flow.

In regime 4, which extends over all values of the back pressure lower than the supersonic design pressure, the adjustment from the exit pressure to the lower back pressure occurs as an oblique expansion shock wave outside the nozzle; this is the case for condition (h).

It is only in regime 1 that the flow rate depends on the back pressure. It will be noticed that this is only a small part of the nozzle's range of operation. Once the sonic speed has been reached at the throat (at the pressure P_*), the flow becomes choked and the flow rate remains constant, for constant supply conditions, and is independent of the back pressure.

Before deriving expressions for the flow rate in a nozzle, it will be instructive to see how the constraints of continuity, energy and the equation of state allow both subsonic and supersonic flow in a nozzle.

From equation 1.7, continuity can be expressed as

$$\rho u S = \text{constant} \tag{6.93}$$

Writing this in differential form and dividing throughout by $\rho u S$ gives

$$\frac{\mathrm{d}S}{S} + \frac{\mathrm{d}u}{u} + \frac{\mathrm{d}\rho}{\rho} = 0 \tag{6.94}$$

For isentropic flow with negligible change of elevation and no shaft work, equation 6.11 reduces to

$$V\mathrm{d}P + u\mathrm{d}u = 0 \tag{6.95}$$

Thus

$$\frac{\mathrm{d}u}{u} = -\frac{V\mathrm{d}P}{u^2} = -\frac{\mathrm{d}P}{\rho u^2} \tag{6.96}$$

For an isentropic change, the equation of state is

$$PV^{\gamma} = \text{constant} \tag{6.32}$$

Putting $k = \gamma$ in equation 6.68 and using equation 6.70 gives

$$\frac{\partial P}{\partial \rho} = \gamma P V = c^2 \tag{6.97}$$

where c is the speed of sound. Thus

$$\frac{\mathrm{d}\rho}{\rho} = \frac{\mathrm{d}P}{\rho c^2} \tag{6.98}$$

Substituting for $\mathrm{d}u/u$ and $\mathrm{d}\rho/\rho$ in equation 6.94 using equations 6.96 and 6.98 gives

$$\frac{\mathrm{d}S}{S} - \frac{\mathrm{d}P}{\rho u^2} + \frac{\mathrm{d}P}{\rho c^2} = 0 \tag{6.99}$$

Equation 6.99 may be written as

$$\frac{\mathrm{d}S}{S} = \frac{\mathrm{d}P}{\rho u^2}(1 - Ma^2) = \frac{\mathrm{d}u}{u}(Ma^2 - 1) \tag{6.100}$$

where $Ma = u/c$ is the Mach number.

Equation 6.100 shows that in a diverging section ($\mathrm{d}S/S > 0$) the velocity must decrease for subsonic flow ($Ma < 1$) and increase for supersonic flow

($Ma > 1$). If flow in a converging section is subsonic, it accelerates but if it were then to become supersonic, equation 6.100 shows that it would decelerate. Thus the maximum speed in a converging section is the sonic speed and this is reached at the throat where $\mathrm{d}S/S = 0$.

As equation 6.95 shows, for isentropic flow with negligible change in elevation (potential energy) and no shaft work, there is an interchange between only two forms of energy: pressure energy and kinetic energy. This is reflected in equation 6.100, which shows that for a given change in flow area the pressure changes in the opposite way to the velocity.

6.10.1 *Flow rate through a nozzle*

Integrating equation 6.95 from state 1 to state 2 gives

$$\int_1^2 V\mathrm{d}P + \frac{u_2^2 - u_1^2}{2} = 0 \tag{6.101}$$

Using the equation of state, equation 6.32, to evaluate the integral in equation 6.101:

$$\int_1^2 V\mathrm{d}P = \left(\frac{\gamma}{\gamma - 1}\right)(P_2V_2 - P_2V_1) \tag{6.102}$$

Thus

$$u_2^2 - u_1^2 = \left(\frac{2\gamma}{\gamma - 1}\right)(P_1V_1 - P_2V_2) \tag{6.103}$$

It is convenient to refer the conditions in the nozzle to a basis corresponding to a stationary gas with pressure P_0, specific volume V_0, and temperature T_0. Thus, equation 6.103 can be written in the form

$$u^2 = \left(\frac{2\gamma}{\gamma - 1}\right)(P_0V_0 - PV) \tag{6.104}$$

where the variables without subscripts are at some general location. In most cases the upstream gas speed will be relatively low and the upstream values of P and V may be taken as P_0 and V_0. When this is not the case, the corresponding value of P_0V_0 can be calculated from equation 6.104 by inserting the upstream values for PV.

The mass flow rate M is given by

$$M = \rho u S = uS/V \tag{6.105}$$

Eliminating u between equations 6.104 and 6.105 gives

$$M^2 = \frac{S^2}{V^2}\left(\frac{2\gamma}{\gamma-1}\right)(P_0V_0 - PV) \tag{6.106}$$

Using the equation of state and rearranging the result enables the mass flow rate to be given as

$$M = (\gamma P_0/V_0)^{1/2} S\psi \tag{6.107}$$

where

$$\psi = \left\{\left(\frac{2}{\gamma-1}\right)\left[\left(\frac{P}{P_0}\right)^{2/\gamma} - \left(\frac{P}{P_0}\right)^{(\gamma+1)/\gamma}\right]\right\}^{1/2} \tag{6.108}$$

The pressure-dependence of the flow is contained entirely within the term ψ. The quantity $\sqrt{\gamma P_0/V_0}$ is constant for specified upstream conditions. (It is equal to the mass flux the gas would have if flowing at the sonic speed $c_0 = \sqrt{\gamma P_0 V_0}$ corresponding to the reservoir conditions P_0, V_0, T_0.)

6.10.2 *Critical pressure ratio*

The maximum attainable mass flux for given supply conditions must occur when ψ is a maximum. Therefore the pressure P_* causing the maximum flux can be found by differentiating ψ^2 with respect to P and equating the result to zero:

$$\left[\frac{2}{\gamma}\left(\frac{P_*}{P_0}\right)^{2/\gamma} - \left(\frac{\gamma+1}{\gamma}\right)\left(\frac{P_*}{P_0}\right)^{(\gamma+1)/\gamma}\right]\frac{1}{P_*} = 0 \tag{6.109}$$

from which the critical pressure ratio P_*/P_0 is given by

$$\frac{P_*}{P_0} = \left(\frac{2}{\gamma+1}\right)^{\gamma/(\gamma-1)} \tag{6.110}$$

For $\gamma = 1.40$, $P_*/P_0 = 0.528$. For other values of γ, the value of the critical pressure ratio lies in the approximate range 0.5 to 0.6.

Inserting P_* from equation 6.110 into equation 6.104 gives

$$u_*^2 = \gamma P_* V_* \tag{6.111}$$

showing that the sonic speed is achieved at the critical pressure.

Thus, if the pressure ratio at the minimum flow area is equal to the critical value given by equation 6.110, the flow there will be at the sonic speed. If the pressure ratio is higher than this value the flow will be subsonic and will depend on the back pressure. In both convergent and convergent–divergent nozzles, the pressure at the throat cannot be lower than P_* but if a low back pressure is imposed a shock wave will occur somewhere downstream of the throat.

Example 6.3

Nitrogen contained in a large tank at a pressure $P = 200000$ Pa and a temperature of 300 K flows steadily under adiabatic conditions into a second tank through a converging nozzle with a throat diameter of 15 mm. The pressure in the second tank and at the throat of the nozzle is $P_t = 140000$ Pa. Calculate the mass flow rate, M, of nitrogen assuming frictionless flow and ideal gas behaviour. Also calculate the gas speed at the nozzle and establish that the flow is subsonic. The relative molecular mass of nitrogen is 28.02 and the ratio of the specific heat capacities γ is 1.39.

Calculations

$$M = (\gamma P_0/V_0)^{1/2} S \psi \tag{6.107}$$

where

$$\psi = \left\{\left(\frac{2}{\gamma - 1}\right)\left[\left(\frac{P}{P_0}\right)^{2/\gamma} - \left(\frac{P}{P_0}\right)^{(\gamma+1)/\gamma}\right]\right\}^{1/2} \tag{6.108}$$

The gas is initially at rest so that the stagnation pressure P_0 is equal to the pressure in the first tank, ie 200000 Pa. The pressure at the throat $P_t = 140000$ Pa so that

$$P_t/P_0 = 0.7$$

Evaluating ψ at the throat

$$\psi = \left\{\left(\frac{2}{0.39}\right)[0.7^{1.4388} - 0.7^{1.7194}]\right\}^{1/2}$$

$$= 0.5407$$

The throat area is given by

$$S_t = \frac{\pi(0.015 \text{ m})^2}{4} = 1.767 \times 10^{-4} \text{ m}^2$$

Equation of state

$$PV = R'T = \frac{RT}{\text{RMM}}$$

$$= \frac{(8314.3\ \text{J/kmol K})(300\ \text{K})}{28.02\ \text{kg/kmol}}$$

$$= 8.904 \times 10^4\ \text{J/kg}$$

Thus, for $P_0 = 2 \times 10^5$ Pa, $V_0 = 0.4451\ \text{m}^3/\text{kg}$. Substituting these values into equation 6.107

$$M = \left(\frac{1.39 \times 2 \times 10^5\ \text{Pa}}{0.4451\ \text{m}^3/\text{kg}}\right)^{1/2} (1.767 \times 10^{-4}\ \text{m}^2)(0.5407)$$

$$M = \underline{0.0755\ \text{kg/s}}$$

At the throat, $P_t = 1.4 \times 10^5$ Pa and the specific volume there is given by

$$V_t = V_0 \left(\frac{P_0}{P_t}\right)^{1/\gamma} = 0.4451 \left(\frac{2.0}{1.4}\right)^{\frac{1}{1.39}} = 0.5753\ \text{m}^3/\text{kg}$$

The gas speed at the throat is given by

$$\frac{MV_t}{S_t} = \frac{(0.0755\ \text{kg/s})(0.5753\ \text{m}^3/\text{kg})}{1.767 \times 10^{-4}\ \text{m}^2}$$

$$= \underline{245.8\ \text{m/s}}$$

Sonic speed at throat conditions

$$c = \sqrt{\gamma P_t V_t} = \sqrt{(1.39)(1.4 \times 10^5\ \text{Pa})(0.5753\ \text{m}^3/\text{kg})}$$

$$= \underline{334.6\ \text{m/s}}$$

Thus, the gas speed at the throat is less than the sonic speed there. The flow is subsonic throughout the nozzle. This result is to be expected because the pressure ratio is 0.7 and the critical pressure ratio (for $\gamma = 1.39$) is 0.53.

6.10.3 *Shock waves*

Properties of the gas, such as the velocity, pressure, density and temperature, change by large amounts across the narrow shock wave. Although mass, energy and momentum are conserved across a shock wave, entropy is not. Entropy is created by a shock from supersonic to subsonic flow.

The above analysis, comprising equations 6.95 to 6.100 and 6.101 to

6.111, applies to isentropic, ie constant entropy flow. It therefore applies to the smooth curves but not across a shock wave. When a normal shock wave occurs in regime 2, the flow is isentropic up to the shock wave, then isentropic but at a different entropy downstream of the shock wave. The value of the product $S\psi$ is constant throughout the flow if no shock wave occurs but is different upstream and downstream of a shock when one occurs.

A shock wave from subsonic to supersonic flow would require a decrease in the entropy so what would be an alarming phenomenon is thermodynamically impossible.

References

Barna, P.S., *Fluid Mechanics for Engineers*, London: Butterworths, p. 248 (1969).

Perry, J.H., *Chemical Engineers' Handbook*. Sixth edition, New York, McGraw-Hill Book Company Inc, p. 3–268 (1984).

Shapiro, A.H., *The Dynamics and Thermodynamics of Compressible Flow*, Volume I, New York, The Ronald Press Company (1953).

Smith, J.M. and Van Ness, H.C., *Introduction to Chemical Engineering Thermodynamics*, International edition, Singapore, McGraw-Hill Book Company Inc, pp. 88–89 (1987).

7 Gas–liquid two-phase flow

The flow of gas–liquid mixtures in pipes and other items of process equipment is common and extremely important. In some cases the quality, that is the mass fraction of gas in the two-phase flow, will vary very little over a large distance. An example of this is the flow in many gas–oil pipelines. In other cases, boiling or condensation occurs and the quality may change very significantly although the total mass flow rate remains constant.

It is important to appreciate that different flow regimes occur at different gas and liquid flow rates and differences also occur for different materials. In order to have any confidence when calculating pressure losses in two-phase flow it is necessary to be able to predict the flow regime and then to use an appropriate pressure drop calculation procedure.

7.1 Flow patterns and flow regime maps

7.1.1 *Flow patterns*

The flow regimes that are obtained in vertical, upward, cocurrent flow at different gas and liquid flow rates are shown in Figure 7.1. The sequence shown is that which would normally be seen as the ratio of gas to liquid flow rates is increased. In the *bubbly* regime there is a distribution of bubbles of various sizes throughout the liquid. As the gas flow rate increases, the average bubble size increases. The next regime occurs when the gas flow rate is increased to the point when many bubbles coalesce to produce *slugs* of gas. The gas slugs have spherical noses and occupy almost the entire cross section of the tube, being separated from the wall by a thin liquid film. Between slugs of gas there are slugs of liquid in which there may be small bubbles entrained in the wakes of the gas slugs.

This well-defined flow pattern is destroyed at higher flow rates and a chaotic type of flow, generally known as *churn flow*, is established. Over

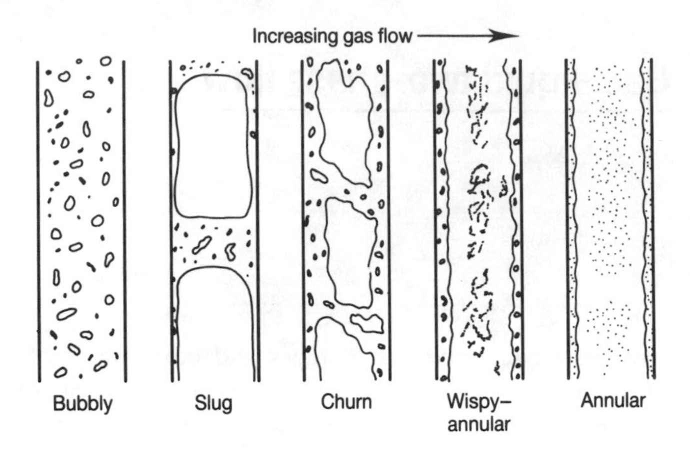

Figure 7.1
Flow regimes in vertical gas–liquid flow

most of the cross section there is a churning motion of irregularly shaped portions of gas and liquid. Further increase in the gas flow rate causes a degree of separation of the phases, the liquid flowing mainly on the wall of the tube and the gas in the core. Liquid drops or droplets are carried in the core: it is the competing tendencies for drops to impinge on the liquid film and for droplets to be entrained in the core by break-up of waves on the surface of the film that determine the flow regime. The main differences between the *wispy-annular* and the *annular* flow regimes are that in the former the entrained liquid is present as relatively large drops and the liquid film contains gas bubbles, while in the annular flow regime the entrained droplets do not coalesce to form larger drops.

Cocurrent gas–liquid flow in horizontal pipes displays similar patterns to those for vertical flow; however, asymmetry is caused by the effect of gravity, which is most significant at low flow rates. The sequence of flow regimes identified by Alves (1954) is shown in Figure 7.2. In the *bubbly* regime the bubbles are confined to a region near the top of the pipe. On increasing the gas flow rate, the bubbles become larger and coalesce to form long bubbles giving what is known as the *plug* flow regime. At still higher gas flow rates the gas plugs join to form a continuous gas layer in the upper part of the pipe. This type of flow, in which the interface between the gas and the liquid is smooth, is known as the *stratified* flow regime. Owing to the lower viscosity and lower density of the gas it will flow faster than the liquid.

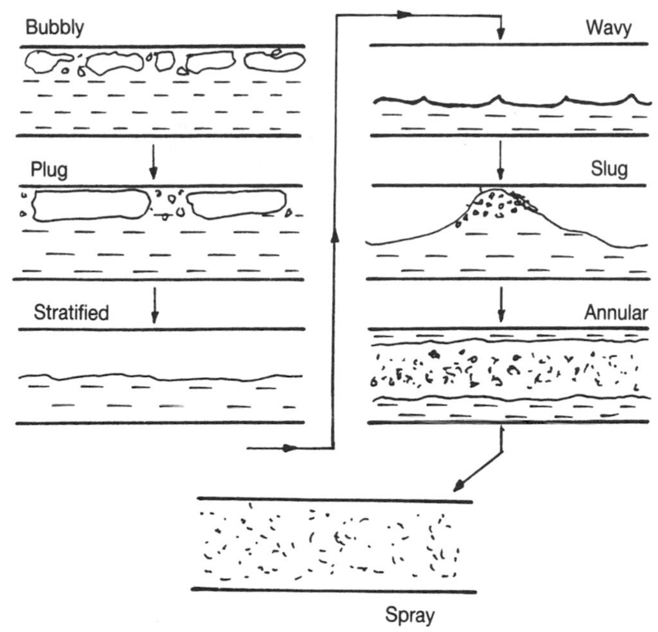

Figure 7.2
Flow regimes in horizontal gas–liquid flow

As the gas flow rate is increased further, the interfacial shear stress becomes sufficient to generate waves on the surface of the liquid producing the *wavy* flow regime. As the gas flow rate continues to rise, the waves, which travel in the direction of flow, grow until their crests approach the top of the pipe and, as the gas breaks through, liquid is distributed over the wall of the pipe. This is known as the *slug* regime and should not be confused with the regime of the same name for vertical flow.

At higher gas flow rates an *annular* regime is found as in vertical flow. At very high flow rates the liquid film may be very thin, the majority of the liquid being dispersed as droplets in the gas core. This type of flow may be called the *spray* or *mist* flow regime.

It may be noted that similar flow regimes can be seen with immiscible liquid systems. If the densities of the two liquids are close the flow regimes for horizontal flow will more nearly resemble those for vertical flow.

7.1.2 *Flow regime maps*

The prediction of the flow regime in gas–liquid two-phase flow is rather uncertain partly because the transitions between the flow regimes are

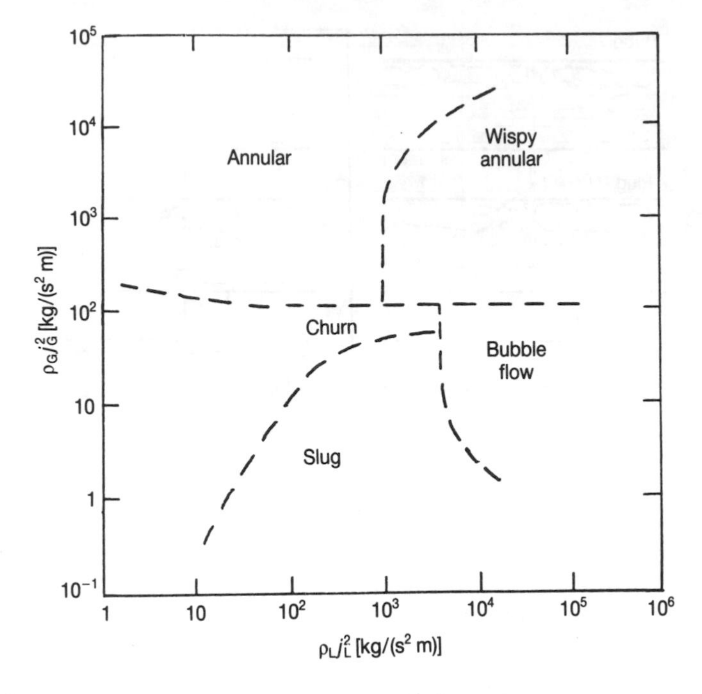

Figure 7.3
Flow regime map for vertical gas–liquid flow
Source: G. F. Hewitt and D. N. Roberts, *Studies of two-phase flow patterns by simultaneous X-ray and flash photography*, Report AERE-M 2159 (London: HMSO, 1969)

gradual and the classification of a particular flow is subjective. There are various flow regime maps in the literature, two of which are given in Figures 7.3 and 7.4. For vertical flow of low pressure air–water and high pressure steam–water mixtures, Hewitt and Roberts (1969) have determined a flow regime map shown in Figure 7.3. Here, j_G and j_L denote the volumetric fluxes of the gas and liquid. For the gas

$$j_G = Q_G/S \tag{7.1}$$

and for the liquid

$$j_L = Q_L/S \tag{7.2}$$

where Q_G, Q_L are the volumetric flow rates of the gas and the liquid, and S is the cross-sectional area of the pipe. The axes of Figure 7.3 represent the superficial momentum fluxes of the gas and liquid. (The volumetric flux is

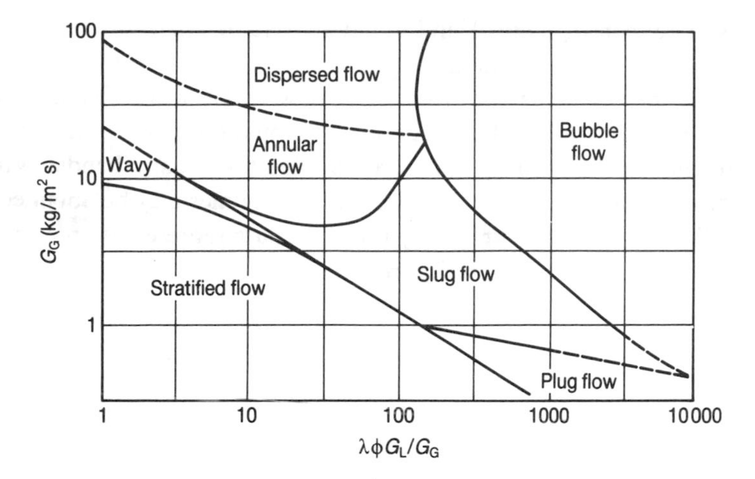

Figure 7.4
Flow regime map for horizontal gas–liquid flow
Source: O. Baker, *Oil and Gas Journal* **53**, pp. 185–95 (26 July, 1954)

the same as the superficial velocity.) In addition to allowing the flow regime for a specified combination of gas and liquid flow rates to be determined, the diagram shows how changes of operating conditions change the flow regime. In particular it can be seen that the sequence of flow regimes described above is produced by increasing the gas momentum flux and/or reducing the liquid momentum flux.

The best known flow regime map for horizontal gas–liquid flow was given by Baker (1954) and is shown in Figure 7.4. Here G_G, G_L denote the superficial mass fluxes of the gas and the liquid. For the gas

$$G_G = M_G/S = j_G\rho_G \tag{7.3}$$

and for the liquid

$$G_L = M_L/S = j_L\rho_L \tag{7.4}$$

The quantities λ and Φ are physical property correction factors defined by the expressions

$$\lambda = \left(\frac{\rho_G}{\rho_A} \cdot \frac{\rho_L}{\rho_W}\right)^{1/2} \quad \text{and} \quad \Phi = \frac{\sigma_W}{\sigma_L}\left[\frac{\mu_L}{\mu_W}\left(\frac{\rho_W}{\rho_L}\right)^2\right]^{1/3} \tag{7.5}$$

where σ denotes the coefficient of surface tension. The subscripts A and W indicate the values for air and water at 20 °C and a pressure of 1 atmosphere; consequently λ and Φ have the value unity for the air–water system under these conditions.

One of the problems with two-phase flow is that a significant distance may be required for the flow regime to become established and the flow regime may be changed by flow through pipe fittings and bends. When a change of phase occurs several different flow regimes may be obtained in a short distance as demonstrated by the schematic representation of flow in an evaporator tube shown in Figure 7.5.

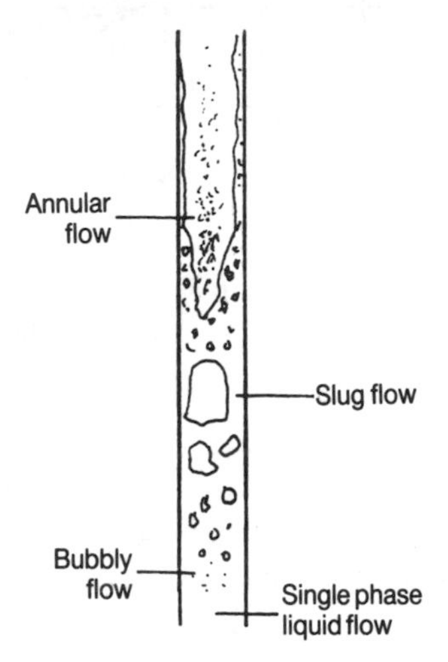

Figure 7.5
Flow regimes in a vertical evaporator or boiler tube

7.2 Momentum equation for two-phase flow

Figure 7.6 illustrates a gas–liquid two-phase flow through an inclined pipe. For clarity the diagram is drawn for stratified flow but the equations to be derived are not limited to that flow regime. A momentum equation can be written for each phase but it will be sufficient for the present purposes to treat the whole flow. In this case the interfacial shear force δF_S makes no direct contribution but it would have to be considered in writing the momentum equation for either of the phases individually. The net force acting in the positive x-direction is

$$-S\delta P - \delta F_G - \delta F_L - (S_G\rho_G + S_L\rho_L)g\sin\theta\,\delta x \tag{7.6}$$

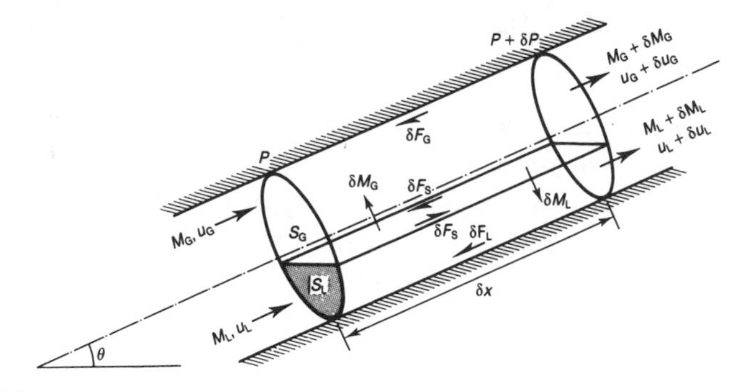

Figure 7.6
Element of two–phase flow

The terms represent the contributions from the total pressure gradient, the frictional drag of the pipe wall and the hydrostatic head of the two-phase mixture.

The rate of change of momentum is

$$(M_G+\delta M_G)(u_G+\delta u_G)+(M_L+\delta M_L)(u_L+\delta u_L)-M_G u_G-M_L u_L \quad (7.7)$$

where the first two terms are the momentum flow rate of the fluid leaving the element and the third and fourth terms are the momentum flow rates into the element.

For steady flow the net force acting on the fluid in the element is equal to the change of momentum flow rate:

$$-S\delta P-\delta F_G-\delta F_L-(S_G\rho_G+S_L\rho_L)g\sin\theta\,\delta x = \delta(M_G u_G+M_L u_L) \quad (7.8)$$

where second order terms have been neglected. Consequently,

$$-S\frac{\mathrm{d}P}{\mathrm{d}x}-\frac{\mathrm{d}F}{\mathrm{d}x}-(S_G\rho_G+S_L\rho_L)g\sin\theta=\frac{\mathrm{d}}{\mathrm{d}x}(M_G u_G+M_L u_L) \quad (7.9)$$

or

$$\frac{\mathrm{d}P}{\mathrm{d}x}=-\frac{1}{S}\frac{\mathrm{d}F}{\mathrm{d}x}-\frac{1}{S}\frac{\mathrm{d}}{\mathrm{d}x}(M_G u_G+M_L u_L)-\left(\frac{S_G\rho_G}{S}+\frac{S_L\rho_L}{S}\right)g\sin\theta \quad (7.10)$$

Equation 7.10 shows that the total pressure gradient comprises three components that are due to fluid friction, the rate of change of momentum and the static head. The momentum term is usually called the accelerative component. Thus

$$\frac{dP}{dx} = \left(\frac{dP}{dx}\right)_f + \left(\frac{dP}{dx}\right)_a + \left(\frac{dP}{dx}\right)_{sh} \tag{7.11}$$

In principle, this is the same as for single-phase flow. For example in steady, fully developed, isothermal flow of an incompressible fluid in a straight pipe of constant cross section, friction has to be overcome as does the static head unless the pipe is horizontal, however there is no change of momentum and consequently the accelerative term is zero. In the case of compressible flow, the gas expands as it flows from high pressure to low pressure and, by continuity, it must accelerate. In Chapter 6 this was noted as an increase in the kinetic energy.

In gas–liquid flow with no change of phase, for example air and water, the gas phase expands and acceleration occurs as in single-phase flow. Where the two-phase flow may be dramatically different from single-phase flow is when boiling or condensation occurs. For example, in a boiling two-phase flow the relatively dense liquid is changed into a vapour having a much lower density (higher specific volume) and a substantial accelerative pressure gradient may be required.

7.2.1 *Two-phase flow terminology*

It is convenient to work in terms of the void fraction α and the mass fraction of gas w, which is also known as the quality. The void fraction is defined as the time average fraction of the cross-sectional area through which the gas flows:

$$\alpha = \frac{S_G}{S} \tag{7.12}$$

and consequently

$$\frac{S_L}{S} = 1 - \alpha \tag{7.13}$$

For the quality

$$w = \frac{M_G}{M_G + M_L} = \frac{M_G}{M} \tag{7.14}$$

and

$$\frac{M_L}{M} = 1 - w \tag{7.15}$$

It is also conventional to work in terms of the mass flux, sometimes called the mass velocity, G:

$$G = \frac{M}{S} \tag{7.16}$$

so, using equations 7.14 and 7.15

$$M_G = wGS \qquad \text{and} \qquad M_L = (1-w)GS \tag{7.17}$$

Using the specific volumes of the gas V_G and liquid V_L, the gas and liquid velocities can be written in the following forms:

$$u_G = \frac{M_G V_G}{\alpha S} = \frac{wGV_G}{\alpha} \tag{7.18}$$

and

$$u_L = \frac{M_L V_L}{(1-\alpha)S} = \frac{(1-w)GV_L}{1-\alpha} \tag{7.19}$$

It is left as an exercise in using this notation for the reader to show that the three components of the pressure gradient may be written as

$$\left(\frac{\mathrm{d}P}{\mathrm{d}x}\right)_f = -\frac{1}{S}\frac{\mathrm{d}F}{\mathrm{d}x} \tag{7.20}$$

$$\begin{aligned}\left(\frac{\mathrm{d}P}{\mathrm{d}x}\right)_a &= -\frac{1}{S}\frac{\mathrm{d}}{\mathrm{d}x}(M_G u_G + M_L u_L) \\ &= -G^2\frac{\mathrm{d}}{\mathrm{d}x}\left(\frac{w^2 V_G}{\alpha} + \frac{(1-w)^2 V_L}{1-\alpha}\right)\end{aligned} \tag{7.21}$$

$$\begin{aligned}\left(\frac{\mathrm{d}P}{\mathrm{d}x}\right)_{sh} &= -\left(\frac{S_G\rho_G}{S} + \frac{S_L\rho_L}{S}\right) g\sin\theta \\ &= -[\alpha\rho_G + (1-\alpha)\rho_L]\, g\sin\theta = -\left(\frac{\alpha}{V_G} + \frac{1-\alpha}{V_L}\right) g\sin\theta\end{aligned} \tag{7.22}$$

7.3 **Flow in bubble columns**

The dispersion of gas bubbles in a liquid is widely used in bubble column reactors and bioreactors. As shown in Figure 7.7, the gas is introduced by some kind of distributor at the bottom of the column. The liquid may be introduced at the bottom of the column and removed at the top, in which

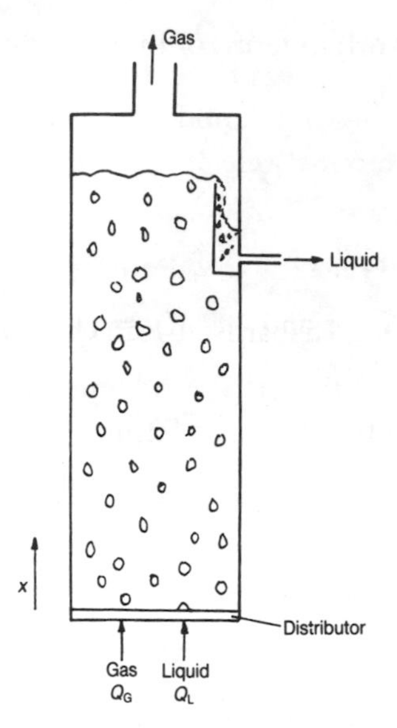

Figure 7.7

Schematic representation of a bubble column with cocurrent flow of gas and liquid

case cocurrent flow occurs. Alternatively, the liquid may be introduced at the top of the column and be removed at the bottom thus providing counter-current flow of the two phases.

In bubble columns the static head of the fluid is the dominant component of the pressure drop and consequently it is important to determine the void fraction of the dispersion. All quantities will be measured as positive in the upward direction, this being the direction of flow of the dispersed phase. Assuming that the gas bubbles are of uniform size and are uniformly distributed over any cross section of the column, the gas and liquid velocities relative to the column are

$$u_G = \frac{Q_G}{\alpha S} \quad \text{and} \quad u_L = \frac{Q_L}{(1-\alpha)S} \tag{7.23}$$

where Q_G, Q_L are the volumetric flow rates of the gas and the liquid.

The relative velocity of the gas with respect to the liquid is known as the *slip velocity*:

$$\text{slip velocity} = u_G - u_L = \frac{Q_G}{\alpha S} - \frac{Q_L}{(1-\alpha)S} \tag{7.24}$$

For counter-current flow the value of Q_L must be taken as negative.

For many dispersed systems (gas bubbles in liquids, liquid droplets in another liquid, solid particles in a liquid), it has been found that the slip velocity is related to the terminal velocity u_t of a single bubble, droplet or particle by the equation

$$\text{slip velocity} = u_t(1-\alpha)^{n-1} \tag{7.25}$$

This result follows from the Richardson–Zaki equation. In their original work, Richardson and Zaki (1954) studied batch sedimentation, in particular the settling of coarse solid particles through a liquid in a vertical cylinder with a closed bottom. Richardson and Zaki found that the settling speed u_c of the equal-sized particles in the concentrated suspension was related to the terminal settling speed u_t of a single particle in a large expanse of liquid by the equation

$$u_c = u_t(1-\alpha)^n \tag{7.26}$$

where α is the volume fraction of particles and n is an empirical constant that depends on the value of the Reynolds number. It is important to appreciate that u_c is the particle speed relative to the apparatus. Of greater fundamental importance is the relative velocity of the particles with respect to the liquid and an expression for this can easily be derived from equation 7.26, as follows. In a batch sedimentation process, the net volumetric flux across any horizontal plane must be zero: as the particles settle, they displace liquid upwards. The downwards volumetric flux of the particles is $u_c\alpha$ and the liquid flux *upwards* must be equal to this. Consequently, the average upward speed of the liquid as it is displaced is equal to $u_c\alpha/(1-\alpha)$ because the volume fraction of the liquid is $1-\alpha$. The velocity of the particles *relative to the liquid* is therefore given by

$$\text{relative velocity} = u_c + \frac{u_c\alpha}{(1-\alpha)} = \frac{u_c}{(1-\alpha)} \tag{7.27}$$

Substituting for u_c from equation 7.26 allows equation 7.27 to be written in the form of equation 7.25. This result expresses the interaction between the particles (or bubbles) and the liquid and is therefore applicable also to cases in which the net flux is non-zero, as in bubble columns.

The value of n varies considerably with different types of dispersion but for bubble columns it has been found experimentally that

$$n = 4.6 \quad \text{for } Re<1$$

$$n = 2.4 \quad \text{for } Re>500$$

where the bubble Reynolds number $Re = \rho_L u_t D_e / \mu_L$ is based on the terminal rise velocity of the bubble and its equivalent diameter, that is the diameter of a sphere having the same volume as the bubble. Equation 7.25 shows that the velocity of rise of a bubble swarm relative to the liquid is lower than the terminal rise velocity of a single, isolated bubble of the same size in the same liquid. It follows that the presence of neighbouring bubbles increases the drag on a bubble. In practice it is usually adequate to use the approximation $n = 2$.

It should be noted that equation 7.25 represents the interaction of forces acting on bubbles in a swarm while equation 7.24 represents the principle of conservation of mass (continuity). Both equations must be satisfied simultaneously.

Combining equations 7.24 and 7.25, gives

$$\frac{Q_G}{\alpha S} - \frac{Q_L}{(1-\alpha)S} = u_t(1-\alpha)^{n-1} = \text{slip velocity} \qquad (7.28)$$

Following Wallis (1969), equation 7.28 is multiplied throughout by $\alpha(1-\alpha)$ giving

$$\frac{Q_G}{S}(1-\alpha) - \frac{Q_L}{S}\alpha = u_t\alpha(1-\alpha)^n = (\text{slip velocity})\alpha(1-\alpha) = u_{G,L} \qquad (7.29)$$

The quantity $u_{G,L}$ is called the *characteristic velocity* or *drift flux* and has an important physical significance, which can be seen by making the following manipulations:

$$u_{G,L} = \frac{Q_G}{S}(1-\alpha) - \frac{Q_L}{S}\alpha = \frac{Q_G}{S} - \left(\frac{Q_G + Q_L}{S}\right)\alpha = \left(\frac{Q_G}{\alpha S} - \frac{Q}{S}\right)\alpha \qquad (7.30)$$

where $Q = Q_G + Q_L$. In the last term of equation 7.30, $Q_G/\alpha S$ is the velocity of the gas bubbles and Q/S is the average velocity of the gas–liquid mixture, which is the same as the net volumetric flux. Thus, the drift flux $u_{G,L}$ is equal to the volumetric flux of the dispersed phase (the bubbles) relative to a plane moving at the volumetric average velocity.

Equation 7.29 provides a convenient method of determining the value of the void fraction for specified gas and liquid flow rates. Although this equation can be solved algebraically, it is more convenient and illuminating to use a graphical method. In Figure 7.8 the two parts of equation 7.29 are plotted against the void fraction α for the case of cocurrent flow. The intersection of the straight line representing continuity and the curve representing the forces acting on the bubbles gives the value of the void

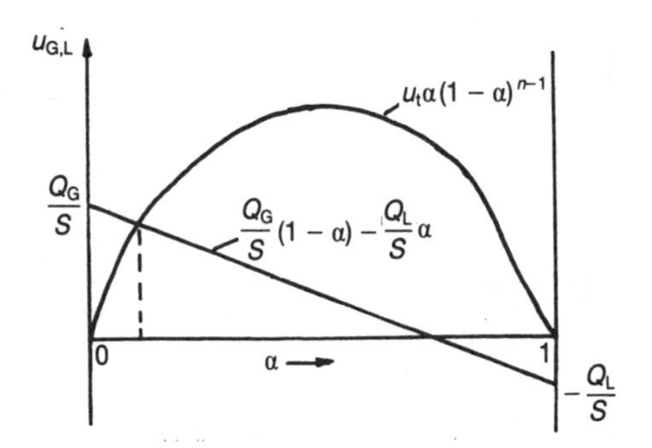

Figure 7.8
Wallis plot for flow in a bubble column

fraction. Using this Wallis plot, it is easy to see how the void fraction must change to accommodate variations of the gas and liquid flow rates. As an example, Figure 7.9 shows the effect of reducing the liquid flow rate (1 to 3) when the gas flow rate is kept constant. The flow is still cocurrent. As the liquid flow rate is reduced the void fraction increases. It will be noted that for cocurrent flow only one value of the void fraction is possible for a given pair of gas and liquid flow rates.

The case of counter-current flow (downward liquid flow) is shown in

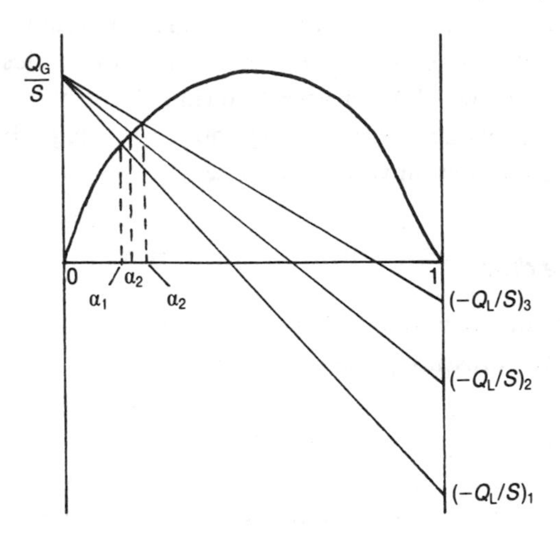

Figure 7.9
Wallis plot for cocurrent flow

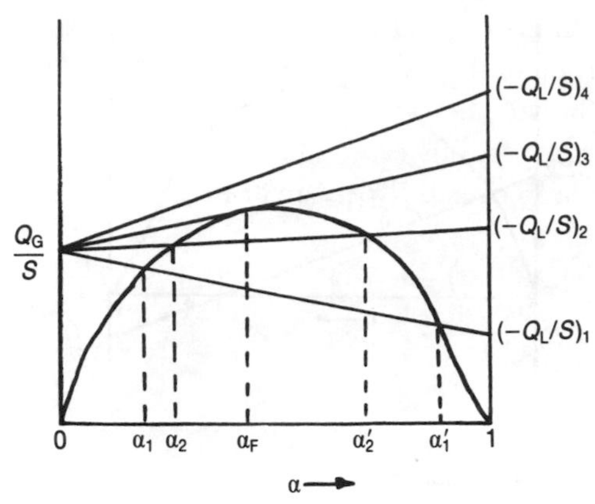

Figure 7.10
Wallis plot for counter–current flow

Figure 7.10. The sequence 1 to 4 corresponds to increasing the magnitude of the liquid flow for a fixed gas flow rate. Cases 1 and 2 are typical and demonstrate the fact that in general two values of the void fraction are possible for counter-current flow, however, in practice the higher value is difficult to obtain owing to coalescence of the bubbles. Condition 3 represents the greatest downward liquid flow rate for which there is a solution for the specified gas flow rate. Consequently, this represents the flooding condition: the drag of the descending liquid is so great that it prevents the gas bubbles rising. The void fraction at flooding is denoted as α_f. If an attempt is made to operate with a greater liquid flow rate, as in case 4, the bubbles are retarded at the distributor and coalesce. The larger bubbles that result will have a greater terminal rise velocity and consequently a greater characteristic velocity thus producing a higher $u_{G,L} - \alpha$ curve, for which a solution will be possible.

7.3.1 *Pressure drop*

Once the value of the void fraction has been determined, the static head component of the pressure gradient can be calculated:

$$\left(\frac{\mathrm{d}P}{\mathrm{d}x}\right)_{sh} = -[\alpha\rho_G + (1-\alpha)\rho_L]\,g \tag{7.31}$$

In most cases the frictional component of the pressure gradient is negligible in bubble columns but if necessary it can be calculated using the homogeneous model discussed in Section 7.5.

For bubble columns of moderate height the static head is sufficiently low for the expansion of the rising bubbles to be negligible. Consequently, the gas density, the volumetric flow rate of the gas, and the void fraction are sensibly constant. In this case, equation 7.31 is readily integrated to give the pressure drop over a dispersion of height H:

$$\Delta P = \int_0^H [\alpha \rho_G + (1-\alpha)\rho_L] g \, dx = [\alpha \rho_G + (1-\alpha)\rho_L] gH \qquad (7.32)$$

Note that the quantity

$$\alpha \rho_G + (1-\alpha)\rho_L = \frac{\alpha}{V_G} + \frac{1-\alpha}{V_L} \qquad (7.33)$$

is simply the average density of the dispersion.

7.3.2 *Bubble rise velocity*

It has been assumed that the terminal rise velocity of a single bubble is known so the characteristic velocity can be calculated. In general the bubble's velocity depends on its size, which in turn depends on the design of the distributor. An excellent survey of bubble and droplet formation has been given by Kumar and Kuloor (1970).

The subject of the rise velocity of a single bubble is a fascinating one. In many cases the curve of bubble rise velocity versus equivalent diameter is very close to that for a rigid sphere of the same diameter and density; however, large bubbles become flattened and their terminal velocities are significantly lower than for the equivalent rigid sphere. A further, interesting phenomenon occurs for bubbles of intermediate diameter in carefully purified liquids: the rise velocity is about 50 per cent greater than that of the equivalent rigid sphere. This is due to the mobility of the gas–liquid interface, allowing circulation inside the bubble, which therefore 'rolls' through the liquid. The presence of a small amount of surface active contaminant, which accumulates at the interface, inhibits this mobility of the interface causing the bubble to behave like a rigid sphere. The higher velocity is not observed with very small bubbles because so little contaminant is required to inhibit circulation, nor with large bubbles where form drag is dominant. Contamination with surface active agents is particularly prevalent in aqueous systems.

An accepted empirical correlation for the terminal rise velocity of single isolated bubbles is that of Peebles and Garber (1953), who identified four regions as shown in Table 7.1.

Table 7.1 *Peebles and Garber correlation*

Terminal velocity	*Range of applicability*
$u_t = \dfrac{2R_{be}^2(\rho_L - \rho_G)g}{9\mu_L}$	$Re < 2$
$u_t = 0.33g^{0.76}\nu_L^{-0.52}R_{be}^{1.28}$	$2 < Re < 4.02G_1^{-0.214}$
$u_t = 1.35\left(\dfrac{\sigma}{\rho_L R_{be}}\right)^{0.5}$	$4.02G_1^{-0.214} < Re < 3.10G_1^{-0.25}$ or $16.32G_1^{0.144} < G_2 < 5.75$
$u_t = 1.18\left(\dfrac{g\sigma}{\rho_L}\right)^{0.25}$	$3.10G_1^{-0.25} < Re$ or $5.75 < G_2$

The bubble Reynolds number *Re* is defined as

$$Re = \frac{\rho_L u_t(2R_{be})}{\mu_L} = \frac{\rho_L u_t D_{be}}{\mu_L} \tag{7.34}$$

where $D_{be} = 2R_{be}$ is the diameter of a sphere having the same volume as the bubble.

The quantity G_1 is the Morton number:

$$G_1 = \frac{g\mu_L^4}{\rho_L\sigma^3} \tag{7.35}$$

and G_2 is defined by

$$G_2 = \frac{gR_{be}^4 u_t^4 \rho_L^3}{\sigma^3} \tag{7.36}$$

As Wallis (1969) points out, the upper limit of region 4 is with very large bubbles when their rise is dominated by inertial forces. Under these conditions, the terminal rise velocity is readily calculated from potential flow theory and is given by

$$u_t = 1.00\sqrt{gR_{be}} \qquad \text{for} \qquad R_{be} > 2\sqrt{\frac{\sigma}{g\rho_L}} \tag{7.37}$$

This may be considered as a fifth region to be added to the Peebles and Garber correlation.

The Peebles and Garber correlation is valid when the gas density is much lower than the liquid density. Grace, Wairegi and Nguyen (1976) have given a correlation that is valid for both bubbles and liquid drops over a very wide range of conditions, and Wallis (1974) has given a comprehensive correlation for the rise velocity of bubbles and drops in both pure and contaminated liquids.

It should be noted that the bubble rise velocity is independent of the bubble diameter over an extremely wide range of bubble size; this corresponds to the fourth region in the Peebles and Garber correlation. This region is likely to encompass almost all conditions in bubble columns so prediction of the bubble diameter is unnecessary in order to calculate the terminal velocity and hence the void fraction. Harmathy (1960) claims that a better value for the coefficient (1.18) in the Peebles and Garber correlation is 1.53. The corresponding expression for this constant velocity region given by Wallis (1974) can be rearranged to the same form:

$$u_t = \sqrt{2}\left(\frac{g\sigma}{\rho_L}\right)^{0.25} \tag{7.38}$$

and the coefficient, $\sqrt{2}$, is recommended as the best value.

If the bubble size and shape are required, for example for mass transfer calculations, the work of Kumar and Kuloor (1970) and that of Grace, Wairegi and Nguyen (1976) may be consulted.

7.4 Slug flow in vertical tubes

When a very large bubble of gas is allowed to rise in a large expanse of liquid it is found that the bubble becomes rather flattened, having a spherical upper surface and a fairly flat lower surface, as shown in Figure 7.11a. This is characteristic of the fact that the bubble's motion through the liquid is dominated by inertial forces. Inviscid flow theory shows that the rise velocity is given by the expression

$$u_t = 0.35\sqrt{gD} \tag{7.39}$$

where D is the actual diameter of the bubble as shown in Figure 7.11a. (The same result was given in equation 7.37 in terms of the volumetric equivalent radius.)

If a large bubble is constrained to rise through a tube of liquid a slug is formed as shown in Figure 7.11b. The slug's motion is dominated by the potential flow over its nose in the same way as with a large bubble in an expanse of liquid and the slug's velocity of rise u_s is given by

$$u_s = 0.35\sqrt{gD_p} \tag{7.40}$$

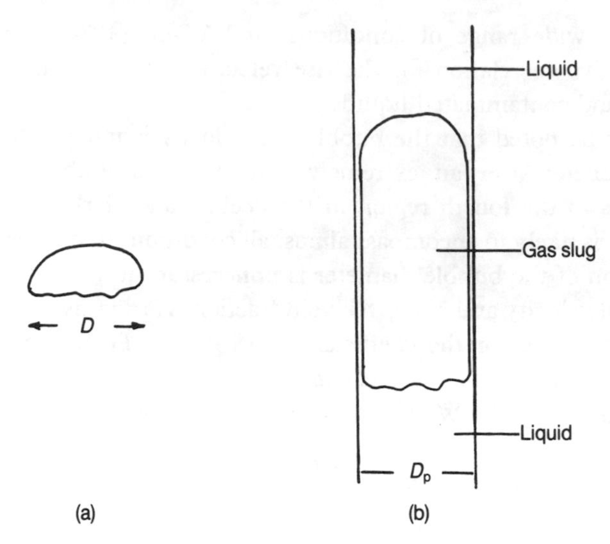

Figure 7.11
Definition of bubble diameters
(a) *For spherical cap bubble* (b) *For a gas slug*

where D_p is the diameter of the pipe or tube. The diameter of the slug is only slightly smaller than the tube diameter, the liquid film between the slug and the tube wall being very thin. It may appear surprising that the shear force in this film is negligible but this must be so because the gas has such a low viscosity and density, and the relative motion is fairly slow.

In a continuously slugging system, in which both gas and liquid flow up through the tube, the slugs rise relative to the total flow of the gas–liquid mixture. If the volumetric fluxes of the gas and the liquid are j_G and j_L, the average velocity of rise of the mixture is $j_G + j_L$ and it might be thought that the slug's velocity of rise relative to the tube would be given by the equation

$$u_s = (j_G + j_L) + 0.35\sqrt{gD_p} \qquad (7.41)$$

This is not true because the slug is found to rise with a velocity $0.35\sqrt{(gD_p)}$ relative to the centre-line velocity of the liquid. The liquid flow will be turbulent and its centre-line velocity therefore approximately 20 per cent greater than its average velocity. Thus the correct expression for the slug's velocity is

$$u_s = 1.2(j_G + j_L) + 0.35\sqrt{gD_p} \qquad (7.42)$$

This equation has been verified by many investigators for well-spaced slugs rising through Newtonian liquids of low viscosity. Corrections for the effects of surface tension and higher viscosities have been given by White and Beardmore (1962). The constant 0.35 is replaced by a constant k_1:

$$u_s = 1.2(j_G + j_L) + k_1\sqrt{gD_p} \tag{7.43}$$

The value of k_1 depends on the Eötvos number, *Eö*, and the Morton number, G_I, as shown in Figure 7.12.

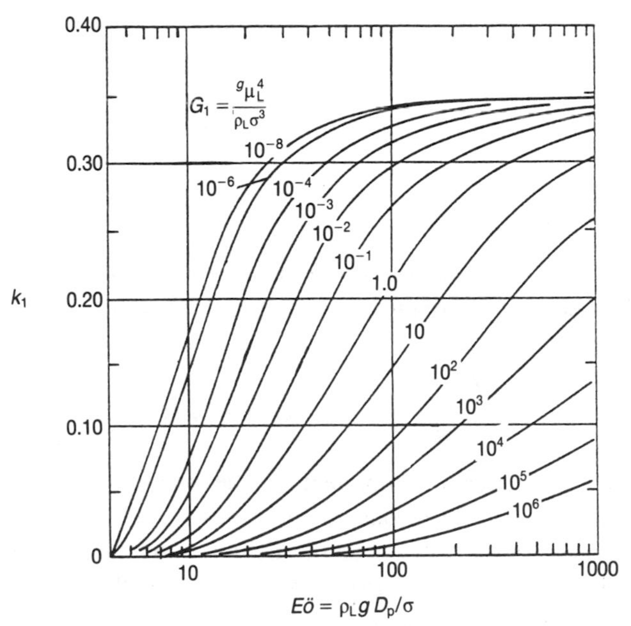

Figure 7.12
Plot of k_1 against Eötvos number for slug flow
Source: E. T. White and R. H. Beardmore, *Chemical Engineering Progress* **17**, pp. 351–361 (1962)

When the slugs are not well spaced, the slug following in the wake of another slug rises more quickly and after some distance the two slugs will coalesce. Moissis and Griffith (1962) have given a correlation to account for this:

$$k_1 = k_{1ss}(1 + 8e^{-1.06L/D_p}) \tag{7.44}$$

where k_{1ss} is the value of k_1 for a single slug and L is the spacing of the slugs. Unfortunately, L is not determined solely by the gas and liquid flow rates and therefore equation 7.44 is difficult to use in practice.

7.4.1 *Void fraction*

For the gas flow, continuity gives

$$Q_G = u_s \alpha S \tag{7.45}$$

or

$$\alpha = \frac{Q_G}{u_s S} = \frac{j_G}{u_s} \tag{7.46}$$

Substituting for u_s from equation 7.43, the void fraction is obtained as

$$\alpha = \frac{j_G}{1.2(j_G + j_L) + k_1\sqrt{gD_p}} \tag{7.47}$$

7.4.2 *Pressure drop*

In many cases the frictional and accelerative components of the pressure gradient are negligible so only the static head component need be considered:

$$\left(\frac{dP}{dx}\right)_{sh} = -[\alpha\rho_G + (1-\alpha)\rho_L]g \tag{7.48}$$

If necessary the frictional component of the pressure gradient can be calculated using the homogeneous model discussed in Section 7.5.

Provided the pressure drop over the height H of the section through which the slugs rise is small, expansion of the gas will be negligible and equation 7.48 can be integrated with ρ_G and α constant:

$$\Delta P = \int_0^H [\alpha\rho_G + (1-\alpha)\rho_L]\, g\, dx = [\alpha\rho_G + (1-\alpha)\rho_L]gH \tag{7.49}$$

Figure 7.12 shows that the maximum value of k_1 is 0.35. If this value is used in cases where a lower value of k_1 is appropriate, the void fraction calculated from equation 7.47 will be slightly underestimated and the pressure drop correspondingly overestimated, resulting in a conservative design.

Wallis (1969) discusses slug flow in horizontal and inclined pipes.

7.5 The homogeneous model for two-phase flow

The complex nature of the churn flow regime does not lend itself to the relatively simple analyses used for the bubble flow and slug flow regimes. Friction will certainly be significant and acceleration may be, particularly when a change of phase occurs. Although the annular flow regime appears to be simpler and Wallis (1969) has suggested an approach using an interfacial friction factor, the details of annular flow are complex. A significant part of the liquid flow is in the form of droplets in the gas core and there is a large interchange of liquid between the core and the film. This is a topic of continuing research and in this introductory text a separate analysis for the annular flow regime will not be presented.

The homogeneous flow model and the separated flow model may be used to estimate the pressure drop for the churn regime but the former is not recommended for use with annular flow. The separated flow model of Martinelli and Nelson (1948), and developments thereof, may be used for annular flow.

The approaches adopted in the homogeneous model and the separated flow model are opposites: in the former it is assumed that the two-phase flow can be treated as a hypothetical single-phase flow having some kind of average properties, while in the separated flow model it is assumed that distinct parts of the flow cross section can be assigned to the two phases, reflecting what occurs to a large extent in annular flow.

7.5.1 *Momentum equation for the homogeneous flow model*

Recall that in Section 7.2 it was shown that the total pressure gradient comprises three components due to friction, acceleration and the static head of the mixture. In the homogeneous flow model the two-phase flow is treated as a hypothetical single-phase flow with a single, uniform velocity over a given cross section. It is assumed that the frictional component of the pressure gradient can be described by the use of a single friction factor and the model is sometimes known as the Friction Factor Model. It would be expected that the model's predictions would be most accurate for flows in which one phase is well dispersed in the other, that is the spray regime, the bubbly flow regime and, possibly, the wispy-annular regime.

The continuity equation can be written as

$$M = \bar{\rho} u S \tag{7.50}$$

where u is the velocity of both phases and $\bar{\rho}$ is the average density of the two-phase mixture.

By definition, the average density is the reciprocal of the average specific volume of the mixture, so that

$$\frac{1}{\bar{\rho}} = \bar{V} = wV_G + (1-w)V_L \tag{7.51}$$

The momentum equation takes a simple form, which may be derived from the general form given in equations 7.9 by putting

$$u_G = u_L = u \tag{7.52}$$

and

$$S_G\rho_G + S_L\rho_L = S\bar{\rho} \tag{7.53}$$

Thus the momentum equation becomes

$$-S\frac{dP}{dx} - \frac{dF}{dx} - S\bar{\rho}g\sin\theta = M\frac{du}{dx} \tag{7.54}$$

or

$$\frac{dP}{dx} = -\frac{1}{S}\frac{dF}{dx} - G\frac{du}{dx} - \bar{\rho}g\sin\theta \tag{7.55}$$

Again, the three components of the total pressure gradient, friction, acceleration and the static head, are clear.

It is necessary to write each of these terms in a convenient form. The treatment of the frictional component is the same as in Chapter 2.

Frictional component

The frictional force δF over length δx of the wall gives rise to a balancing frictional component of the pressure gradient:

$$\delta F = \pi d_i \tau_w \delta x = -\frac{\pi d_i^2}{4}(\delta P)_f \tag{7.56}$$

where τ_w is the shear stress at the wall. Therefore

$$\frac{dF}{dx} = \pi d_i \tau_w = -\frac{\pi d_i^2}{4}\left(\frac{dP}{dx}\right)_f \tag{7.57}$$

or

$$-\left(\frac{dP}{dx}\right)_f = \frac{4\tau_w}{d_i} \tag{7.58}$$

which is equivalent to equation 2.7, the latter being written for a finite length of pipe.

Using the Fanning friction factor, the wall shear stress is given by

$$\tau_w = \tfrac{1}{2}\bar{\rho}u^2 f$$

and the frictional component of the pressure gradient is then

$$-\left(\frac{dP}{dx}\right)_f = \frac{2f\bar{\rho}u^2}{d_i} = \frac{2fG^2\bar{V}}{d_i} \tag{7.59}$$

This equation is equivalent to equation 2.13. The velocity, which generally varies with position along the pipe, has been rewritten in terms of the constant mass flux G and the average specific volume of the mixture. Of course $\bar{V}$ is a function of pressure and therefore varies along the pipe.

Accelerative component

Substituting for the velocity in terms of G and $\bar{V}$, the accelerative component of the pressure gradient in equation 7.55 can be written as

$$-\left(\frac{dP}{dx}\right)_a = G\frac{d}{dx}(G\bar{V}) = G^2\frac{d\bar{V}}{dx} \tag{7.60}$$

The average specific volume $\bar{V}$ is a function of the gas and liquid specific volumes and the mass fraction of gas. The liquid may be treated as incompressible but, in general, the specific volume of the gas and the mass fraction will change along the length of the pipe. Differentiating equation 7.51

$$\frac{d\bar{V}}{dx} = w\frac{dV_G}{dx} + (V_G - V_L)\frac{dw}{dx} \tag{7.61}$$

$$= w\frac{dV_G}{dP}\frac{dP}{dx} + V_{LG}\frac{dw}{dx} \tag{7.61}$$

where $V_{LG} = V_G - V_L$. The first term in equation 7.61 originates from the compressibility of the gas phase and represents the effect of acceleration due to expansion of the gas at constant mass fraction. The second term, of which there is no equivalent in single-phase flow, represents the effect of acceleration due to a change of phase. For example, in evaporation or boiling in a tube, part of the liquid with specific volume V_L is changed into vapour with a much higher specific volume V_G so the mean specific volume increases and by continuity the flow must accelerate.

Substituting equation 7.61 into equation 7.60 allows the accelerative component of the pressure gradient to be written as

$$-\left(\frac{dP}{dx}\right)_a = G^2\left(w\frac{dV_G}{dP}\frac{dP}{dx} + V_{LG}\frac{dw}{dx}\right) \tag{7.62}$$

Static head component

The static head component is simply

$$-\left(\frac{dP}{dx}\right)_{sh} = \bar{\rho}g\sin\theta = \frac{g\sin\theta}{\bar{V}} \tag{7.63}$$

Total pressure gradient

Summing the three components of the pressure gradient in equation 7.55 gives the total pressure gradient:

$$-\frac{dP}{dx} = \frac{2fG^2\bar{V}}{d_i} + G^2\left(w\frac{dV_G}{dP}\frac{dP}{dx} + V_{LG}\frac{dw}{dx}\right) + \frac{g\sin\theta}{\bar{V}} \tag{7.64}$$

The total pressure gradient is implicit because it appears in the accelerative term as well as on the left of the equation. Rewriting equation 7.64 to provide the total pressure gradient explicitly gives

$$-\frac{dP}{dx} = \left(\frac{2fG^2\bar{V}}{d_i} + G^2V_{LG}\frac{dw}{dx} + \frac{g\sin\theta}{\bar{V}}\right)\bigg/\left(1 + G^2w\frac{dV_G}{dP}\right) \tag{7.65}$$

Simplifications

In general, integration of equation 7.65 has to be done by a step by step procedure. Assuming that the conditions at the pipe entrance are known, all the quantities on the right of equation 7.65 are evaluated at those conditions and the pressure gradient calculated. Making the approximation that the pressure gradient is constant over the short length, the pressure at the end of that length is calculated and used to evaluate the quantities in equation 7.65 for the next step.

In some cases the change of quality will be negligible so the second term in the numerator of equation 7.65 can be neglected. The other feature leading to acceleration, namely the compressibility of the gas phase, is responsible for the form of the denominator. Sometimes gas compressibility will be negligible, particularly if the overall pressure is high: this is the case in gas–oil pipelines.

Simplifications that may be applicable can be summarized as follows.

1 $\left|G^2 w \dfrac{dV_G}{dP}\right| \ll 1$ ie gas compressibility negligible.

2 f, $\bar{V}$ remain constant over the length of integration.
3 With the above simplifications it is possible to integrate equation 7.65 analytically for the special case of evaporation with dw/dx constant. This condition occurs when the heat flux is uniform along the tube length. If the liquid is saturated at $x=0$ and has a quality w_e at the tube exit $x=L$, then the pressure drop is given by

$$\Delta P = \frac{2fG^2V_L L}{d_i}\left[1+\frac{w_e}{2}\left(\frac{V_{LG}}{V_L}\right)\right] + G^2V_L\left(\frac{V_{LG}}{V_L}\right)w_e + \frac{Lg\sin\theta}{V_{LG}w_e}\ln\left[1+w_e\left(\frac{V_{LG}}{V_L}\right)\right] \tag{7.66}$$

For equation 7.66 to be strictly valid it is necessary that V_{LG} be sensibly constant. This condition will be satisfied sufficiently closely provided the pressure drop is not large.

7.5.2 *Friction factors for the homogeneous model*

All quantities in equation 7.64 can be readily evaluated with the exception of the friction factor. There are several approaches to estimating the friction factor for two-phase flow.

1 Use a constant value of f irrespective of conditions: a suitable choice would be $f = 0.007$. This is the simplest but least satisfactory approach.
2 Calculate the friction factor in the normal way for single-phase flow but evaluating the Reynolds number using a mean viscosity, $\bar{\mu}$:

$$Re = \frac{Gd_i}{\bar{\mu}} \tag{7.67}$$

Expressions for the mean viscosity that can be used are

(i) $\bar{\mu} = w\mu_G + (1-w)\mu_L$ Cicchitti *et al* (1960)

(ii) $\dfrac{1}{\bar{\mu}} = \dfrac{w}{\mu_G} + \dfrac{(1-w)}{\mu_L}$ McAdams *et al* (1942)

(iii) $\bar{\mu} = \bar{\rho}[wV_G\mu_G + (1-w)V_L\mu_L]$

$= (j_G\mu_G + j_L\mu_L)/j$ Dukler *et al* (1964)

3 Use a friction factor for a corresponding single-phase flow.

As an example of method 3, in bubbly flow with a low quality it would be appropriate to calculate the friction factor based on the properties of the liquid. The frictional component of the pressure gradient for the actual two-phase flow is given by

$$-\left(\frac{\mathrm{d}P}{\mathrm{d}x}\right)_f = \frac{2fG^2\bar{V}}{d_i} \tag{7.59}$$

while if the *whole* of the two-phase flow were liquid the frictional component of the pressure gradient would be given by

$$-\left(\frac{\mathrm{d}P}{\mathrm{d}x}\right)_{LO} = \frac{2f_{LO}G^2V_L}{d_i} \tag{7.68}$$

Here, the subscripts LO have been used to denote the hypothetical single-phase flow that corresponds to the whole of the two-phase flow being liquid.

Dividing equation 7.59 by equation 7.68

$$\left(\frac{\mathrm{d}P}{\mathrm{d}x}\right)_f \Big/ \left(\frac{\mathrm{d}P}{\mathrm{d}x}\right)_{LO} = \frac{f\bar{V}}{f_{LO}V_L} \tag{7.69}$$

If the approximation is made that the friction factor for the two-phase flow f is equal to that for the hypothetical liquid flow f_{LO}, a very simple relationship is obtained between the frictional pressure gradients for the two flows:

$$\left(\frac{\mathrm{d}P}{\mathrm{d}x}\right)_f = \frac{\bar{V}}{V_L}\left(\frac{\mathrm{d}P}{\mathrm{d}x}\right)_{LO} = \left[1 + w\left(\frac{V_G}{V_L} - 1\right)\right]\left(\frac{\mathrm{d}P}{\mathrm{d}x}\right)_{LO} \tag{7.70}$$

The pressure gradient for the 'wholly liquid' flow can be calculated from equation 7.68, in which the friction factor is evaluated for the Reynolds number given by

$$Re_{LO} = \frac{Gd_i}{\mu_L} \tag{7.71}$$

It is important to appreciate that, although the notation here is slightly different from that in Chapter 2, the calculation procedure for the single-phase flow is identical to that in the earlier chapter.

For a two-phase flow with a high quality, for example the spray regime, it would be more appropriate to use a gas flow as the reference single-phase flow. Again this reference flow has the same total flow rate. Writing the equivalents of equations 7.68, 7.69 and 7.70 will give

$$\left(\frac{\mathrm{d}P}{\mathrm{d}x}\right)_f = \frac{\bar{V}}{V_G}\left(\frac{\mathrm{d}P}{\mathrm{d}x}\right)_{GO} = \left[w + (1-w)\frac{V_L}{V_G}\right]\left(\frac{\mathrm{d}P}{\mathrm{d}x}\right)_{GO} \tag{7.72}$$

The concept in this method is to replace the actual two-phase flow by a corresponding single-phase flow, for which the frictional pressure gradient is readily calculated. The relationship between the frictional pressure gradient for the two-phase flow and that for the reference single-phase flow is then derived from the basic frictional pressure gradient equations (eg equations 7.59 and 7.68). The form of this relationship is given in equation 7.69. If the value of the right hand side of equation 7.69 can be estimated then it is possible to calculate the two-phase frictional pressure gradient from the frictional pressure gradient for the reference single-phase flow. This is an example of the Two-Phase Multiplier. In general

$$\left(\frac{\mathrm{d}P}{\mathrm{d}x}\right)_f = \phi_R^2\left(\frac{\mathrm{d}P}{\mathrm{d}x}\right)_R$$

where ${\phi_R}^2$ is known as the Two-Phase Multiplier; its value depends on which single-phase flow is chosen as the reference flow.

Example 7.1

Air and water flow at 8×10^{-3} kg/s and 0.4 kg/s upwards in a vertical, smooth-wall tube of internal diameter $d_i = 20$ mm and length $L = 1.3$ m. Using the homogeneous flow model, calculate the pressure drop across the tube (neglecting end effects). The fluids are at a temperature of 20 °C and the expansion of the air may be assumed to be isothermal. The exit pressure is 1 bar.

Calculations

The required equation for the pressure gradient is

$$-\frac{\mathrm{d}P}{\mathrm{d}x} = \left(\frac{2fG^2\bar{V}}{d_i} + G^2V_{LG}\frac{\mathrm{d}w}{\mathrm{d}x} + \frac{g\sin\theta}{\bar{V}}\right) \Big/ \left(1 + G^2w\frac{\mathrm{d}V_G}{\mathrm{d}P}\right) \tag{7.65}$$

There is no change of quality, so $\mathrm{d}w/\mathrm{d}x = 0$.

The following values are readily calculated:

$$S = \frac{\pi d_i^2}{4} = 3.142 \times 10^{-4}\ \text{m}^2$$

$$G = \frac{M_G + M_L}{S} = \frac{(0.008 + 0.4)\ \text{kg/s}}{3.142 \times 10^{-4}\ \text{m}^2} = 1299\ \text{kg/(m}^2\text{s)} \qquad (7.16)$$

$$w = \frac{M_G}{M_G + M_L} = 0.0196 \qquad (7.14)$$

For isothermal expansion, PV_G = constant. Therefore

$$\frac{dV_G}{dP} = -\frac{V_G}{P}$$

From thermodynamic tables, ρ_{air} = 1.1984 kg/m^3 at 20 °C, 1 atm (1.01325 bar). Therefore

$$\frac{dV_G}{dP} = -8.455 \times 10^{-6}\ \text{m}^3\text{/(kg Pa) at 1 bar}$$

and

$$G^2 w \frac{dV_G}{dP} = 1299^2 \times 0.0196 \times (-8.455 \times 10^{-6}) = -0.2796$$

Thus the denominator of equation 7.65 has the value 0.7204. This accounts for acceleration.

Frictional term

$$-\left(\frac{dP}{dx}\right)_f = \frac{2fG^2\bar{V}}{d_i} \qquad (7.59)$$

It is necessary to estimate f for the two-phase flow. Using method 3 outlined above, it is appropriate to use the liquid as the reference flow because the quality is low (0.0196). The pressure gradient for the whole flow as liquid is

$$-\left(\frac{dP}{dx}\right)_{LO} = \frac{2f_{LO}G^2V_L}{d_i}$$

$$= \frac{2f_{LO}[1299\ \text{kg/(m}^2\text{s)}]^2\ (1 \times 10^{-3}\ \text{m}^3\text{/kg})}{20 \times 10^{-3}\ \text{m}}$$

$$= 1.687 \times 10^5 f_{LO} \quad \text{Pa/m}$$

f_{LO} is determined for the whole flow as liquid, for which the Reynolds number is

$$Re_{LO} = \frac{Gd_i}{\mu_L} = \frac{[1299\ \text{kg/(m}^2\text{s)}]\ (20 \times 10^{-3}\ \text{m})}{1 \times 10^{-3}\ \text{Pa s}} = 2.598 \times 10^4$$

From the friction factor chart, Figure 2.1, for a smooth tube and this value of the Reynolds number, $f_{LO} = 0.0058$. Therefore

$$-\left(\frac{dP}{dx}\right)_{LO} = (1.687 \times 10^5\ \text{Pa/m})(0.0058) = 978.5\ \text{Pa/m}$$

Making the approximation that the value of the friction factor for the two-phase flow is equal to f_{LO}, the frictional component of the two-phase pressure gradient is given by

$$\left(\frac{dP}{dx}\right)_f = \frac{\bar{V}}{V_L}\left(\frac{dP}{dx}\right)_{LO} \qquad (7.70)$$

Evaluating the specific volume at the (known) outlet pressure

$$\begin{aligned} \bar{V} &= wV_G + (1-w)V_L \qquad (7.51) \\ &= (0.0196)(0.8455\ \text{m}^3/\text{kg}) + (0.9804)(1.00 \times 10^{-3}\ \text{m}^3/\text{kg}) \\ &= 0.01755\ \text{m}^3/\text{kg} \end{aligned}$$

Therefore

$$\begin{aligned} -\left(\frac{dP}{dx}\right)_f &= \frac{0.01755\ \text{m}^3/\text{kg}}{1.00 \times 10^{-3}\ \text{m}^3/\text{kg}} \times 978.5\ \text{Pa/m} \\ &= 1.717 \times 10^4\ \text{Pa/m} \end{aligned}$$

Static head

$$-\left(\frac{dP}{dx}\right)_{sh} = \frac{g}{\bar{V}} = \frac{9.81\ \text{m/s}^2}{0.01755\ \text{m}^3/\text{kg}} = 559.0\ \text{Pa/m} \qquad \text{from (7.63)}$$

Total pressure drop

Total pressure gradient, from equation 7.65

$$-\frac{dP}{dx} = \frac{(1.717 \times 10^4 + 559.0)\ \text{Pa/m}}{0.7204} = 2.461 \times 10^4\ \text{Pa/m}$$

This is the value of the pressure gradient at the exit of the tube because conditions (particularly the value of $\bar{V}$) at that location have been used in the above calculations. If the pressure gradient were constant over the length of the tube, the pressure drop over the 1.3 m length would be

$$\Delta P = (2.461 \times 10^4 \text{ Pa/m})(1.3 \text{ m}) = 3.199 \times 10^4 \text{ Pa} = 0.3199 \text{ bar}$$

This estimated pressure drop is a significant fraction of the outlet pressure (1 bar) so various quantities calculated above will vary through the tube; this is particularly true for V_G and consequently $\bar{V}$.

The calculation accuracy can be increased by splitting the tube length into a number of increments and making the calculation of the pressure drop for each.

Incremental calculation

As an illustration, the tube length is divided into five equal increments. From above, the pressure gradient at the exit is -2.461×10^4 Pa/m so the pressure drop over the fifth increment (counting from the inlet) is

$$\Delta P_5 = (2.461 \times 10^4 \text{ Pa/m})\frac{1.3 \text{ m}}{5} = 6398.6 \text{ Pa} = 0.06399 \text{ bar}$$

and the pressure at the end of the fourth increment is therefore 1.06399 bar. This pressure can now be used to evaluate the properties for Increment 4, allowing the pressure drop calculation to be made for that increment. These calculations are then made for each increment. The results are given in Table 7.2.

Table 7.2 *Incremental calculation of pressure drop*

Increment number	P (bar)	$-G^2 w \dfrac{dV_G}{dP}$	$\bar{V}$ (m^3/kg)	$-\left(\dfrac{dP}{dx}\right)_f$ (Pa/m)	$-\left(\dfrac{dP}{dx}\right)_{sh}$ (Pa/m)	ΔP (bar)
5	1.0000	0.2796	0.01755	1.717×10^4	559	0.06399
4	1.0640	0.2470	0.01656	1.620×10^4	592	0.05798
3	1.1220	0.2221	0.01575	1.541×10^4	623	0.05359
2	1.1756	0.2023	0.01508	1.475×10^4	651	0.05020
1	1.2233	0.1861	0.01450	1.419×10^4	677	0.04749
						0.27325

This calculation with five increments gives the total pressure drop over the 1.3 m of the tube as 0.273 bar compared with the estimate of 0.320 bar found using the exit value of the pressure gradient. The error in making this simplification compared with the value using five increments is 17 per cent. In high pressure systems the pressure drop will generally be a very small fraction of the average pressure and pressure-dependent variations will be less significant.

An example of the use of the homogeneous flow model for a case in which boiling occurs, and equation 7.66 is used, is given as part of Example 7.2.

7.6 Two-phase multiplier

In the separated flow models presented in Sections 7.7 and 7.8, the method of calculating the frictional component of the pressure gradient involves use of the two-phase multiplier ϕ^2 defined by

$$\left(\frac{dP}{dx}\right)_f = \phi_R^2 \left(\frac{dP}{dx}\right)_R \qquad (7.73)$$

That is, the two-phase frictional pressure gradient is calculated from a reference single-phase frictional pressure gradient $(dP/dx)_R$ by multiplying by the two-phase multiplier, the value of which is determined from empirical correlations. In equation 7.73 the two-phase multiplier is written as ϕ_R^2 to denote that it corresponds to the reference single-phase flow denoted by R.

For a gas–liquid two-phase flow there are four possible reference flows:

1 whole flow liquid, denoted by subscripts LO
2 whole flow gas, denoted by subscripts GO
3 only the liquid in the two-phase flow, denoted by subscript L
4 only the gas in the two-phase flow, denoted by subscript G.

When the reference flow is the whole of the two-phase flow as liquid, then the two-phase frictional pressure gradient is given by

$$\left(\frac{dP}{dx}\right)_f = \phi_{LO}^2 \left(\frac{dP}{dx}\right)_{LO} \qquad (7.74)$$

The frictional pressure gradient for this 'wholly liquid' reference flow is given by

$$-\left(\frac{dP}{dx}\right)_{LO} = \frac{2f_{LO}G^2V_L}{d_i} \qquad (7.75)$$

where the friction factor f_{LO} is evaluated for the Reynolds number $Re_{LO} = Gd_i/\mu_L$. Thus if the value of the two-phase multiplier ϕ_{LO}^2 can be determined using a suitable correlation, the two-phase frictional pressure gradient is readily calculated from equation 7.74. The case of a wholly gas reference flow is similar.

When the reference flow is only the liquid in the two-phase flow, the equations are slightly different because the liquid flow rate and not the total flow rate must be used for the reference.

The 'only liquid' frictional pressure gradient is given by

$$-\left(\frac{dP}{dx}\right)_L = \frac{2f_L(1-w)^2G^2V_L}{d_i} \tag{7.76}$$

with the friction factor f_L evaluated for the Reynolds number $Re_L = (1-w)Gd_i/\mu_L$. The two-phase frictional pressure gradient is then calculated from the defining equation

$$\left(\frac{dP}{dx}\right)_f = \phi_L^2\left(\frac{dP}{dx}\right)_L \tag{7.77}$$

For an 'only gas' reference flow, the reference frictional pressure gradient is given by

$$-\left(\frac{dP}{dx}\right)_G = \frac{2f_Gw^2G^2V_G}{d_i} \tag{7.78}$$

with the friction factor f_G evaluated for the Reynolds number $Re_G = wGd_i/\mu_G$. The two-phase frictional pressure gradient is given by the equation

$$\left(\frac{dP}{dx}\right)_f = \phi_G^2\left(\frac{dP}{dx}\right)_G \tag{7.79}$$

Warning about notation

The notation used here and in Section 7.7 is standard in the literature on this subject and originates in the pioneering work of Martinelli and co-workers. There are two aspects of the notation that may lead to confusion and error. First, note that LO and GO do not denote 'liquid only' and 'gas only' reference flows, as might be expected. On the contrary, they denote flows in which the whole of the flow rate is liquid or gas. It may help to remember them as 'liquid overall' and 'gas overall'. The second point to note is that ϕ^2 denotes the two-phase multiplier. Correlations may present values of ϕ but it must be remembered that this is the square root of the two-phase multiplier.

7.7 Separated flow models

In these models the phases are treated as if they are separate and flow in well defined but unspecified parts of the cross section. Only the simplest case, in which the phases are allowed to have different but uniform velocities, will be considered here. An overall momentum equation will be given and it will be seen that merely allowing the gas and liquid velocities to differ leads to considerable complexity. Two empirical correlations from the pioneering work of Martinelli and co-workers will then be described. These methods can be used for the churn and annular flow regimes.

7.7.1 *Momentum equation*

The basic momentum equation derived in Section 7.2 is

$$-\frac{\mathrm{d}P}{\mathrm{d}x} = -\left(\frac{\mathrm{d}P}{\mathrm{d}x}\right)_f + G^2\frac{\mathrm{d}}{\mathrm{d}x}\left[\frac{w^2V_G}{\alpha} + \frac{(1-w)^2V_L}{(1-\alpha)}\right] + \left(\frac{\alpha}{V_G} + \frac{1-\alpha}{V_L}\right)g\sin\theta \tag{7.80}$$

In contrast to the case of the homogeneous model, the accelerative term cannot be put in a simpler form because the phase velocities differ. It is therefore necessary to carry out the differentiation in the accelerative term. When this is done and the frictional component of the pressure gradient is represented using the 'wholly liquid' two-phase multiplier, the resulting form of the momentum equation is

$$-\frac{\mathrm{d}P}{\mathrm{d}x} = \frac{\dfrac{2f_{LO}G^2V_L\phi_{LO}^2}{d_i} + G^2\dfrac{\mathrm{d}w}{\mathrm{d}x}A(\alpha,w) + \left(\dfrac{\alpha}{V_G} + \dfrac{1-\alpha}{V_L}\right)g\sin\theta}{1 + G^2\left\{\dfrac{w^2}{\alpha}\dfrac{\mathrm{d}V_G}{\mathrm{d}P} + \left(\dfrac{\partial\alpha}{\partial P}\right)_w\left[\dfrac{(1-w)^2V_L}{(1-\alpha)^2} - \dfrac{w^2V_G}{\alpha^2}\right]\right\}} \tag{7.81}$$

where

$$A(\alpha,w) = \left[\frac{2wV_G}{\alpha} - \frac{2(1-w)V_L}{1-\alpha}\right] + \left(\frac{\partial\alpha}{\partial w}\right)_P\left[\frac{(1-w)^2V_L}{(1-\alpha)^2} - \frac{w^2V_G}{\alpha^2}\right] \tag{7.82}$$

If the 'only liquid' reference flow had been used the frictional term in equation 7.81 would be $2f_L(1-w)^2G^2V_L\phi_L^2/d_i$.

Comparing equation 7.81 with its homogeneous model equivalent, equation 7.65, it is clear that merely allowing the phases to have different velocities leads to a considerable increase in complexity. In both equations, the middle term in the numerator derives from the accelerative component of the pressure gradient and represents acceleration due to a change of phase. Also, in both equations, the term involving G^2 in the denominator originates in the accelerative component. In the homogenous model, both phases must accelerate equally, so there is only a term including dV_G/dP. In the separated flow model, equation 7.81, there is a further term, that in square brackets multiplied by $(\partial\alpha/\partial P)_w$ and resulting from the fact that the phases are not constrained to have the same velocity.

Integration of equation 7.81 to determine the pressure drop over a length of pipe generally requires a stepwise procedure. As with the homogeneous model, in some cases simplifications may be possible:

1 the denominator of equation 7.81 differs little from unity;
2 f_L, V_G, V_L remain sensibly constant over the length of integration;
3 if conditions 1 and 2 are satisfied and evaporation takes place from saturation at $x = 0$ with a constant value of dw/dx then

$$\Delta P = \frac{2f_{LO}G^2V_LL}{d_i}\left[\frac{1}{w_e}\int_0^{w_e}\phi_{LO}^2\,dw\right] + G^2V_L\left[\frac{w_e^2}{\alpha}\left(\frac{V_G}{V_L}\right) + \frac{(1-w_e)^2}{1-\alpha} - 1\right]$$
$$+\frac{Lg\sin\theta}{w_e}\int_0^{w_e}\left(\frac{\alpha}{V_G}+\frac{1-\alpha}{V_L}\right)dw \qquad (7.83)$$

where w_e is the quality at the tube exit $x = L$.

Equations 7.81 and 7.83 are not easy to evaluate. In the following sections the Lockhart–Martinelli and Martinelli–Nelson correlations will be considered. The Lockhart–Martinelli correlation is valid when there is no change of phase, so $dw/dx=0$ in equation 7.81 and the second term in the numerator vanishes. In the Martinelli–Nelson correlation, values are given for the quantities in square brackets in equation 7.83.

7.7.2 *Lockhart–Martinelli correlation*

The experimental work on which the correlation is based was done for horizontal flow of air–liquid mixtures at near-atmospheric pressures and with no change of phase. It is inadvisable to use the correlation for other conditions. For the conditions employed, the accelerative component of the pressure gradient was assumed to be negligible, while the static head component vanishes. Consequently the frictional pressure gradients in the two phases were assumed to be equal.

Lockhart and Martinelli (1949) used 'only liquid' and 'only gas' reference flows and, having derived equations for the frictional pressure gradient in the two-phase flow in terms of shape factors and equivalent diameters of the portions of the pipe through which the phases are assumed to flow, argued that the two-phase multipliers ϕ_L^2 and ϕ_G^2 could be uniquely correlated against the ratio X^2 of the pressure gradients of the two reference flows:

$$X^2 = \left(\frac{\mathrm{d}P}{\mathrm{d}x}\right)_L \Big/ \left(\frac{\mathrm{d}P}{\mathrm{d}x}\right)_G \tag{7.84}$$

This was confirmed by their experimental results.

It was assumed that four flow regimes could occur depending on whether each phase was in turbulent or laminar (viscous) flow. Their empirical correlation is shown in Figure 7.13. The second and third subscripts denote the type of flow of the liquid and gas respectively. Note that ϕ and X are the *square roots* of the two-phase multiplier and the ratio of reference flow pressure gradients.

The four curves of Figure 7.13 can be well represented [Collier (1972)] by equations of the form:

$$\phi_L^2 = 1 + \frac{C}{X} + \frac{1}{X^2} \tag{7.85}$$

and

$$\phi_G^2 = 1 + CX + X^2 \tag{7.86}$$

where the values of C for the various flow combinations are shown in Table 7.3.

Table 7.3

	liquid	*gas*	C
(*tt*)	turbulent	turbulent	20
(*vt*)	viscous	turbulent	12
(*tv*)	turbulent	viscous	10
(*vv*)	viscous	viscous	5

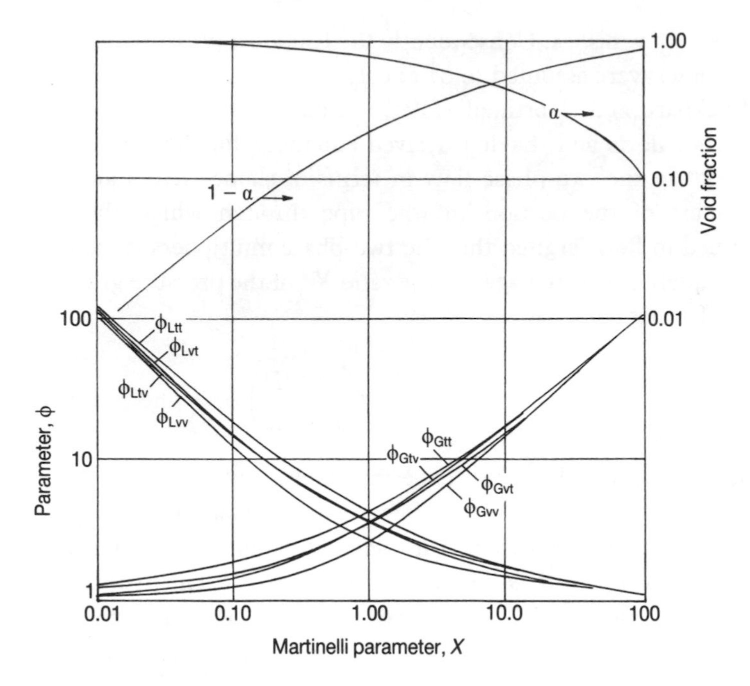

Figure 7.13
Void fraction and square root of two–phase multiplier against Martinelli parameter X
Source: R. W. Lockhart and R. C. Martinelli, *Chemical Engineering Progress* **45**, pp. 39–46 (1949)

Use of the correlation is very simple. First the frictional pressure gradients are calculated for only the liquid flowing in the pipe, and for only the gas:

$$-\left(\frac{\mathrm{d}P}{\mathrm{d}x}\right)_L = \frac{2f_L(1-w)^2G^2V_L}{d_i} \tag{7.76}$$

and

$$-\left(\frac{\mathrm{d}P}{\mathrm{d}x}\right)_G = \frac{2f_Gw^2G^2V_G}{d_i} \tag{7.78}$$

The ratio of these pressure gradients gives X^2.

In order to determine whether each phase is in laminar or turbulent flow, Lockhart and Martinelli suggested tentatively that the Reynolds number for the appropriate reference flow should be greater than 2000 for turbulent flow and less than 1000 for laminar flow. At intermediate values the flow was thought to be transitional.

The value of the square root of the two-phase multiplier is read from Figure 7.13, or calculated from equation 7.85 or 7.86, and the two-phase frictional pressure gradient calculated from

$$\left(\frac{dP}{dx}\right)_f = \phi_L^2\left(\frac{dP}{dx}\right)_L$$

or

$$\left(\frac{dP}{dx}\right)_f = \phi_G^2\left(\frac{dP}{dx}\right)_G$$

Obviously, it is better to use the smaller multiplier. The curves for each combination of flow regimes cross at $X = 1$, as they must from the definition of X, so the 'only gas' reference is preferable for $X < 1$ and the 'only liquid' reference for $X > 1$. The reader is reminded that this correlation should be used only when the frictional component of the pressure gradient is dominant.

7.7.3 *Martinelli parameter*

In developing their correlation, Lockhart and Martinelli assumed that the friction factors could be determined from equations of the same form as the Blasius equation:

$$f_L = K_L Re_L^{-n} = K_L\left[\frac{d_i(1-w)G}{\mu_L}\right]^{-n} \tag{7.87}$$

and

$$f_G = K_G Re_G^{-m} = K_G\left(\frac{d_i wG}{\mu_G}\right)^{-m} \tag{7.88}$$

For laminar flow $K = 16$ and $m,n = 1$. For turbulent flow the values $K = 0.046$ and $m,n = 0.20$–0.25 are recommended.

From the definition of X, equation 7.84, and equations 7.76 and 7.78,

$$X^2 = \left(\frac{dP}{dx}\right)_L \bigg/ \left(\frac{dP}{dx}\right)_G = \frac{f_L}{f_G}\left(\frac{1-w}{w}\right)^2 \frac{V_L}{V_G} \tag{7.89}$$

Substituting for the friction factors from equations 7.87 and 7.88,

$$X^2 = \frac{K_L}{K_G}\frac{[(1-w)/\mu_L]^{-n}}{(w/\mu_G)^{-m}}\left(\frac{1-w}{w}\right)^2 \frac{\rho_G}{\rho_L}(d_i G)^{m-n} \tag{7.90}$$

If both the gas and the liquid are in laminar flow or both in turbulent flow, $K_L = K_G$ and $m = n$. Consequently

$$X^2 = \left(\frac{1-w}{w}\right)^{2-n}\left(\frac{\mu_L}{\mu_G}\right)^n \frac{\rho_G}{\rho_L} \tag{7.91}$$

In practice, the flow regimes of both phases will be turbulent in most cases. Using the value $n = 0.20$

$$X_{tt} = \left(\frac{1-w}{w}\right)^{0.9}\left(\frac{\mu_L}{\mu_G}\right)^{0.1}\left(\frac{\rho_G}{\rho_L}\right)^{0.5} \tag{7.92}$$

This provides a simple method of determining the ratio of the 'only liquid' and 'only gas' frictional pressure gradients without evaluating both pressure gradients.

At high values of the Martinelli parameter X, the gas–liquid flow behaves more like the liquid; at low values of X it behaves more like the gas.

7.7.4 *Martinelli–Nelson correlation*

This correlation is an extension of the Lockhart–Martinelli correlation. The earlier correlation is limited to low pressures and systems in which no change of phase occurs. Although Lockhart and Martinelli provided for four flow regimes, it is unusual in industrial processes for either phase to be in laminar flow. The Martinelli–Nelson (1948) correlation is specifically for forced circulation boiling of water in which it is assumed that both phases are in turbulent flow.

When a change of phase occurs, as in boiling, it is necessary to use the 'wholly liquid' reference flow (an 'only liquid' basis would change as the liquid flow rate decreases during boiling). At low pressures, the results of the Lockhart–Martinelli correlation can be used for the frictional component of the pressure gradient but it is necessary to convert the 'only liquid' basis used in the earlier correlation to the 'wholly liquid' basis. It is assumed that the frictional pressure gradients for the two reference flows are related by the expression

$$\left(\frac{dP}{dx}\right)_{LO} = \left(\frac{dP}{dx}\right)_L \left(\frac{M_L + M_G}{M_L}\right)^{2-n} \tag{7.93}$$

This is consistent with the Blasius type of expression used for the friction factors in deriving the Martinelli parameter. Using the value $n = 0.20$ and expressing the ratio of flow rates in terms of the quality

$$\left(\frac{dP}{dx}\right)_{LO} = \left(\frac{dP}{dx}\right)_L (1-w)^{-1.8} \tag{7.94}$$

Consequently, from the definitions of the Two-Phase Multipliers, equations 7.74 and 7.77

$$\phi_{LO}^2 = \phi_L^2 \left(\frac{dP}{dx}\right)_L \bigg/ \left(\frac{dP}{dx}\right)_{LO} = \phi_L^2 (1-w)^{1.8} \tag{7.95}$$

The Lockhart–Martinelli correlation provides the relationship between ϕ_L^2 and the Martinelli parameter X_{tt}. Therefore, use of equation 7.95 enables the relationship between ϕ_{LO}^2 and X_{tt} at low pressures to be found.

At the other pressure extreme, namely the critical pressure, the phases are indistinguishable so it follows that

$$\left.\begin{aligned} \phi_{LO}^2 &= 1 \\ X_{tt} &= \left(\frac{1-w}{w}\right)^{0.9} \end{aligned}\right\} \quad \text{at the critical pressure}$$

Thus, at the critical pressure ϕ_{LO}^2 has the value unity at all values of the quality and Martinelli parameter.

Martinelli and Nelson found the relationship between ϕ_{LO}^2 and X_{tt} at intermediate pressures by trial and error using experimental results as a guide. The resulting correlation is shown in Figure 7.14 where ϕ_{LO}^2 is presented as a function of quality. The relationship between void fraction and quality is given in Figure 7.15. The use of these graphs and appropriate thermodynamic data for steam enable equation 7.81 to be integrated numerically.

The special case of evaporation from saturation at $x = 0$ with a constant value of dw/dx, represented by equation 7.83 can be treated much more easily because Martinelli and Nelson have presented correlations for the quantities

$$\overline{\phi_{LO}^2} = \frac{1}{w_e} \int_0^{w_e} \phi_{LO}^2 \, dw \tag{7.96}$$

and

$$r_2 = \frac{w_e^2}{\alpha}\left(\frac{V_G}{V_L}\right) + \frac{(1-w_e)^2}{1-\alpha} - 1 \tag{7.97}$$

The correlations for these quantities are shown in Figures 7.16 and 7.17.

These quantities enable the frictional and accelerative terms in equation 7.83 to be calculated. The static head term can be evaluated by numerical integration using the α–w relationship given in Figure 7.15.

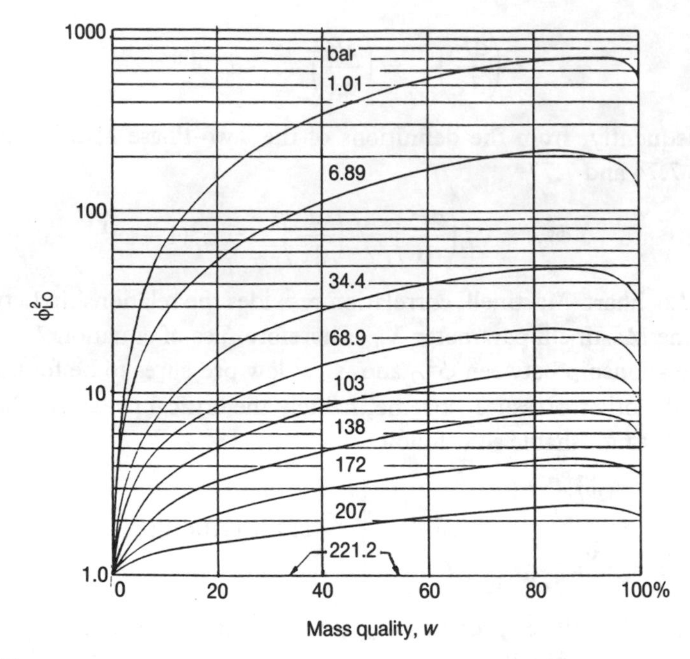

Figure 7.14
Two-phase multiplier as a function of mass quality
Source: R. C. Martinelli and D. B. Nelson, *Transactions of ASME*, **70**, No. 6, pp. 695–702 (1948)

The definition of r_2 used by Martinelli and Nelson is different from that used here: theirs is the present quantity multiplied by V_L. The advantage of the present definition is that it makes r_2 dimensionless. The ordinate of Figure 7.17 has been scaled accordingly.

7.7.5 *Comparison with measurements*

Over the range of Martinelli parameter of practical importance $(0.1 < X_{tt} < 0.7)$ the Martinelli–Nelson correlation predicts a frictional pressure gradient approximately twice that predicted by the homogeneous model.

When these predictions are compared with measurements it is found that there is a significant dependence on the total mass flux G: the Martinelli–Nelson correlation is accurate in the mass flux range 500–1000 kg/(m^{2}s) while the homogeneous model gives good agreement when the mass flux is greater than 2000 kg/(m^{2}s).

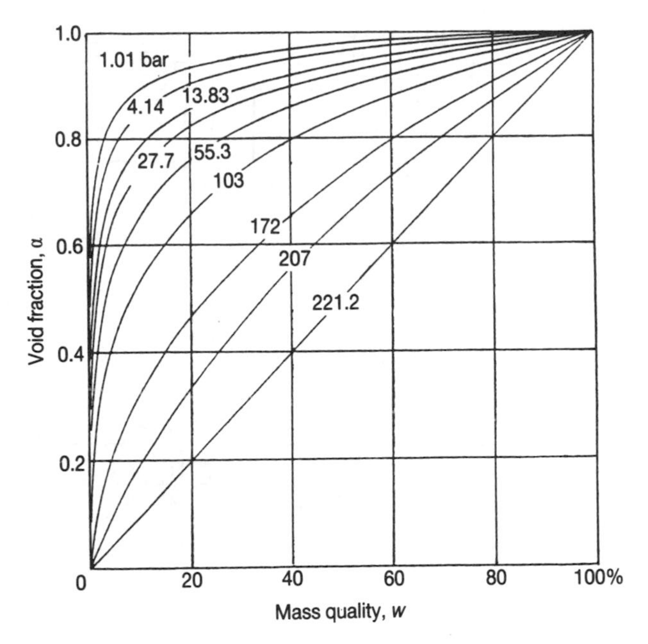

Figure 7.15

Void fraction as a function of mass quality

Source: R. C. Martinelli and D. B. Nelson, *Transactions of ASME*, **70**, No. 6, pp. 695–702 (1948)

The original Martinelli–Nelson correlation was based on relatively limited data. Thom (1964) has derived revised values of the quantities shown in Figures 7.14 to 7.17 from extensive measurements.

One of the disadvantages of the Martinelli–Nelson and Thom correlations is that they are for the water–steam system only. Although this system is of great importance, there is a need for information on other materials. A correlation with the dual aims of correcting the mass flux dependency missing from the Martinelli–Nelson correlation and allowing predictions to be made for other systems has been given by Baroczy (1965). Whilst Baroczy's correlation may be resorted to in the absence of anything better, the wild excursions of the graphs suggest that this is fundamentally not the correct approach.

Both Baroczy (1965) and Chisholm (1968) have modified the Martinelli–Nelson correlation to take into account the influence of the mass flux. Chisholm's modification is recommended by Collier (1972). Subsequently,

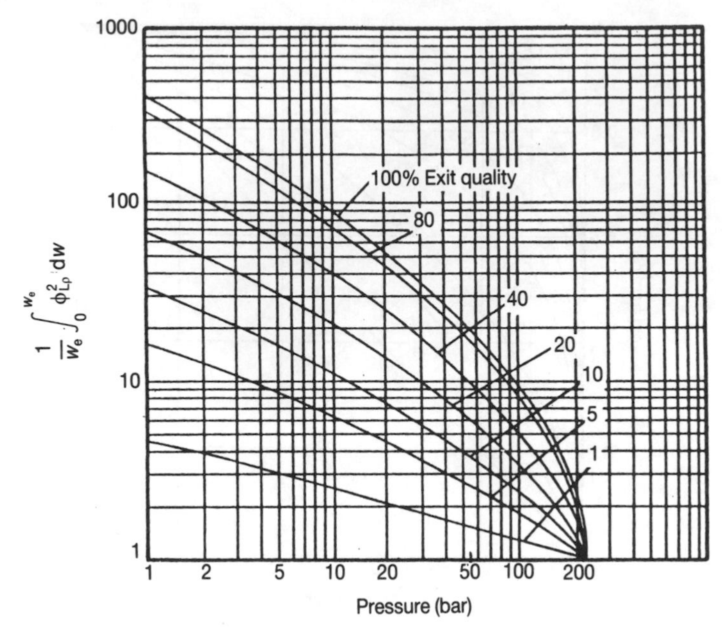

Figure 7.16

Mean value of the two–phase multiplier as a function of absolute pressure

Source: R. C. Martinelli and D. B. Nelson, *Transactions of ASME*, **70**, No. 6, pp. 695–702 (1948)

Chisholm (1973) presented a convenient form of correlation incorporating his own and Baroczy's modifications.

Example 7.2

Steam is generated in a high pressure boiler containing tubes 2.5 m long and 12.5 mm internal diameter. The wall roughness is 0.005 mm. Water enters the tubes at a pressure of 55.05 bar and a temperature of 270 °C, and the water flow rate through each tube is 500 kg/h. Each tube is heated uniformly at a rate of 50 kW.

(a) Estimate the pressure drop across each tube (neglecting end effects) using (i) the homogeneous flow model and (ii) the Martinelli–Nelson correlation.

(b) How should the calculation be modified if the inlet temperature were 230 °C at the same pressure?

Calculations

(a) To avoid excessive calculation, the whole tube length will be treated as

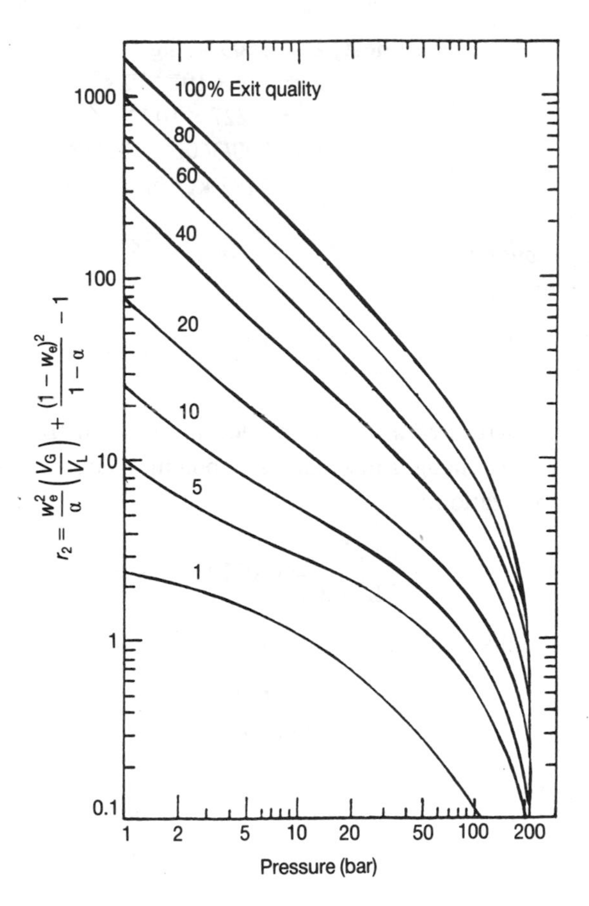

Figure 7.17
Factor r_2 as a function of absolute pressure
Source: R. C. Martinelli and D. B. Nelson, *Transactions of ASME*, **70**, No. 6, pp. 695–702 (1948)

a single increment. For the conditions specified, the water entering the tubes is at its boiling temperature. From the data and using steam tables, the following values are found:

$$V_L = 1.302 \times 10^{-3}\ \text{m}^3/\text{kg},\ V_G = 3.563 \times 10^{-2}\ \text{m}^3/\text{kg}$$
(at inlet conditions)

Therefore

$$V_{LG} = 3.433 \times 10^{-2}\ \text{m}^3/\text{kg and } V_{LG}/V_L = 26.37$$

Also

$$\begin{aligned}\text{latent heat, } \lambda &= 1605 \text{ kJ/kg}\\ \mu_L &= 9.9\times 10^{-5} \text{ Pa s}\\ S &= 1.227\times 10^{-4} \text{ m}^2\\ M &= 500/3600 = 0.1389 \text{ kg/s}\\ G &= 1132 \text{ kg/(m}^2\text{s)}\end{aligned}$$

(i) Homogeneous model

From steam tables

$$-\frac{dV_G}{dP} \approx \frac{(0.03944 - 0.03244) \text{ m}^3\text{/kg}}{10^6 \text{ Pa}} = 7.0\times 10^{-9} \text{ m}^3\text{/(kg Pa)}$$

As the water is saturated at the tube inlet, all the heat transfer results in boiling (neglecting changes in kinetic and potential energy), consequently the rate of vaporization is

$$\frac{50 \text{ kW}}{1605 \text{ kJ/kg}} = 0.03115 \text{ kg/s}$$

and the exit quality w_e is therefore given by

$$w_e = \frac{0.03115 \text{ kg/s}}{0.1389 \text{ kg/s}} = 0.2243$$

Thus, in equation 7.65

$$\left| G^2 w_e \frac{dV_G}{dP} \right| \approx 2.0\times 10^{-3} \ll 1$$

In view of this inequality and the fact that the inlet water is saturated, equation 7.66 may be used:

$$\Delta P = \frac{2fG^2V_LL}{d_i}\left[1+\frac{w_e}{2}\left(\frac{V_{LG}}{V_L}\right)\right] + G^2V_L\left(\frac{V_{LG}}{V_L}\right)w_e + \frac{Lg\sin\theta}{V_{LG}w_e}\ln\left[1+w_e\left(\frac{V_{LG}}{V_L}\right)\right]$$

Frictional term

As in Example 7.1, f_{LO} may be used as an approximation to f for the two-phase flow.

$$Re_{LO} = \frac{Gd_i}{\mu_L} = \frac{[1132 \text{ kg/(m}^2\text{s)}](12.5\times 10^{-3} \text{ m})}{9.9\times 10^{-5} \text{ Pa s}} = 1.429\times 10^5$$

Relative roughness = 0.005 mm/12.5 mm = 0.0004.

Hence, from the friction factor chart, Figure 2.1, $f_{LO} = 0.0044 \approx f$. The frictional pressure drop is given by

$$\Delta P_f = \frac{2fG^2 V_L L}{d_i}\left[1 + \frac{w_e}{2}\left(\frac{V_{LG}}{V_L}\right)\right]$$
$$= \frac{(2)(0.0044)(1132^2)(1.302\times 10^{-3})(2.5)}{12.5\times 10^{-3}}\left[1 + \frac{0.2243}{2}(26.37)\right] \text{Pa}$$
$$= 11.62 \text{ kPa}$$

Accelerative term

$$\Delta P_a = G^2 V_L\left(\frac{V_{LG}}{V_L}\right)w_e = G^2 V_{LG} w_e$$
$$= [1132 \text{ kg/(m}^2\text{s)}]^2(3.433\times 10^{-2} \text{ m}^3\text{/kg})\times 0.2243 \text{ Pa}$$
$$= 9.867 \text{ kPa}$$

Static head term

$$\Delta P_{sh} = \frac{Lg}{V_{LG} w_e}\ln\left[1 + w_e\left(\frac{V_{LG}}{V_L}\right)\right]$$
$$= \frac{(2.5)(9.81)}{(3.433\times 10^{-2})(0.2243)}\ln[1 + (0.2243)(26.37)] \text{ Pa}$$
$$= 6.159 \text{ kPa}$$

Total pressure drop
Summing the three terms

$$\underline{\underline{\Delta P = 27.6 \text{ kPa}}}$$

(ii) Martinelli–Nelson correlation
It is necessary first to check that the denominator of equation 7.81 differs little from unity. From the values in part (i)

$$-\frac{dV_G}{dP} \approx 7\times 10^{-9} \text{ m}^3\text{/(kg Pa)}$$

From Figure 7.15, at 55 bar and $w \approx 0.2$

$$\left(\frac{\partial \alpha}{\partial P}\right)_w \approx 0.003 \text{ bar}^{-1} = 3\times 10^{-8} \text{ Pa}^{-1}$$

Using these values shows that

$$G^2\left\{\frac{w^2}{\alpha}\frac{dV_G}{dP}+\left(\frac{\partial\alpha}{\partial P}\right)_w\left[\frac{(1-w)^2V_L}{(1-\alpha)^2}-\frac{w^2V_G}{\alpha^2}\right]\right\}\ll 1$$

It is therefore possible to use equation 7.83:

$$\Delta P=\frac{2f_{LO}G^2V_LL}{d_i}\left[\frac{1}{w_e}\int_0^{w_e}\phi_{LO}^2\,dw\right]+G^2V_L\left[\frac{w_e^2}{\alpha}\left(\frac{V_G}{V_L}\right)+\frac{(1-w_e)^2}{1-\alpha}-1\right]$$
$$+\frac{Lg\sin\theta}{w_e}\int_0^{w_e}\left(\frac{\alpha}{V_G}+\frac{1-\alpha}{V_L}\right)dw$$

Frictional term

From the values in part (i)

$$\frac{2f_{LO}G^2V_LL}{d_i}=\frac{(2)(0.0044)(1132^2)(1.302\times10^{-3})(2.5)}{12.5\times10^{-3}}\ \text{Pa}$$
$$=2.936\times10^3\ \text{Pa}$$

From Figure 7.16, for $P = 55.05$ bar and $w_e = 0.2243$

$$\frac{1}{w_e}\int_0^{w_e}\phi_{LO}^2\,dw\approx 6.2$$

Therefore

$$\Delta P_f=(2.936\times10^3\ \text{Pa})(6.2)\ =\ 1.821\times10^4\ \text{Pa}\ =\ 18.21\ \text{kPa}$$

Accelerative term

$$\Delta P_a\ =\ G^2V_Lr_2$$

From Figure 7.17

$$r_2\ \approx\ 3.2$$

Therefore

$$\Delta P_a\ =\ [1132\ \text{kg/(m}^2\text{s)}]^2(1.302\times10^{-3}\ \text{m}^3\text{/kg})(3.2)\ =\ 5339\ \text{Pa}\ =\ 5.34\ \text{kPa}$$

Static head term

$$\Delta P_{sh}=\frac{Lg}{w_e}\int_0^{w_e}\left(\frac{\alpha}{V_G}+\frac{1-\alpha}{V_L}\right)dw$$

Using Figure 7.15, a curve of α against w at 55 bar can be estimated. Reading off values of α at various values of w (Table 7.4) allows the above integrand to be evaluated as a function of w:

Table 7.4

w	α	$\frac{\alpha}{V_G}+\frac{1-\alpha}{V_L}$
–	–	(kg/m^3)
0	0	768
0.02	0.40	472
0.04	0.53	376
0.06	0.56	354
0.08	0.61	317
0.10	0.63	302
0.15	0.72	235
0.20	0.76	206
0.2243	0.77	198

Using these values to evaluate the integral graphically,

$$\int_0^{0.2243}\left(\frac{\alpha}{V_G}+\frac{1-\alpha}{V_L}\right)\mathrm{d}w \approx 76$$

and

$$\Delta P_{sh} = \frac{(2.5)(9.81)}{(0.2243)}(76) = 8310 \text{ Pa} = 8.31 \text{ kPa}$$

Total pressure drop
Summing the three terms

$$\underline{\Delta P = 31.9 \text{ kPa}}$$

(b) If the inlet water were at a temperature of 230 °C and the same pressure of 55.05 bar, it would be unsaturated. It is necessary first to calculate the heat transfer required to bring the water to the boiling point and the length of tube required to do this. The pressure drop for the single-phase flow in this part of the tube can be calculated easily. The two-phase flow calculations for the remainder of the tube can then be done as in part (a) but it must be noted that the exit quality will be lower in this case.

Further reading

The presentation of the material in Sections 7.3 and 7.4 has been greatly influenced by the work of Wallis (1969), while the remainder of this chapter closely follows the treatment of Collier (1972). These two books represent excellent starting points for anyone seeking further reading. It should be noted that it is customary in the two-phase flow literature to use the symbol x to denote mass quality and z to denote the axial coordinate. In this chapter, x has been used to denote the coordinate and w the mass quality in order to be consistent with the rest of the book.

References

Alves, G.E., Co-current liquid–gas flow in a pipeline contactor, *Chemical and Process Engineering*, **50**, No. 9, pp. 449–56 (1954).

Baker, O., Design of pipe lines for simultaneous flow of oil and gas, *Oil and Gas Journal*, **53**, pp. 185–95 (26 July 1954).

Baroczy, C.J., A systematic correlation for two-phase pressure drop. AIChE. reprint no. 37, paper presented at 8th National Heat Transfer Conference, Los Angeles, August 1965, *Chemical Engineering Progress Symposium Series*, **62**, No. 64, pp. 232–49 (August 1965).

Chisholm, D., The influence of mass velocity on frictional pressure gradients during steam–water flow. Paper 35 presented at 1968 Thermodynamics and Fluid Mechanics Convention, Institution of Mechanical Engineers, Bristol, March (1968).

Chisholm, D., Pressure gradient due to friction during the flow of evaporating two-phase mixtures in smooth tubes and channels, *International Journal of Heat and Mass Transfer*, **16**, pp. 347–58 (1973).

Cicchitti, A., Lombardi, C., Silvestri, M., Soldaini, G., and Zavattarelli, R., Two-phase cooling experiments—pressure drop, heat transfer and burnout measurements, *Energia Nucleare*, **7**, No. 6, pp. 407–25 (1960).

Collier, J.G., *Convective boiling and condensation*, London, McGraw-Hill Book Company (UK) Ltd (1972).

Dukler, A.E., Wicks, M. and Cleveland, R., Pressure drop and hold-up in two-phase flow, *AIChE Journal*, **10**, No. 1, pp. 38–51 (1964).

Grace, J.R., Wairegi, T. and Nguyen, T.H., Shapes and velocities of single drops and bubbles moving freely through immiscible liquids, *Transactions of the Institution of Chemical Engineers*, **54**, pp. 167–73 (1976).

Harmathy, T.Z., Velocity of large drops and bubbles in media of infinite and restricted extent, *AIChE Journal*, **6**, pp. 281–88 (1960).

Hewitt, G.F. and Roberts, D.N., Studies of two-phase flow patterns by simultaneous X-ray and flash photography. Report AERE-M 2159. London, HMSO (1969).

Kumar, R. and Kuloor, N.R., The formation of bubbles and drops. In *Advances in Chemical Engineering*, Volume 8, 255–368. Eds. Drew, T. B., Cokelet, G. R., Hoopes, J. W., Jr. and Vermeulen, T. (1970).

Lockhart, R.W. and Martinelli, R.C., Proposed correlation of data for isothermal two-phase, two-component flow in pipes, *Chemical Engineering Progress*, **45**, pp. 39–46 (1949).

McAdams, W.H., Woods, W.K. and Heroman, L.C. Jr., Vaporization inside horizontal tubes—II—Benzene–oil mixtures, *Transactions of ASME*, **64**, pp. 193–200 (1942).

Martinelli, R.C. and Nelson, D.B., Prediction of pressure drop during forced circulation boiling of water, *Transactions of ASME*, **70**, No. 6, pp. 695–702 (1948).

Moissis, R. and Griffith, P., Entrance effects in two-phase slug flow, *Transactions of ASME, Journal of Heat Transfer*, Series C **84**, pp. 29–38 (1962).

Peebles, F.N. and Garber, H.J., Studies on motion of gas bubbles in liquids, *Chemical Engineering Progress*, **49**, No. 2, pp. 88–97 (1953).

Richardson, J.F. and Zaki, W.N., Sedimentation and fluidisation: Part I, *Transactions of the Institution of Chemical Engineers*, **32**, pp. 35–52 (1954).

Thom, J.R.S., Prediction of pressure drop during forced circulation boiling of water, *International Journal of Heat and Mass Transfer*, **7**, pp. 709–24 (1964).

Wallis, G.B., *One-dimensional two-phase flow*, New York, McGraw-Hill Book Company Inc. (1969).

Wallis, G.B., The terminal speed of liquid drops and bubbles in an infinite medium, *International Journal of Multiphase Flow*, **1**, pp. 491–511 (1974).

White, E.T. and Beardmore, R.H., The velocity of rise of single cylindrical air bubbles through liquids contained in vertical tubes, *Chemical Engineering Science*, **17**, pp. 351–61 (1962).

8 Flow measurement

8.1 Flowmeters and flow measurement

The flow of fluids is most commonly measured using head flowmeters. The operation of these flowmeters is based on the Bernoulli equation. A constriction in the flow path is used to increase the flow velocity. This is accompanied by a decrease in pressure head and since the resultant pressure drop is a function of the flow rate of fluid, the latter can be evaluated. The flowmeters for closed conduits can be used for both gases and liquids. The flowmeters for open conduits can only be used for liquids. Head flowmeters include orifice and venturi meters, flow nozzles, Pitot tubes and weirs. They consist of a primary element which causes the pressure or head loss and a secondary element which measures it. The primary element does not contain any moving parts. The most common secondary elements for closed conduit flowmeters are U-tube manometers and differential pressure transducers.

A U-tube manometer is shown schematically in Figure 8.1(a). One arm is connected to the high pressure tap and the other arm to the low pressure tap in the flowing fluid. The fluid in one arm of the manometer is separated from the fluid in the other arm by an immiscible liquid of higher density which is usually mercury. Consider the pressures at levels a and b in the two arms of the manometer shown in Figure 8.1(a) when the system is in equilibrium. Let the pressures at level b be P_1 and P_2 in arm 1 and arm 2, respectively. Let the difference in the heights of immiscible liquid in the two arms of the manometer be Δz_m.

The pressure at level a in the manometer is $(P_1 + \rho \Delta z_m g)$ in arm 1 and $(P_2 + \rho_m \Delta z_m g)$ in arm 2 where ρ and ρ_m are the densities of the flowing fluid and immiscible liquid respectively. These two pressures are equal since the two arms of the manometer are connected by a continuous column of stationary liquid. Therefore

$$P_1 + \rho \Delta z_m g = P_2 + \rho_m \Delta z_m g \tag{8.1}$$

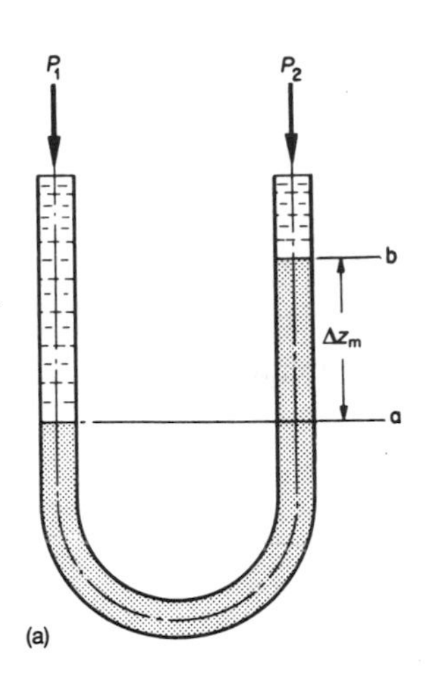

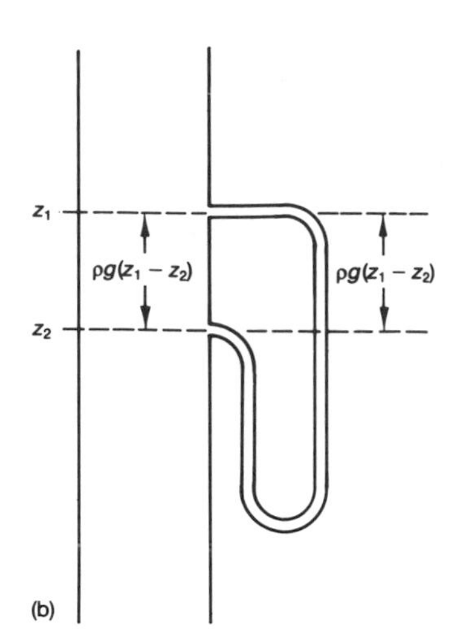

Figure 8.1
U-tube manometer
(a) *Manometer reading Δz_m when $P_1 > P_2$* (b) *A vertical difference in the location of the manometer taps does not affect the reading*

which can be written as

$$P_1 - P_2 = (\rho_m - \rho)\Delta z_m g \tag{8.2}$$

If ρ and ρ_m are in kg/m^3, Δz_m is in m, and g is 9.81 m/s^2, the pressure differential across the primary element $P_1 - P_2$ is in N/m^2 or Pa. The head differential across the primary element Δh based on the flowing fluid is

$$\Delta h = \frac{P_1 - P_2}{\rho g} \tag{8.3}$$

Combining equations 8.2 and 8.3 gives

$$\Delta h = \frac{(\rho_m - \rho)\Delta z_m}{\rho} \tag{8.4}$$

which relates the head differential across the primary element to the difference in height of immiscible liquid in the two arms of the manometer.

Other flowmeters are in common use which operate on principles differing from head flowmeters. Mechanical flowmeters have primary elements which contain moving parts. These flowmeters include rotameters, positive displacement meters and velocity meters. Electromagnetic flowmeters have the advantages of no restriction in a conduit and no moving parts.

8.2 Head flowmeters in closed conduits

The primary element of an orifice meter is simply a flat plate containing a drilled hole located in a pipe perpendicular to the direction of fluid flow as shown in Figure 8.2.

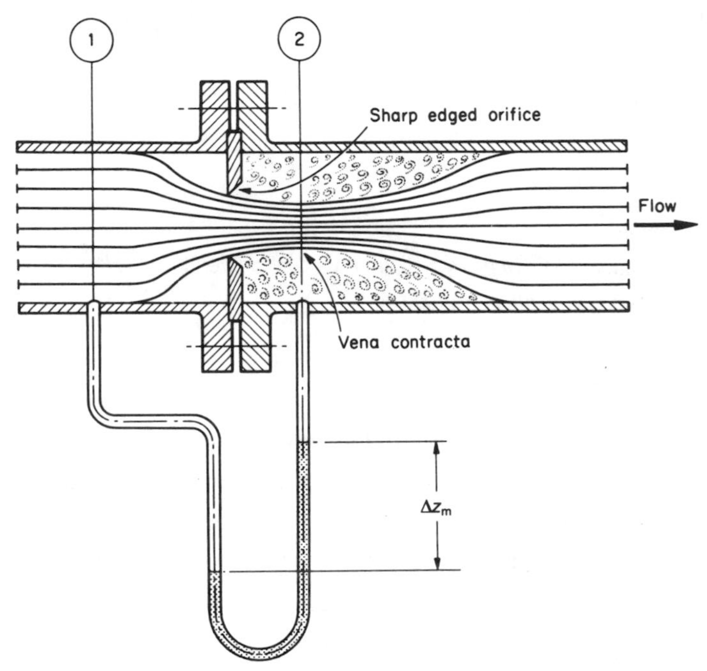

Figure 8.2
Orifice meter

Equation 1.15 is the modified Bernoulli equation for steady flow in a pipe with no pump in the section.

$$\left(z_2 + \frac{P_2}{\rho_2 g} + \frac{u_2^2}{2g\alpha_2}\right) = \left(z_1 + \frac{P_1}{\rho_1 g} + \frac{u_1^2}{2g\alpha_1}\right) - h_f \qquad (1.15)$$

For the steady horizontal flow of an incompressible fluid of density ρ between points 1 and 2 in a pipe with no pump and no friction, equation 1.15 can be written as

$$\frac{u_2^2}{2\alpha_2}\left(1-\frac{\alpha_2 u_1^2}{\alpha_1 u_2^2}\right)=\frac{(P_1-P_2)}{\rho} \tag{8.5}$$

Consider points 1 and 2 in Figure 8.2. At point 1 in the pipe, the fluid flow is undisturbed by the orifice plate. The fluid at this point has a mean velocity u_1 and a cross-sectional flow area S_1. At point 2 in the pipe the fluid attains its maximum mean velocity u_2 and its smallest cross-sectional flow area S_2. This point is known as the *vena contracta*. It occurs at about one half to two pipe diameters downstream from the orifice plate. The location is a function of the flow rate and the size of the orifice relative to the size of the pipe. Let the mean velocity in the orifice be u_o and let the diameter and cross-sectional flow area of the orifice be d_o and S_o respectively.

For this case the principle of continuity can be expressed by any of the following three equations.

$$M=\rho S_1 u_1=\rho S_2 u_2=\rho S_o u_o \tag{8.6}$$

$$Q=S_1 u_1=S_2 u_2=S_o u_o \tag{8.7a}$$

or

$$Q=\frac{\pi}{4}d_1^2 u_1=\frac{\pi}{4}d_2^2 u_2=\frac{\pi}{4}d_o^2 u_o \tag{8.7b}$$

where M is the flow rate of fluid and Q is the volumetric flow rate.

Using equation 8.7 to substitute for u_1 and u_2 in equation 8.5 gives

$$\frac{u_o^2}{2\alpha_2}\left(\frac{d_o}{d_2}\right)^4\left[1-\frac{\alpha_2}{\alpha_1}\left(\frac{d_2}{d_1}\right)^4\right]=\frac{(P_1-P_2)}{\rho} \tag{8.8}$$

which can be rearranged in the form

$$u_o=\left(\frac{d_2}{d_o}\right)^2\sqrt{\frac{2(P_1-P_2)\alpha_2}{\rho[1-(\alpha_2/\alpha_1)(d_2/d_1)^4]}} \tag{8.9}$$

giving the mean velocity through the orifice.

Using equation 8.7, the volumetric flow rate is given by

$$Q=S_o\left(\frac{d_2}{d_o}\right)^2\sqrt{\frac{2(P_1-P_2)\alpha_2}{\rho[1-(\alpha_2/\alpha_1)(d_2/d_1)^4]}} \tag{8.10}$$

Equation 8.10 gives the volumetric flow rate Q when there is no friction in the system.

In practice, the measured volumetric flow rate is always less than Q

given by equation 8.10. Viscous frictional effects retard the flowing fluid. In addition, boundary layer separation occurs on the downstream side of the orifice plate resulting in a substantial pressure or head loss from form friction. This effect is a function of the geometry of the system.

In practice, the volumetric flow rate Q is given by equation 8.11

$$Q = S_o C_d \sqrt{\frac{2(P_1 - P_2)}{\rho[1 - (d_o/d_1)^4]}} \tag{8.11}$$

In equation 8.11, which is analogous to equation 8.10, C_d is the dimensionless discharge coefficient which accounts for geometry and friction; d_o/d_1 is the ratio of the diameter of the orifice to the inside diameter of the pipe. This ratio does not vary as does the ratio d_2/d_1 in equation 8.10 for frictionless flow.

Using equation 8.3 to substitute for the pressure difference in equation 8.11 gives

$$Q = S_o C_d \sqrt{\frac{2g\Delta h}{[1 - (d_o/d_1)^4]}} \tag{8.12}$$

Equation 8.12 gives the volumetric flow rate Q in terms of the head differential across the orifice plate Δh. The latter is based on the flowing fluid.

Both equations 8.11 and 8.12 refer to horizontal pipes. When the pipe is not horizontal, the total pressure difference $(P_1 - P_2)$ must be corrected for the pressure difference due to the static head between the two pressure taps. Thus, equation 8.11 should be replaced by

$$Q = S_o C_d \sqrt{\frac{2[(P_1 - P_2) + \rho g(z_1 - z_2)]}{\rho[1 - (d_o/d_1)^4]}} \tag{8.13}$$

and equation 8.12 by

$$Q = S_o C_d \sqrt{\frac{2g[\Delta h + (z_1 - z_2)]}{[1 - (d_o/d_1)^4]}} \tag{8.14}$$

It must be remembered that, in equation 8.14, Δh is still defined by equation 8.3.

Provided that location 1 is always the upstream pressure tap and location 2 the downstream tap, these equations are applicable for both upward and downward flow, but note that the sign of $(z_1 - z_2)$ will change. The value of ΔP, and consequently Δh, will be negative for downward flow if the pressure drop due to flow is smaller than the static pressure difference. Equations 8.13 and 8.14 reduce to equations 8.11 and 8.12, respectively, when $z_1 = z_2$.

It is essential to appreciate that the pressure difference measured by a manometer automatically eliminates the static head difference. This is shown in Figure 8.1(b). The static head $\rho g(z_1 - z_2)$ in the pipe is exactly balanced by the extra static head above the right hand limb of the manometer. Consequently, if Δh is calculated from Δz_m using equation 8.4, *no further correction for the static head should be made.*

The holes in orifice plates may be concentric, eccentric or segmental as shown in Figure 8.3. Orifice plates are prone to damage by erosion.

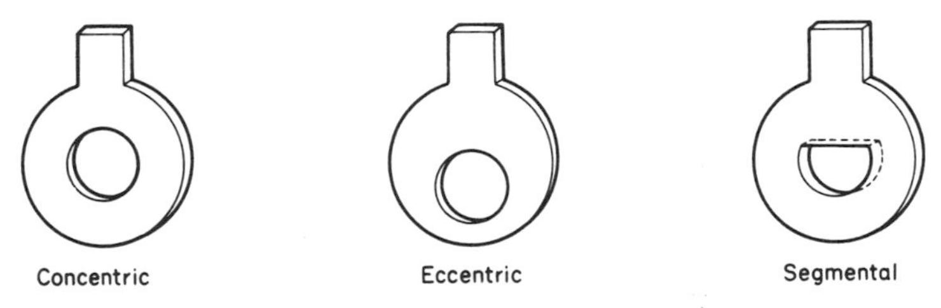

Figure 8.3
Concentric, eccentric and segmental orifice plates

The coefficient of discharge C_d for a particular orifice meter is a function of the location of the pressure taps, the ratio of the diameter of the orifice to the inside diameter of the pipe d_o/d_1, the Reynolds number in the pipeline Re, and the thickness of the orifice plate.

Most orifices used for flow measurement are sharp-edged as shown in Figure 8.2; this produces well-defined separation of the flow at the orifice and consequently consistent values of the discharge coefficient. Figure 8.4 shows how the discharge coefficient for a circular, sharp-edged orifice depends on the Reynolds number and the ratio of the orifice diameter to the internal diameter of the pipe. The Reynolds number is based on the orifice diameter and the fluid speed through the orifice. The discharge coefficient varies greatly with Reynolds number for relatively large orifices making them unsuitable for flow measurement. At Reynolds numbers above about 2×10^4 the discharge coefficient has a constant value of about 0.62: it is preferable to use orifice meters in this constant C_d region.

Orifice meters suffer from high frictional pressure or head losses. Thus, most of the pressure drop is not recoverable. The pressure loss is given by the equation [Barna (1969)]

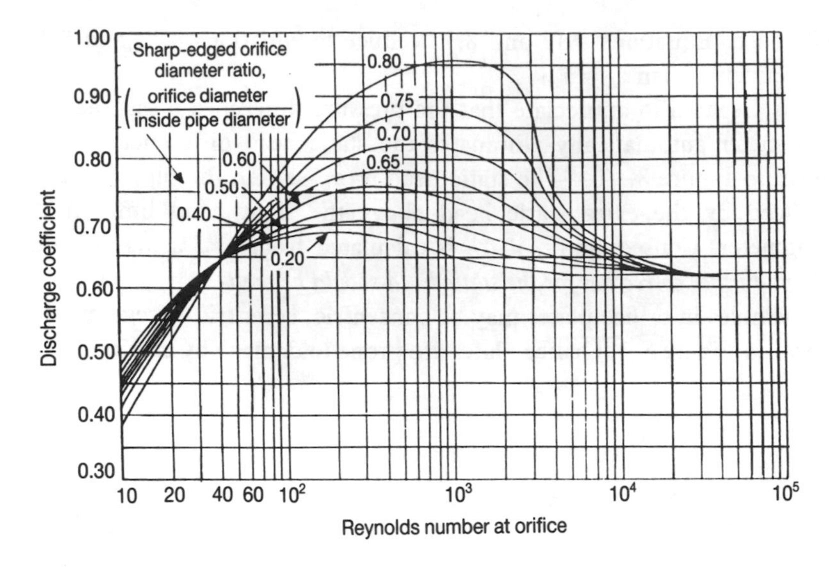

Figure 8.4
Orifice discharge coefficients
Source: J. H. Perry, *Chemical Engineers' Handbook* (Sixth edition, New York: McGraw-Hill, 1984) p. 5–15

$$\Delta P = \left[1 - \left(\frac{d_o}{d_1}\right)^2\right](P_1 - P_2) \tag{8.15}$$

Orifice plates are inexpensive and easy to install since they can readily be inserted at a flanged joint.

Figure 8.5 shows a Venturi meter. The theory is the same as for the orifice meter but a much higher proportion of the pressure drop is recoverable than is the case with orifice meters. The gradual approach to and the gradual exit from the orifice substantially eliminates boundary layer separation. Thus, form drag and eddy formation are reduced to a minimum.

A series of tap connections in an annular pressure ring gives a mean value for the pressure at point 1 in the approach section and also at point 2 in the throat. Although Venturi meters are relatively expensive and tend to be bulky, they can meter up to 60 per cent more flow than orifice plates for the same inside pipe diameter and differential pressure [Foust *et al.* (1964)]. The coefficient of discharge C_d for a Venturi meter is in the region of 0.98. Venturies are more suitable than orifice plates for metering liquids containing solids.

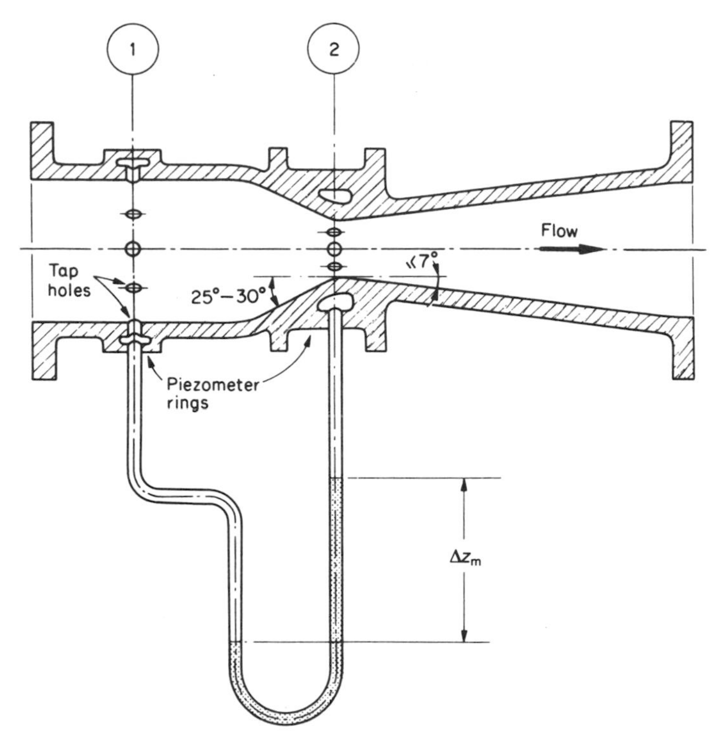

Figure 8.5
Venturi meter

Figure 8.6 shows a flow nozzle. This is a modified and less expensive type of Venturi meter.

The theoretical treatment of head flowmeters in this section is for incompressible fluids. The flow of compressible fluids through a constriction in a pipe is treated in Chapter 6.

Orifice meters, Venturi meters and flow nozzles measure volumetric flow rate Q or mean velocity u. In contrast the Pitot tube shown in a horizontal pipe in Figure 8.7 measures a point velocity v. Thus Pitot tubes can be used to obtain velocity profiles in either open or closed conduits. At point 2 in Figure 8.7 a small amount of fluid is brought to a standstill. Thus the combined head at point 2 is the pressure head $P/(\rho g)$ plus the velocity head $v^2/(2g)$ if the potential head z at the centre of the horizontal pipe is arbitrarily taken to be zero. Since at point 3 fluid is not brought to a standstill, the head at point 3 is the pressure head only if points 2 and 3 are sufficiently close for them to be considered to have the same potential head z.

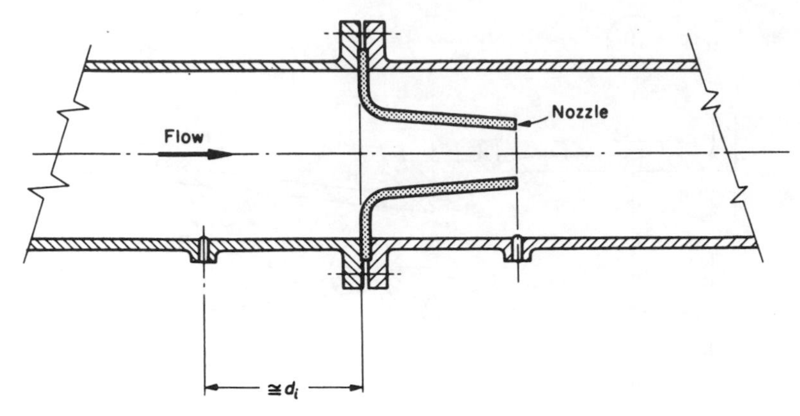

Figure 8.6
Flow nozzle

Thus the difference in head Δh between points 2 and 3 neglecting friction is the velocity head $v^2/(2g)$. Therefore the point velocity v is given by the equation

$$v = \sqrt{2g\Delta h} \tag{8.16}$$

The difference in heads between points 2 and 3, Δh, is usually measured with a manometer.

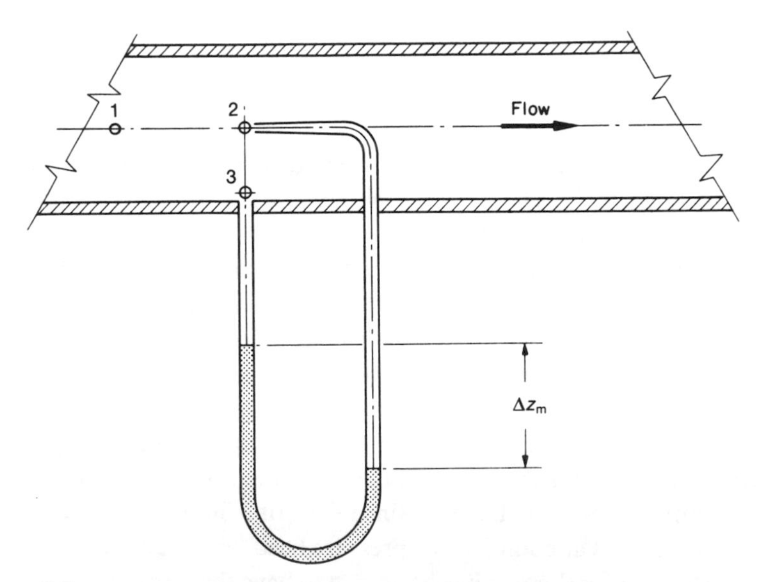

Figure 8.7
Pitot tube

Combining equations 8.4 and 8.16 gives

$$v = \sqrt{\frac{2g(\rho_m - \rho)\Delta z_m}{\rho}} \tag{8.17}$$

Equation 8.17 gives the point velocity v in terms of the difference in level between the two arms of the manometer Δz_m, the density of the flowing fluid ρ, the density of the immiscible manometer liquid ρ_m and the gravitational acceleration g.

Most Pitot tubes consist of two concentric tubes parallel to the direction of fluid flow. The inner tube points into the flow and the outer tube is perforated with small holes which are perpendicular to the direction of flow. The inner tube transmits the combined pressure and velocity heads and the outer tube only the pressure head.

Although Pitot tubes are inexpensive and have negligible permanent head losses they are not widely used. They are highly sensitive to fouling, their required alignment is critical and they cannot measure volumetric flow rate Q or mean velocity u. The latter can be calculated from a single measurement only if the velocity distribution is known: this can be found if the Pitot tube can be traversed across the flow.

Example 8.1

Calculate the volumetric flow rate of water through a pipe with an inside diameter of 0.15 m fitted with an orifice plate containing a concentric hole of diameter 0.10 m given the following data.

Difference in level on a mercury manometer connected across the orifice plate

$$\Delta z_m = 0.254 \text{ m}$$

Mercury specific gravity

$$= 13.6$$

Discharge coefficient

$$C_d = 0.60$$

Calculations

The head differential across the orifice is given by

$$\Delta h = \frac{(\rho_m - \rho)\Delta z_m}{\rho} \tag{8.4}$$

$$= \left(\frac{\rho_m}{\rho} - 1\right)\Delta z_m$$

$$= (13.6 - 1)(0.254 \text{ m})$$

$$= 3.20 \text{ m}$$

The volumetric flow rate is given by

$$Q = S_o C_d \sqrt{\frac{2g\Delta h}{[1 - (d_o/d_1)^4]}} \tag{8.12}$$

Given that

$$S_o = \frac{\pi d_o^2}{4} = \frac{(3.142)(0.10 \text{ m})^2}{4} = 0.007\,855 \text{ m}^2$$

$$g = 9.81 \text{ m/s}^2$$

$$\Delta h = 3.20 \text{ m}$$

$$\frac{d_o}{d_1} = \frac{0.10}{0.15} = \frac{1}{1.5}$$

$$1 - \left(\frac{d_o}{d_1}\right)^4 = 0.8025$$

$$\sqrt{\frac{2g\Delta h}{[1 - (d_o/d_1)^4]}} = \sqrt{\frac{(2)(9.81 \text{ m/s}^2)(3.20 \text{ m})}{0.8025}} = 8.845 \text{ m/s}$$

$$C_d = 0.60$$

it follows that

$$Q = (0.007\,855 \text{ m}^2)(0.60)(8.845 \text{ m/s})$$

$$= \underline{0.0417 \text{ m}^3\text{/s}}$$

8.3 Head flowmeters in open conduits

Weirs are commonly used to measure the flow rate of liquids in open conduits. The theory is based on the Bernoulli equation for frictionless flow. From equation 1.13 with $\Delta h = 0$ and $h_f = 0$

$$\left(z_2 g + \frac{P_2}{\rho_2} + \frac{v_2^2}{2}\right) - \left(z_1 g + \frac{P_1}{\rho_1} + \frac{v_1^2}{2}\right) = 0 \tag{8.18}$$

Consider a liquid flowing over a sharp crested weir as shown in Figure 8.8. Let the upstream level of the liquid be z_o above the level of the weir crest. As the liquid approaches the weir, the liquid level gradually drops and the flow velocity increases. Downstream from the weir, a jet is formed. This is called the nappe and it is ventilated underneath to enable it to spring free from the weir crest [Barna (1969)].

Consider any point in the liquid at a height z vertically above the weir

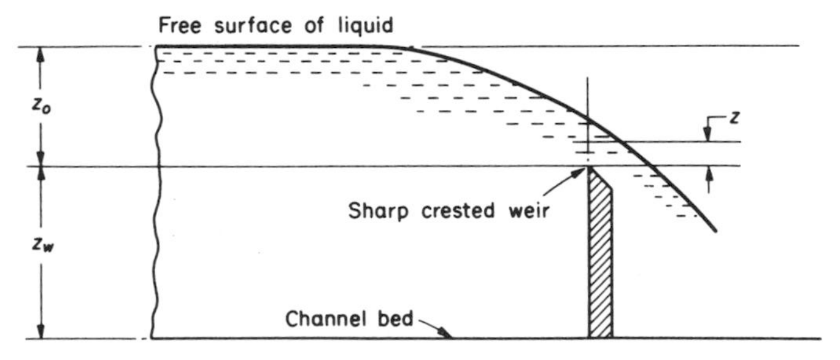

Figure 8.8
Flow of liquid over a sharp–crested weir

crest in Figure 8.8. Let the liquid at this point have a velocity v. For this case the Bernoulli equation can be written in the form

$$\left(z_o g + \frac{v_o^2}{2}\right) - \left(z g + \frac{v^2}{2}\right) = 0 \tag{8.19}$$

Equation 8.19 is based on the following assumptions: the approach velocity v_o is uniform and parallel, the streamlines are horizontal above the weir, there is atmospheric pressure under the nappe, and the flow is frictionless.

In order to evaluate the flow rate as a function of z_o it is necessary to neglect the fall of the liquid surface above the weir. It should be noted that the fall shown in Figure 8.8 is greatly exaggerated; in practice the fall is very small.

Equation 8.19 can be rewritten in the form

$$v = \sqrt{(2g)}\sqrt{\left(z_o + \frac{v_o^2}{2g} - z\right)} \tag{8.20}$$

Equation 8.20 gives the point velocity v at a height z above the weir crest. If the width of the conduit is b at this point, the volumetric flow rate through an element of cross-sectional flow area of height dz is

$$dQ = bv\,dz \tag{8.21}$$

Substituting equation 8.20 into equation 8.21 gives

$$dQ = b\sqrt{(2g)}\sqrt{\left(z_o + \frac{v_o^2}{2g} - z\right)}\,dz \tag{8.22}$$

Let

$$h = z_o + \frac{v_o^2}{2g} - z \tag{8.23}$$

so that

$$dh = -dz \tag{8.24}$$

On substituting equations 8.23 and 8.24 into equation 8.22, the volumetric flow rate through the element is given by

$$dQ = -b\sqrt{(2g)}h^{1/2}\,dh \tag{8.25}$$

For a rectangular conduit, b is a constant and equation 8.25 can be integrated to

$$Q = -\tfrac{2}{3}b\sqrt{(2g)}h^{3/2} + c \tag{8.26}$$

Substituting for h in terms of z from equation 8.23, and noting that $Q = 0$ for $z = 0$, allows the constant to be evaluated. The flow rate for $z = z_o$ is then given by

$$Q = \frac{2}{3}b\sqrt{(2g)}\left[\left(z_o + \frac{v_o^2}{2g}\right)^{3/2} - \left(\frac{v_o^2}{2g}\right)^{3/2}\right] \tag{8.27}$$

If the approach velocity v_o is neglected, equation 8.27 becomes

$$Q = \tfrac{2}{3}b\sqrt{(2g)}z_o^{3/2} \tag{8.28}$$

Equations 8.27 and 8.28 give the volumetric flow rate Q through a rectangular weir when there is no friction in the system.

In practice the volumetric flow rate Q is given by equation 8.29

$$Q = \tfrac{2}{3}C_d b\sqrt{(2g)}z_o^{3/2} \tag{8.29}$$

C_d is the dimensionless discharge coefficient which is a function of z_o and the ratio z_w/z_o where z_w is the height of the weir crest above the channel bed. A typical value is $C_d = 0.65$ for $z_w/z_o = 2$ and $z_o = 0.3$ m. Viscosity has a negligible influence on C_d. Thus the flow rate can be determined from a measurement of z_o.

In addition to rectangular weirs, V-notch or triangular weirs are commonly used with a cross-sectional flow area as shown in Figure 8.9. In

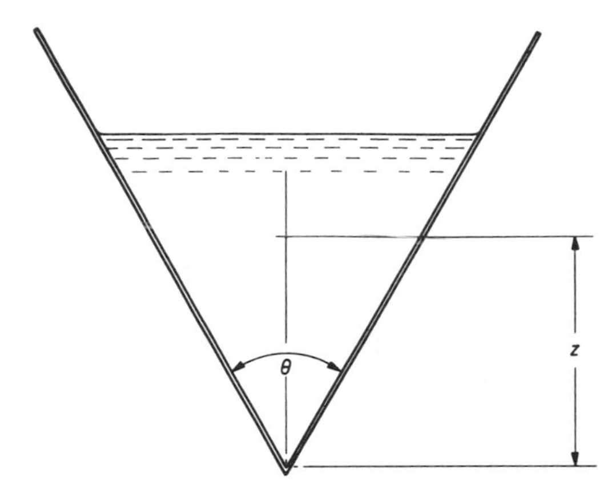

Figure 8.9
Cross-sectional flow area in a V-notch weir

this case the width of the conduit b is variable and at any height z above the bottom of the weir is

$$b = 2z \tan \frac{\theta}{2} \tag{8.30}$$

If the approach velocity v_o is neglected, equation 8.22 can be written

$$dQ = b\sqrt{(2g)}\sqrt{(z_o - z)}\, dz \tag{8.31}$$

where dQ is the volumetric flow rate through an element of cross-sectional flow area of height dz when friction is neglected. Substituting for b from equation 8.30 allows equation 8.31 to be written as

$$dQ = 2 \tan \frac{\theta}{2} \sqrt{(2g)}\sqrt{(z_o - z)} z \, dz \tag{8.32}$$

Let

$$h = z_o - z \tag{8.33}$$

then

$$dQ = -2 \tan \frac{\theta}{2} \sqrt{(2g)}\, h^{1/2}(z_o - h)\, dh \tag{8.34}$$

Integrating equation 8.34 gives

$$Q = -2 \tan \frac{\theta}{2} \sqrt{(2g)} \left(\frac{2}{3} z_o h^{3/2} - \frac{2}{5} h^{5/2} \right) + C \tag{8.35}$$

Noting that $Q = 0$ for $z = 0$ (ie $h = z_o$), then

$$C = \frac{8}{15} \tan \frac{\theta}{2} \sqrt{(2g)} z_o^{5/2} \tag{8.36}$$

The full flow rate occurs for $z = z_o$, ie $h = 0$, when the first term on the right hand side of equation 8.35 vanishes. Thus the flow rate is given by

$$Q = \frac{8}{15} \tan \frac{\theta}{2} \sqrt{(2g)} z_o^{5/2} \tag{8.37}$$

Equation 8.37 gives the volumetric flow rate Q through a V-notch weir when there is no friction in the system.

In practice the volumetric flow rate Q is given by Equation 8.38

$$Q = \frac{8}{15} C_d \tan \frac{\theta}{2} \sqrt{(2g)} z_o^{5/2} \tag{8.38}$$

where C_d is the dimensionless discharge coefficient which is mainly a function of z_o and θ. A typical value is $C_d = 0.62$ for $z_o = 0.15$ m and $\theta = 20°$.

V-notch weirs are particularly useful for measuring flow rates that vary considerably.

8.4 Mechanical and electromagnetic flowmeters

In the head flowmeters discussed so far, the primary element has no moving parts. The constriction area is fixed and the pressure drop varies as the flow rate changes. In the rotameter shown in Figure 8.10, the pressure drop is held constant and the constriction area varies as the flow rate changes. A float is free to move up and down in a tapered tube. The float remains steady when the upward force of the flowing fluid exactly balances the weight of the float in the fluid. The tapered tube is marked with a scale which is calibrated for a given fluid to give flow speed at each scale reading. As the fluid flow rate is increased, the float moves to a higher position in the tube.

Other commonly used mechanical flowmeters are velocity meters.

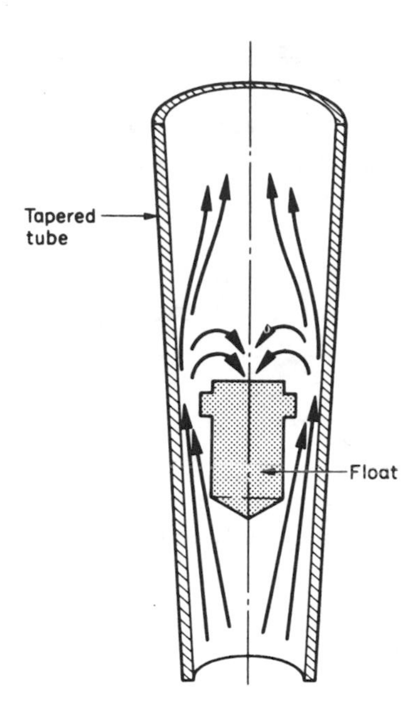

Figure 8.10
Rotameter

These usually consist of a non-magnetic casing, a rotor, and an electromagnetic pickup. The rotor is either a propeller or turbine freely suspended on ball bearings in the path of the flowing fluid with the axis of rotation in line with the flow. The rotor turns in the fluid flow stream at a rate proportional to the flow rate. As the rotor turns it cuts through the lines of force of an electric field produced by an adjacent induction coil. The electrical pulse output from the induction coil pickup is amplified and fed to readout instruments or recorders to give either total flow or flow rate [Holland and Chapman (1966)].

The head flowmeters and mechanical flowmeters considered so far all involve some kind of restriction in the flow line, which in turn produces additional frictional head losses. In contrast, the electromagnetic flowmeter consists of a straight length of non-magnetic pipe containing no restrictions through which the fluid flows. The pipe is normally lined with electrically insulating material. Two small electrodes located diametrically opposite each other are sealed in flush with the interior surface. Coil windings on the outside of the non-magnetic pipe provide a magnetic field. The flowing fluid acts as a moving conductor which cuts the magnetic lines of force. A voltage is induced in the fluid which is directly proportional to the fluid velocity. This voltage is detected by the electrodes and is then amplified and transmitted to either readout instruments or recorders. The magnetic field is usually alternating so that the induced voltage is also alternating. Electromagnetic flowmeters can only be used to meter fluids which have some electrical conductivity. They cannot be used to meter hydrocarbons.

Although electromagnetic flowmeters are expensive they are especially suitable for metering liquids containing suspended solids. Furthermore, unlike head flowmeters, they are unaffected by variations in fluid viscosity, density or temperature. Since they are also unaffected by turbulence or variations in velocity profile, they can be installed close to valves, bends, fittings, etc.

The flowmeters discussed above are used either to measure velocity or volumetric flow rate. They can only be used to measure the mass flow rate if the fluid density is also measured and the volumetric flow rate and density signals are coordinated.

Direct reading mass flowmeters have now been developed. These operate on the angular momentum principle. The primary element consists of a cylindrical impeller and turbine both of which contain tubes through which the fluid flows. Angular momentum is given to the fluid by driving the impeller at a constant speed via a magnetic coupling using a synchronous motor. Fluid from the impeller discharges into the cylindrical turbine where all the angular momentum is removed. The torque on the turbine is proportional to the mass flow rate. This torque is transferred to a gyro-integrating mechanism via a magnetic coupling. A cyclometer registers the rotation of the gyroscope and totalizes the fluid mass flow rate. Direct reading mass flowmeters are expensive.

8.5 Scale errors in flow measurement

Rotameters, velocity flowmeters, and electromagnetic flowmeters have the advantage that they can be used with a linear scale in which the volumetric flow rate Q is directly proportional to the scale reading s.

$$Q = k_1 s \tag{8.39}$$

where k_1 is a constant. Differentiating equation 8.39 gives

$$\frac{dQ}{ds} = k_1 \tag{8.40}$$

All head flowmeters are used with a square root scale in which the volumetric flow rate Q is proportional to the square root of the scale reading s.

$$Q = k_2 s^{1/2} \tag{8.41}$$

where k_2 is a constant. Differentiating equation 8.41 gives

$$\frac{dQ}{ds} = \frac{k_2}{2s^{1/2}} = \frac{k_2^2}{2Q} \tag{8.42}$$

The per cent flow rate error Δe can be defined as

$$\Delta e = 100\frac{\Delta Q}{Q} \tag{8.43}$$

where ΔQ is the absolute error in the volumetric flow rate.

Let Δs be the indicator or recorder error. Then, for small errors

$$\frac{\Delta Q}{\Delta s} \cong \frac{dQ}{ds} \tag{8.44}$$

Combining equations 8.43 and 8.44 gives

$$\Delta e = \left(\frac{100\,\Delta s}{Q}\right)\left(\frac{dQ}{ds}\right) \tag{8.45}$$

Substituting equation 8.40 for a linear scale into equation 8.45 gives

$$\Delta e = \frac{100\,\Delta s\, k_1}{Q} \tag{8.46}$$

If the maximum volumetric flow rate $Q_{max} = 100$ and the maximum scale reading $s_{max} = 100$, equation 8.46 can be written as

$$\Delta e = \frac{100\,\Delta s}{Q} \tag{8.47}$$

Substituting equation 8.42 for a square root scale into equation 8.45 gives

$$\Delta e = \frac{50\,\Delta s\, k_2^2}{Q^2} \tag{8.48}$$

If the maximum volumetric flow rate $Q_{max} = 100$ and the maximum scale reading $s_{max} = 100$, equation 8.48 can be written as

$$\Delta e = \frac{5000\ \Delta s}{Q^2} \tag{8.49}$$

The square root scale is more accurate than the linear scale at flow rates near the maximum of the scale. The linear scale is more accurate than the square root scale at flow rates much less than the maximum of the scale. Thus head flowmeters are unsuitable for measuring flow rates which vary widely.

Example 8.2

A flowmeter is inherently accurate at all points to 0.5 per cent of the full range. Calculate the per cent flow rate error using first a linear and then a square root scale for flow rates of 10, 25, 50 and 100 per cent of maximum flow.

Calculations

The indicator error is given by

$$\Delta s = 0.5 \qquad s_{max} = 100$$

For a linear scale, per cent flow rate error

$$\Delta e = \frac{100\ \Delta s}{Q} \qquad Q_{max} = 100 \text{ and } s_{max} = 100 \tag{8.47}$$

Therefore

$$\Delta e = \frac{(100)(0.5)}{10} = 5 \text{ per cent for a flow rate 10 per cent of } Q_{max}$$

$$\Delta e = \frac{(100)(0.5)}{25} = 2 \text{ per cent for a flow rate 25 per cent of } Q_{max}$$

$$\Delta e = \frac{(100)(0.5)}{50} = 1 \text{ per cent for a flow rate 50 per cent of } Q_{max}$$

$$\Delta e = \frac{(100)(0.5)}{100} = 0.5 \text{ per cent for } Q_{max}$$

For a square root scale, per cent flow rate error

$$\Delta e = \frac{5000\ \Delta s}{Q^2} \qquad Q_{max} = 100 \text{ and } s_{max} = 100 \tag{8.49}$$

Therefore

$$\Delta e = \frac{(5000)(0.5)}{(10)^2} = 25 \text{ per cent for a flow rate 10 per cent of } Q_{max}$$

$$\Delta e = \frac{(5000)(0.5)}{(25)^2} = 4 \text{ per cent for a flow rate 25 per cent of } Q_{max}$$

$$\Delta e = \frac{(5000)(0.5)}{(50)^2} = 1 \text{ per cent for a flow rate 50 per cent of } Q_{max}$$

$$\Delta e = \frac{(5000)(0.5)}{(100)^2} = 0.25 \text{ per cent for } Q_{max}$$

References

Barna, P.S., *Fluid Mechanics for Engineers*, London, Butterworths (1969).

Buzzard, W., Instrument scale error study throws new light on flowmeter accuracy, *Chemical Engineering*, **66**, pp. 147–50 (9 Mar 1959).

Foust, A.S., Wenzel, L.A., Clump, C.W., Maus, L. and Anderson, L.B., *Principles of Unit Operations*, New York, John Wiley and Sons, Inc (1964).

Holland, F.A. and Chapman, F.S., *Pumping of Liquids*, New York, Reinhold Publishing Corporation (1966).

Perry, J.H., *Chemical Engineers' Handbook*, Sixth edition, New York, McGraw-Hill Book Company Inc, p. 5–15 (1984).

Ziemke, M.C. and McCallie, B.G., Design of orifice-type flow reducers, *Chemical Engineering*, **71**, pp. 195–8 (14 Sept 1964).

9 Fluid motion in the presence of solid particles

9.1 Relative motion between a fluid and a single particle

Consider the relative motion between a particle and an infinitely large volume of fluid. Since only the relative motion is considered the following cases are covered:

1 a stationary particle in a moving fluid;
2 a moving particle in a stationary fluid;
3 a particle and a fluid moving in opposite directions;
4 a particle and a fluid both moving in the same direction but at different velocities.

In contrast to single-phase flow in a pipe of constant cross section, flow around a sphere or other bluff object exhibits several different flow regimes at different values of the Reynolds number.

For flow around a spherical particle of diameter d_p, the appropriate definition of the Reynolds number is

$$Re_p = \frac{\rho u_p d_p}{\mu} \tag{9.1}$$

where u_p is the speed of the particle relative to the fluid.

Except at very low values of the particle Reynolds number, a wake forms behind the sphere as shown in Figure 9.1. The upper half of this composite diagram shows the streamlines for flow at an intermediate value of Re_p, while the lower half shows the streamlines for a higher value of Re_p. In general, separation of the flow from the surface of the sphere occurs over the rear part creating a large low pressure wake shown by the recirculating flow. The presence of this low pressure wake is responsible for most of the drag when flow separation occurs.

At all values of the particle Reynolds number Re_p, the fluid is brought to rest relative to the particle at A, which is therefore a stagnation point where the pressure is higher than in the flowing fluid (see equation 1.19 in

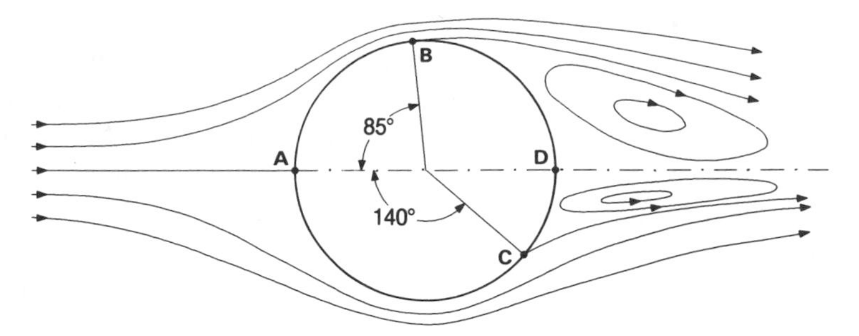

Figure 9.1
Streamlines for flow around a sphere
Upper half: intermediate Re_p giving a laminar boundary layer
Lower half: high Re_p giving a turbulent boundary layer

Section 1.5). In flowing round the sphere, the fluid has to accelerate and therefore, by Bernoulli's equation, the pressure falls towards the mid-point of the sphere's surface.

At very low values of Re_p, when the flow is dominated by viscous stresses, the fluid creeps round the rear of the sphere and no separation occurs: the flow is symmetrical fore and aft. In this case the drag force on the particle is due solely to the shear stress generated by the fluid's velocity gradient normal to the surface. This force is often known as skin friction. When no separation occurs, the highest velocity of the fluid occurs over the mid-point of the sphere's surface and consequently the pressure here is a minimum. As the fluid decelerates over the rear half of the sphere, pressure recovery occurs. Thus, the flow in this region is in the direction of increasing pressure: this is known as an adverse pressure gradient and it tends to make the flow unstable at higher values of the Reynolds number.

As the Reynolds number is increased, fluid inertia becomes more significant. In addition, the pressure at point D increases and, instead of flowing round the rear surface of the sphere, the fluid is pushed away so that flow separation occurs as shown in the upper half of Figure 9.1. Separation occurs at point B. The whole of the wake is a region of relatively low pressure, very close to that at the point of separation, and much lower than the pressure near point A. The force arising from this pressure difference is known as form drag because it is due to the (bluff) shape of the particle. The total drag force is a combination of skin friction and form drag. In this flow regime a laminar boundary layer is formed over the surface of the sphere from A to B.

On increasing the Reynolds number further, a point is reached when the boundary layer becomes turbulent and the point of separation moves further back on the surface of the sphere. This is the case illustrated in the lower half of Figure 9.1 with separation occurring at point C. Although there is still a low pressure wake, it covers a smaller fraction of the sphere's surface and the drag force is lower than it would be if the boundary layer were laminar at the same value of Re_p.

Roughening the surface of a sphere causes the transition to a turbulent boundary layer to occur at a lower value of the Reynolds number. This explains the apparent anomaly that, at certain values of the Reynolds number, the drag will be lower for a sphere with a rough surface than for a similar sphere with a smooth surface. It is for the same reason that golf balls are made with a dimpled surface.

Consider a spherical particle of diameter d_p and density ρ_p falling with a velocity u_p under the influence of gravity in a fluid of density ρ. The net gravitational force F_1 on the particle is given by the equation

$$F_1 = \frac{\pi d_p^3}{6}(\rho_p - \rho)g \tag{9.2}$$

where $\pi d_p^3/6$ is the volume of the spherical particle.

The retarding force F_2 on the particle from the fluid is given by the equation

$$F_2 = C_d S_p \frac{\rho u_p^2}{2} \tag{9.3}$$

where C_d is a dimensionless drag coefficient and S_p is the projected area of the particle in a plane perpendicular to the direction of the fluid stream. Equation 9.3 may be obtained by dimensional analysis. The drag coefficient has a role similar to the friction factor for flow in pipes.

For steady flow the forces F_1 and F_2 are equal and opposite and the particle reaches a constant speed u_t. Equations 9.2 and 9.3 can be combined and written as

$$\frac{\pi d_p^3}{6}(\rho_p - \rho)g = C_d S_p \frac{\rho u_t^2}{2} \tag{9.4}$$

For a spherical particle $S_p = \pi d_p^2/4$ and for this case equation 9.4 can be rewritten in the form

$$u_t = \sqrt{\frac{4d_p(\rho_p - \rho)g}{3C_d\rho}} \tag{9.5}$$

where u_t is known as the terminal settling or falling velocity.

As a result of the changing flow patterns described above, the drag coefficient C_d is a function of the Reynolds number. For the streamline flow range of Reynolds numbers, $Re_p < 0.2$, the drag force F_2 is given by

$$F_2 = 3\pi d_p \mu u_t \tag{9.6}$$

and consequently, from equation 9.3, the drag coefficient is

$$C_d = \frac{24}{Re_p} \tag{9.7}$$

From equations 9.5 and 9.7, the terminal falling velocity u_t for the streamline flow range of Reynolds numbers is given by

$$u_t = \frac{d_p^2(\rho_p - \rho)g}{18\mu} \tag{9.8}$$

Equations 9.6 and 9.8 were derived by Stokes and are known as Stokes's equations for steady creeping flow round a sphere.

For the Reynolds number range $0.2 < Re_p < 500$, it has been shown that

$$C_d = \frac{24}{Re_p}(1 + 0.15 Re_p^{0.687}) \tag{9.9}$$

Equation 9.8 is an empirical equation which is only approximately true. Over the more limited range $2 < Re_p < 500$ a simpler equation is adequate:

$$C_d = \frac{18.5}{Re_p^{0.6}} \tag{9.10}$$

This is particularly useful because equation 9.7 for the Stokes regime can be extended to $Re = 2$ with negligible error.

For the Reynolds number range $500 < Re_p < 200\,000$

$$C_d = 0.44 \tag{9.11}$$

When the Reynolds number Re_p reaches a value of about 300 000, transition from a laminar to a turbulent boundary layer occurs and the point of separation moves towards the rear of the sphere as discussed above. As a result, the drag coefficient suddenly falls to a value of 0.10 and remains constant at this value at higher values of Re_p.

For the most part, solid particles in fluid streams have Reynolds numbers which are much lower than 500.

Pettyjohn and Christiansen (1948) gave equations for the terminal settling velocities of particles which deviate from a spherical shape.

Lapple and Shepherd (1940) presented plots of the dimensional drag coefficient C_d against the particle Reynolds number for spheres, discs and cylinders.

Equation 9.5 gives the terminal settling velocity for a spherical particle. For a non-spherical particle, equation 9.5 can be written in the modified form

$$u_t = \sqrt{\frac{4d_p\psi(\rho_p - \rho)g}{3C_d\rho}} \tag{9.12}$$

For a spherical particle the dimensionless correction factor $\psi = 1$ and equations 9.5 and 9.12 become identical.

9.2 Relative motion between a fluid and a concentration of particles

So far the relative motion between a fluid and a single particle has been considered. This process is called free settling. When a fluid contains a concentration of particles in a vessel, the settling of an individual particle may be hindered by the other particles and by the walls. When this is the case, the process is called hindered settling. Interference is negligible if the particles are at least 10 to 20 diameters away from each other and the vessel wall [Larian (1958)]. In this case the particles can be considered to be free settling.

Hindered settling results from collisions between particles and also between particles and the wall. In addition high particle concentrations reduce the flow area and increase the velocity of the fluid with a consequent decrease in settling rate. Furthermore particle concentrations increase the apparent density and dynamic viscosity of the fluid.

Richardson and Zaki (1954) showed that in the Reynolds number range $Re_p < 0.2$, the velocity u_c of a suspension of coarse spherical particles in water relative to a fixed horizontal plane is given by the equation

$$\frac{u_c}{u_t} = \varepsilon^{4.6} \tag{9.13}$$

where u_t is the terminal settling velocity for a single particle and ε is the voidage fraction of the suspension which is unity for a single particle in an infinite amount of fluid. The velocity of the particles relative to the liquid can be derived from equation 9.13, as explained in Section 7.3.

Settling can be used to classify or separate particles since different sized particles settle at different velocities. Similarly elutriation can also be used to classify particles where small particles are carried upwards with the fluid and large particles sink. Particles of the same size but of different densities can also be separated by settling or elutriation.

Consider two spherical particles 1 and 2 of the same diameter but of different densities settling freely in a fluid of density ρ in the streamline Reynolds number range $Re_p < 0.2$. The ratio of the terminal settling velocities u_{t1}/u_{t2} is given by equation 9.8 rewritten in the form

$$\frac{u_{t1}}{u_{t2}} = \frac{\rho_{p1} - \rho}{\rho_{p2} - \rho} \tag{9.14}$$

The greater the ratio u_{t1}/u_{t2} the greater the ease of separation. Thus the fluid density ρ can be chosen to give a high ratio of terminal settling velocities.

Similarly, for two spherical particles 1 and 2 of the same density ρ_p but of different diameters settling freely in a fluid of density ρ in the streamline Reynolds number range $Re_p < 0.2$, the ratio of the terminal settling velocities u_{t1}/u_{t2} is given by equation 9.8 rewritten in the form

$$\frac{u_{t1}}{u_{t2}} = \left(\frac{d_{p1}}{d_{p2}}\right)^2 \tag{9.15}$$

where d_{p1}/d_{p2} is the ratio of the particle diameters.

Similarly, from equation 9.8, two particles will settle at the same speed in the same fluid in the streamline flow regime if their densities and diameters are related by equation 9.16:

$$\frac{d_{p1}}{d_{p2}} = \sqrt{\left(\frac{\rho_{p2} - \rho}{\rho_{p1} - \rho}\right)} \tag{9.16}$$

The classification or separation of particles can be carried out more rapidly in centrifugal separators than in gravity settlers. In gravity settlers, the particles travel vertically downwards, whereas in centrifugal separators the particles travel radially outwards. A particle of mass m rotating at a radius r with an angular velocity ω is subject to a centripetal force $mr\omega^2$ which can be made very much greater than the vertically directed gravity force mg.

The terminal settling velocity u_t for a single spherical particle in a centrifugal separator can be calculated from equation 9.5 with the centripetal acceleration $r\omega^2$ replacing the gravitational acceleration g to give

$$u_t = \sqrt{\frac{4d_p(\rho_p - \rho)r\omega^2}{3C_d\rho}} \tag{9.17}$$

A very small particle may still be in laminar flow in a centrifugal separator. In this case $r\omega^2$ may be written in place of g in equation 9.8 to give

$$u_t = \frac{d_p^2(\rho_p - \rho)r\omega^2}{18\mu} \quad (9.18)$$

In addition to hydrodynamic interactions between solid particles and a fluid, physico-chemical forces may act between pairs of particles. These forces tend to form a structure which prevents the particles from settling out [Cheng (1970)]. If the forces are sufficiently strong a homogeneous slurry results which usually has non-Newtonian rheological characteristics. If the structure is weak, the slurry is shear thinning. Slurries with a high proportion of solids tend to be shear thickening.

Einstein studied homogeneous slurries of spherical particles in a liquid of the same density. He showed that the distortion of the streamlines around the particles caused the dynamic viscosity of the slurry to increase according to the equation

$$\mu = \mu_L(1 + 2.5\alpha) \quad (9.19)$$

where μ and μ_L are the dynamic viscosities of the slurry and liquid respectively and α is the volume fraction of the solids. Equation 9.19 holds for low concentrations up to $\alpha = 0.02$.

Note that $\alpha = 1 - \varepsilon$.

9.3 Fluid flow through packed beds

In a packed bed of unit volume, the volumes occupied by the voids and the solid particles are ε and $(1 - \varepsilon)$ respectively where ε is the voidage fraction or porosity of the bed. Let S_o be the surface area per unit volume of the solid material in the bed. Thus the total surface area in a packed bed of unit volume is $(1 - \varepsilon)S_o$.

For a spherical particle of diameter d_p the value of $S_o = 6/d_p$. For a non-spherical particle with an average particle diameter d_p, the value $S_o = 6/(d_p\psi)$ where $\psi = 1$ for a spherical particle. Values of ψ for other shapes are readily available [Perry (1984)].

An equivalent diameter d_e for flow through the bed can be defined as four times the cross-sectional flow area divided by the appropriate flow perimeter. For a random packing, this is equal to four times the volume occupied by the fluid divided by the surface area of particles in contact with the fluid.

Thus, the equivalent diameter is

$$d_e = \frac{4\varepsilon}{(1-\varepsilon)S_o} \tag{9.20}$$

The velocity calculated by dividing the volumetric flow rate by the whole cross-sectional area of the bed is known as the superficial velocity u. The mean velocity within the interstices of the bed is then $u_b = u/\varepsilon$.

A Reynolds number for flow through a packed bed can be defined as

$$Re_b = \frac{\rho u_b d_e}{\mu} \tag{9.21}$$

which when combined with equation 9.20 can be written as

$$Re_b = \frac{4\rho u}{\mu(1-\varepsilon)S_o} \tag{9.22}$$

An alternative Reynolds number has been used to correlate data and is defined as

$$Re'_b = \frac{\rho u}{\mu(1-\varepsilon)S_o} \tag{9.23}$$

For a packed bed consisting of spherical particles, equation 9.23 can be written in the form

$$Re'_b = \frac{\rho u d_p}{6\mu(1-\varepsilon)} \tag{9.24}$$

The corresponding equation for non-spherical particles is

$$Re'_b = \frac{\rho u d_p \psi}{6\mu(1-\varepsilon)} \tag{9.25}$$

Consider fluid flowing steadily through a packed bed of height L and unit cross-sectional area. A pressure drop ΔP_f occurs in the bed because of frictional viscous and drag forces. Let the resistance per unit area of surface be τ_b. A force balance across unit cross-sectional area gives

$$\Delta P_f \varepsilon = \tau_b L(1-\varepsilon)S_o \tag{9.26}$$

which can be written either as

$$\frac{f_b}{2} = \frac{\tau_b}{\rho u_b^2} = \left(\frac{\Delta P_f}{L}\right)\left[\frac{\varepsilon}{(1-\varepsilon)S_o \rho u_b^2}\right] \tag{9.27}$$

or since $u_b = u/\varepsilon$ as

$$\frac{f_b}{2} = \frac{\tau_b}{\rho u_b^2} = \left(\frac{\Delta P_f}{L}\right)\left[\frac{\varepsilon^3}{(1-\varepsilon)S_o \rho u^2}\right] \tag{9.28}$$

where f_b is a dimensionless friction factor for flow through a packed bed. Various other definitions of friction factors for flow in packed beds have been used [Longwell (1966)].

For laminar flow where $Re'_b \leqslant 2$

$$\frac{f_b}{2} = \frac{5}{Re'_b} \tag{9.29}$$

The transition to turbulent flow is gradual. Turbulence commences initially in the largest channels and eventually extends to the smaller channels.

For the complete range of Reynolds number Carman (1937) gave the equation

$$\frac{f_b}{2} = \frac{5}{Re'_b} + \frac{0.4}{(Re'_b)^{0.1}} \tag{9.30}$$

Log log plots of f_b against Re'_b are readily available [Perry (1984)] for randomly packed beds.

The Hagen–Poiseuille equation for steady laminar flow of Newtonian fluids in pipes and tubes can be written as

$$u = \left(\frac{\Delta P_f}{L}\right)\frac{d_i^2}{32\mu} \tag{1.65}$$

For a packed bed, substituting the equivalent diameter d_e from equation 9.20 into equation 1.65 gives

$$u_b = \left(\frac{\Delta P_f}{L}\right)\left(\frac{1}{32\mu}\right)\left[\frac{16\varepsilon^2}{(1-\varepsilon)^2 S_o^2}\right] \tag{9.31}$$

or, since $u_b = u/\varepsilon$

$$u = \left(\frac{\Delta P_f}{L}\right)\left(\frac{1}{2\mu}\right)\left[\frac{\varepsilon^3}{(1-\varepsilon)^2 S_o^2}\right] \tag{9.32}$$

Equation 9.32 does not hold for flow through packed beds and should be replaced by the equation

$$u = \left(\frac{\Delta P_f}{L}\right)\left(\frac{1}{K_c \mu}\right)\left[\frac{\varepsilon^3}{(1-\varepsilon)^2 S_o^2}\right] \tag{9.33}$$

which can also be written in the form

$$\Delta P_f = (K_c \mu L)\left[\frac{(1-\varepsilon)^2 S_o^2}{\varepsilon^3}\right] u \tag{9.34}$$

Equation 9.33 is the Carman–Kozeny equation [Carman (1937)]. The parameter K_c has a value which depends on the particle shape, the porosity and particle size range. The value lies in the range 3.5 to 5.5 but the value most commonly used is 5.

For spherical particles $S_o = 6/d_p$ and equation 9.34 can be written as

$$\Delta P_f = (180\,\mu L)\left[\frac{(1-\varepsilon)^2}{\varepsilon^3 d_p^2}\right] u \tag{9.35}$$

where $K_c = 5.0$.

Example 9.1

A gas of density $\rho = 1.25$ kg/m^3 and dynamic viscosity $\mu = 1.5 \times 10^{-5}$ Pa s flows steadily through a bed of spherical particles of diameter $d_p = 0.005$ m. The bed has a height of 3.00 m and a voidage of $\frac{1}{3}$. The superficial velocity $u = 0.03$ m/s. Calculate the Reynolds number and the frictional pressure drop over the bed.

Calculations

$$\text{Reynolds number } Re'_b = \frac{\rho u d_p}{6\mu(1-\varepsilon)} \tag{9.24}$$

Substituting the given values

$$Re'_b = \frac{(1.25\text{ kg/m}^3)(0.03\text{ m/s})(0.005\text{ m})(3)}{(6)(1.50 \times 10^{-5}\text{ Pa s})(2)}$$

$$= 3.125$$

The frictional pressure drop is given by

$$\Delta P_f = (180\mu L)\left[\frac{(1-\varepsilon)^2}{\varepsilon^3 d_p^2}\right] u \tag{9.35}$$

Given that

$$(1-\varepsilon)^2 = \tfrac{4}{9}$$

$$\frac{(1-\varepsilon)^2}{\varepsilon^3} = 12$$

$$d_p^2 = 2.5 \times 10^{-5}\text{ m}^2$$

$$u = 0.03\text{ m/s}$$

$$\mu = 1.5 \times 10^{-5}\text{ Pa s}$$

$$L = 3.0\text{ m}$$

it follows that

$$\Delta P_f = (180)(1.5 \times 10^{-5}\ \text{Pa s})\frac{(3.0\ \text{m})(12)(0.03\ \text{m/s})}{(2.5 \times 10^{-5}\ \text{m}^2)}$$

$$= \underline{116.6\ \text{Pa}}$$

9.4 Fluidization

If a fluid is passed upwards in laminar flow through a packed bed of solid particles the superficial velocity u is related to the pressure drop ΔP by equation 9.33:

$$u = \left(\frac{\Delta P_f}{L}\right)\left(\frac{1}{K_c \mu}\right)\left[\frac{\varepsilon^3}{(1-\varepsilon)^2 S_o^2}\right] \tag{9.33}$$

As the fluid velocity is increased the drag on the particles increases and a point is reached where the pressure drop balances the effective weight of bed per unit cross-sectional area. At this point the fluid drag just supports the solid particles. A small increase in the flow rate causes a slight expansion of the bed from its static, packed state. Further increase in the flow rate allows the bed to expand more and the particles become free to move around and the bed is said to be fluidized. The state when the bed just becomes fluidized is known as incipient, or minimum, fluidization. The fluid velocity required to cause incipient fluidization is called the minimum fluidization velocity and is denoted by u_{mf}.

On increasing the velocity above u_{mf}, two types of fluidization may be observed. With most liquid–solid systems over the whole range of velocities and with gas–solid systems just above u_{mf}, the bed simply expands and remains fairly homogeneous: this is known as particulate fluidization. However, at higher velocities in gas–solid systems gas voids containing few particles form in the bed. This type of fluidization, in which the excess gas passes through the bed as 'bubbles' is known as aggregative fluidization or boiling fluidization. An air-fluidized bed of clear glass ballotini looks remarkably like vigorously boiling water. The bubbles carry particles in their wakes and help to provide the excellent mixing that occurs in fluidized beds. Aggregative fluidization can occur in liquid fluidization of very dense solids.

At higher gas velocities a turbulent bed is formed, in which the gas voids have irregular shapes and channelling may occur. This fluidization regime may be compared with the churn-flow regime in gas–liquid two-phase flow (see Section 7.1).

In tall, narrow beds the gas voids may 'coalesce' producing a slugging bed. This condition is generally undesirable owing to its unsteady nature and the difficulty of scale-up.

At very high gas velocities the particles are carried out of the top of the bed. This is known as fast fluidization and is a type of pneumatic conveying. Fast fluidization has been used in catalytic crackers in order to circulate the catalyst particles; the gas velocity is also high enough to break down any agglomerates of solids thus improving performance.

In a similar way, a high liquid velocity will cause hydraulic conveying in a liquid–solid fluidized bed.

9.4.1 *Determination of the minimum fluidization velocity*

Neglecting the static head component of the pressure drop, a force balance at the point of incipient fluidization can be written as

$$(\Delta P)_{mf} = (1-\varepsilon_{mf})(\rho_p-\rho)L_{mf}g \tag{9.34}$$

Combining equations 9.33 and 9.34, the minimum fluidization velocity u_{mf} is given by

$$u_{mf} = \left[\frac{(\rho_p-\rho)g}{K_c\mu}\right]\left[\frac{\varepsilon_{mf}^3}{(1-\varepsilon_{mf})S_o^2}\right] \tag{9.35}$$

In equations 9.34 and 9.35 ε_{mf} is the value of the void fraction at minimum fluidization. It should be noted that ε_{mf} is not equal to the void fraction in the packed bed and in order to use equation 9.35 to calculate u_{mf} it is necessary to know the value of ε_{mf}.

The best method of determining the minimum fluidization velocity u_{mf} is experimentally, by measuring the pressure drop across the bed over a range of fluid velocities. The pressure drop increases linearly until fluidization occurs and then increases very slowly; indeed up to about twice the minimum fluidization velocity the pressure drop may appear to be constant within experimental error. When a bed is initially fluidized, there is a tendency for the pressure drop across the bed to be rather high and to go through a peak as incipient fluidization occurs. It is possible that this is caused by a need to 'unstick' the particles. If the fluid velocity of an already fluidized bed is reduced, the peak in the pressure drop is not observed and a much clearer transition to the linear pressure drop—flow rate for the packed bed can be seen. The fluid velocity at the transition is taken as u_{mf}. On fluidizing the bed again, the latter type of behaviour with no peak may be observed.

If it is necessary to predict the minimum fluidization velocity the following correlation [Grace (1982)] may be used for gas–solid systems.

$$\frac{\rho u_{mf} d_p}{\mu} = (C^2 + 0.0408 Ar)^{1/2} - C \quad (9.36)$$

where

$$Ar = \rho g d_p^3 (\rho_p - \rho)/\mu^2 \quad (9.37)$$

is the Archimedes (or Galileo) number. Some doubt exists regarding the value of the constant C, with some workers using the value 27.2 and others 33.4 or 33.7. It would seem reasonable to use a mean value of 30.

If it is possible to measure the height of the bed at incipient fluidization, L_{mf}, then ε_{mf} can be calculated from equation 9.34 or simply from the ratio of L_{mf} to the height of the packed bed if the void fraction in the latter is known.

In the absence of pressure drop and void fraction measurements, u_{mf} is calculated from equation 9.36 and ε_{mf} can be estimated from equation 9.35.

Single particles will tend to be carried out of the bed if the fluid velocity exceeds the terminal falling speed u_t of the particles given by equation 9.5. Thus the normal range of fluidization velocity is from u_{mf} to u_t. However, it may be found that the fluid velocity required to bring about fast fluidization is significantly higher than u_t because particles tend to form clusters.

9.5 Slurry transport

A slurry is a liquid containing solid particles in suspension. Slurries can be divided into two classes: settling and non-settling.

Non-settling slurries usually consist of a high concentration of finely divided solid particles suspended in a liquid. The solid particles may also settle so slowly that the slurry may be regarded for all practical purposes as non-settling. Like true liquids, non-settling slurries may exhibit either Newtonian or non-Newtonian flow behaviour. Milk is a non-settling slurry which behaves as a thixotropic liquid. Non-settling homogeneous slurries can be pumped through a pipeline either in laminar or turbulent flow.

Compared with non-settling slurries, settling slurries contain larger solid particles at lower concentrations. Settling slurries are essentially two-phase heterogeneous mixtures. The liquid and the solid particles exhibit their own characteristics. Thus in contrast to non-settling slurries, the solid particles in settling slurries do not alter the viscosity of the conveying liquid.

Settling slurries cannot be pumped in laminar flow. Turbulence must exist to prevent the solid particles from settling. Settling slurries should be pumped through pipelines at velocities which just prevent the solid particles from settling. This results in the minimum pressure drop across the pipeline.

Saltation is also used to transport settling slurries through pipelines [Condolios and Chapus (1963a)]. In this case the solid particles bounce and roll along the bottom of a horizontal pipe.

Below a certain minimum velocity, the turbulence is insufficient to keep all the particles suspended in a settling slurry flowing through a horizontal pipe. At this minimum velocity, there is a concentration gradient from the top to the bottom of the horizontal pipe. At a higher velocity called the standard velocity this gradient disappears and the flow becomes homogeneous. Spells (1955) called the region between the minimum and standard velocities, the heterogeneous flow region.

Empirical equations are available [Durand and Condolios (1955)], which predict values for the minimum and standard velocities for various slurries. Spells analysed the experimental data of a number of investigators for aqueous slurries of sands, boiler ash and lime flowing in horizontal pipes. He obtained the following empirical equations which give the mean minimum liquid velocity u_1 and the mean standard linear liquid velocity u_2 respectively for slurries in horizontal pipes:

$$u_1 = \left[0.0251\, g d_p \left(\frac{\rho_m d_i}{\mu}\right)^{0.775} \left(\frac{\rho_p - \rho}{\rho}\right)\right]^{1/1.225} \tag{9.38}$$

and

$$u_2 = \left[0.0741\, g d_p \left(\frac{\rho_m d_i}{\mu}\right)^{0.775} \left(\frac{\rho_p - \rho}{\rho}\right)\right]^{1/1.225} \tag{9.39}$$

Equations 9.38 and 9.39 are based on experimental data for solid particles with diameters in the range 6×10^{-5} to 6×10^{-4} m and pipe diameters of

2.5×10^{-2} to 3×10^{-1} m. In equations 9.38 and 9.39, μ is the dynamic viscosity of the transporting liquid and ρ, ρ_p, and ρ_m are the densities of the liquid, solid particles and slurry mixture respectively. The last is given by the equation

$$\rho_m = \alpha(\rho_p - \rho) + \rho \tag{9.40}$$

where α is the volume fraction of the solids in the slurry.

Empirical equations are also available to calculate the pressure drop for slurries flowing through pipelines [Condolios and Chapus (1963b)]. Durand and Condolios (1955) found the following equation to fit the experimental data for sand–water mixtures flowing above the minimum velocity in horizontal pipes:

$$\frac{\Delta P - \Delta P_W}{\alpha \Delta P_W} = \frac{180}{\{[u^2/(gd_i)]C_d^{1/2}\}^{3/2}} \tag{9.41}$$

In equation 9.41, ΔP and ΔP_W are the pressure drops for the slurry and for the clear water respectively.

Condolios and Chapus (1963b) found that the presence of fines in a coarse slurry decreases the frictional pressure drop in a horizontal pipe to a much greater extent than might be expected from their relative properties in the solids.

Newitt, Richardson and Gliddon (1961) carried out experiments on aqueous slurries of pebbles, zircon, manganese dioxide, perspex and various kinds of sand in vertical pipes of 2.5×10^{-3} m and 5.0×10^{-3} m diameter. Their pressure drop data were satisfactorily correlated with the following equation:

$$\frac{\Delta P - \Delta P_W}{\alpha \Delta P_W} = 0.0037\left(\frac{gd_i}{u^2}\right)^{1/2}\left(\frac{d_i}{d_p}\right)\left(\frac{\rho_p}{\rho}\right)^2 \tag{9.42}$$

In equation 9.42, the mean velocity u is the volumetric flow rate of the slurry divided by the cross-sectional area of the pipe.

Solid particles hydraulically conveyed in a vertical pipe have a mean velocity which is less than the mean velocity of the liquid. This is because of the tendency of the particles to settle. The volume fraction α in equation 9.42 is the delivered concentration. This is less than the volume fraction in the vertical pipe.

Solid particles hydraulically conveyed in a vertical pipe are subjected to various forces which cause them to rotate and move inwards towards the axis of the pipe. The effect is most pronounced with large velocity gradients.

9.6 **Filtration**

When a slurry flows through a filter, the solid particles become entrapped by the filter medium which is permeable only to the liquid. Either of two mechanisms are used: cake filtration or depth filtration.

In cake filtration, the filter medium acts as a strainer and collects the solid particles on top of the initial layer. A filter cake is formed and the flow obeys the Carman–Kozeny equation for packed beds.

Depth filtration is also called granular filtration. In this case the filter medium is a bed of particulate material through which the slurry flows. Solid particles in the slurry are carried right into and are deposited within the bed. The bed is deep compared to its grain size. The latter is also much larger than the grain size in the slurry. There is virtually no deposition on the surface of the bed. Granular filters are suitable for producing high quality filtrate from large quantities of liquid containing up to 50 parts per million solids. The performance depends not only on the minimum particle size to be removed but also on the affinity of the suspended particles for the granular material. The most commonly used granular material is silica sand.

References

Carman, P.C., Fluid flow through granular beds, *Transactions of the Institution of Chemical Engineers*, **15**, pp. 150–66 (1937).

Cheng, D.C.-H., The flow of non-Newtonian slurries and suspensions in pipeline systems, *Filtration and Separation*, **7**, pp. 434–41 (1970).

Condolios, E. and Chapus, E.E., Transporting solid materials in pipelines, *Chemical Engineering*, **70** pp. 93–8 (24 June 1963a).

Condolios, E. and Chapus, E.E., Designing solids-handling pipelines, *Chemical Engineering*, **70**, pp. 131–8 (8 July 1963b).

Durand, R. and Condolios, E., Centenary Congress, Mineral Industry, France (1955).

Grace, J.R., In Hetsroni, G. (ed.) *Handbook of Multiphase Systems*, Washington; London: Hemisphere Publishing (1982).

Lapple, C.E. and Shepherd, C.B., Calculation of particle trajectories, *Industrial and Engineering Chemistry*, **32**, pp. 605–21 (1940).

Larian, M.G., *Fundamentals of Chemical Engineering Operations*. Englewood Cliffs, N. J. Prentice-Hall Inc., p. 542 (1958).

Longwell, P.A., *Mechanics of Fluid Flow*, New York, McGraw-Hill Book Company Inc, p. 77 (1966).

Newitt, D.M., Richardson, J.F. and Gliddon, B.J., Hydraulic conveying of solids in vertical pipes, *Transactions of the Institution of Chemical Engineers*, **39**, pp. 93–100 (1961).

Perry, J.H., *Chemical Engineers' Handbook*. Sixth edition, New York: McGraw-Hill Book Company Inc, pp. 5–53 and 5–54 (1984).

Pettyjohn, E.S. and Christiansen, E.B., Effect of particle shape on free-settling rates of isomeric particles, *Chemical Engineering Progress*, **44**, pp. 157–72 (1948).

Richardson, J.F. and Zaki, W.N., Sedimentation and fluidisation: Part I, *Transactions of the Institution of Chemical Engineers*, **32**, pp. 35–52 (1954).

Spells, K.E., Correlations for use in transport of aqueous suspensions of fine solids through pipes, *Transactions of the Institution of Chemical Engineers*, **33**, pp. 79–84 (1955).

10 Introduction to unsteady flow

Four aspects of unsteady fluid flow will be considered in this chapter: quasi-steady flow as in the filling or emptying of vessels, incremental calculations, start-up of shearing flow, and pressure surge in pipelines.

10.1 Quasi-steady flow

Very often processing operations change only slowly with time and at any instant may be treated as if conditions were steady. However, in predicting the course of the operation, it is necessary to recognize that conditions drift with time. Processes that change so slowly that they can be treated in this way are described as being quasi-steady. Examples of quasi-steady operations are the emptying of a vessel, the heating of a batch of material and batch distillation.

10.1.1 *Time to empty liquid from a tank*

In practice the resistance of the exit pipe of the tank shown in Figure 10.1 will be sufficiently large that the flow rate will be relatively low and consequently conditions in the tank, in particular the fluid head causing the flow, will change only slowly. In these circumstances the emptying operation can be treated as quasi-steady. In view of this, Bernoulli's equation, which is valid only for steady flow, may be used.

Bernoulli's equation applied between points A and B in the system shown in Figure 10.1 can be written as

$$z+\frac{P_A}{\rho g}+0=0+\frac{P_B}{\rho g}+\frac{u^2}{2g\alpha}+h_f \tag{10.1}$$

or as

$$z+\frac{(P_A-P_B)}{\rho g}=\frac{u^2}{2g\alpha}+h_f \tag{10.2}$$

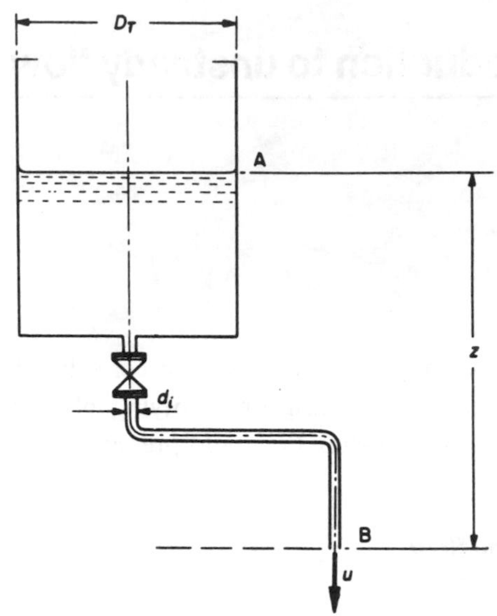

Figure 10.1
Liquid flowing from a tank

In equation 10.2, h_f is the head loss due to friction and is given by the equation

$$h_f = 4f\left(\frac{\Sigma L_e}{d_i}\right)\frac{u^2}{2g} \tag{10.3}$$

In equation 10.3, ΣL_e is the equivalent length of the outlet pipe (including the contraction at its inlet) and d_i is the pipe's inside diameter; u is the mean velocity in the outlet pipe. It has been assumed in writing equations 10.1 and 10.2 that the fluid's velocity in the tank is so low that it can be neglected.

Combining equations 10.2 and 10.3 gives

$$z + \frac{(P_A - P_B)}{\rho g} = \frac{u^2}{2g}\left[\frac{1}{\alpha} + 4f\left(\frac{\Sigma L_e}{d_i}\right)\right] \tag{10.4}$$

from which the mean velocity in the exit pipe is given as

$$u = \left\{\frac{2g[z + (P_A - P_B)/(\rho g)]}{1/\alpha + 4f\Sigma(L_e/d_i)}\right\}^{1/2} \tag{10.5}$$

In an infinitesimal time interval δt the liquid level in the tank changes by an amount δz and the volume of liquid in the tank by δV. Thus, equating the rate of loss of liquid from the tank to the flow rate through the exit pipe

$$-\delta V = -\frac{\pi}{4} D_T^2 \delta z = \frac{\pi}{4} d_i^2 u \delta t \tag{10.6}$$

Note that, for this case of emptying a tank, δz and δV are negative.

Rearranging equation 10.6 and going to the limit $\delta t \to 0$ gives the rate of change of level in the tank as

$$\frac{dz}{dt} = -u \frac{d_i^2}{D_T^2} \tag{10.7}$$

Substituting for u in equation 10.7 using equation 10.5 and integrating from time t_1 to t_2, when the liquid levels are z_1 and z_2, gives

$$\int_{z_1}^{z_2} \frac{dz}{[z + (P_A - P_B)/(\rho g)]^{1/2}} = -(2g)^{1/2} \frac{d_i^2}{D_T^2} \int_{t_1}^{t_2} \frac{dt}{[1/\alpha + 4f(\Sigma L_e/d_i)]^{1/2}} \tag{10.8}$$

Carrying out the integration gives the time interval $\Delta t_{12} = t_2 - t_1$ required for the liquid level to fall from z_1 to z_2:

$$\Delta t_{12} = \{[z_1 + (P_A - P_B)/(\rho g)]^{1/2} - [z_2 + (P_A - P_B)/(\rho g)]^{1/2}\} \times \left(\frac{2D_T^2}{d_i^2}\right) \left[\frac{1/\alpha + 4f(\Sigma L_e/d_i)}{2g}\right]^{1/2} \tag{10.9}$$

When the pressures at points A and B are the same, equation 10.9 takes the simpler form:

$$\Delta t_{12} = (z_1^{1/2} - z_2^{1/2}) \left(\frac{2D_T^2}{d_i^2}\right) \left[\frac{1/\alpha + 4f(\Sigma L_e/d_i)}{2g}\right]^{1/2} \tag{10.10}$$

Example 10.1 shows how slow emptying a tank under gravity will be if the exit pipe is of a small diameter.

Example 10.1

Calculate the time required for the liquid level to fall from a height $z_1 = 9$ m to a height $z_2 = 4$ m above the level of the discharge end of the exit pipe given the following data:

tank diameter $D_T = 2$ m
inside diameter of outlet pipe $d_i = 0.02$ m
Fanning friction factor $f = 0.008$
equivalent length of outlet pipe $\Sigma L_e = 25$ m
liquid viscosity $\mu = 2.1 \times 10^{-3}$ Pa s
liquid density $\rho = 1000$ kg/m^3
tank and discharge at atmospheric pressure

Calculations
Substituting the given values

$$4f\Sigma L_e/d_i = 40$$

In this example, the value of $1/\alpha$ is small compared with $4f\Sigma L_e/d_i$ and could be ignored. When this is not the case it will be necessary to know whether the flow in the outlet pipe is laminar or turbulent, so that the appropriate value of α can be used. The following calculation shows how this can be done.

From equation 10.5, with $P_A = P_B$, the velocity u is given by

$$u = \left[\frac{2g}{1/\alpha + 4f(\Sigma L_e/d_i)}\right]^{1/2} z^{1/2}$$

Assuming that the flow is turbulent, $\alpha = 1$ and u is given by

$$u = \left[\frac{(2)(9.81 \text{ m/s}^2)}{41}\right]^{1/2} z^{1/2}$$

$$= 0.692z^{1/2} \text{ m/s}$$

Therefore, the minimum velocity, when $z = 4\,\text{m}$, is 1.38 m/s and the corresponding value of the Reynolds number is

$$\text{minimum } Re = \frac{ud_i}{\nu} = \frac{(1.38 \text{ m/s})(0.02 \text{ m})}{(2.1 \times 10^{-6} \text{ m}^2/\text{s})}$$

$$= 1.31 \times 10^4$$

Thus, in this example, the flow will always be turbulent justifying the assumption $\alpha = 1$.

Substituting the given values in equation 10.10 (using the quantity evaluated above) shows that the time required is <u>8.04 hours</u>.

10.2 Incremental calculation: time to discharge an ideal gas from a tank

Consider the case in which an ideal gas flows from one tank to another tank at a lower pressure through a convergent nozzle. The second tank is assumed to be at a constant pressure. The flow may be assumed to be isentropic so that the mass flow rate is given by equation 6.107.

$$M = (\gamma P_0/V_0)^{1/2} S\psi \tag{6.107}$$

where

$$\psi = \left\{ \left(\frac{2}{\gamma - 1} \right) \left[\left(\frac{P}{P_0} \right)^{2/\gamma} - \left(\frac{P}{P_0} \right)^{(\gamma+1)/\gamma} \right] \right\}^{1/2} \tag{6.108}$$

As the gas flows from the first tank, to which there is no feed, the pressure P_0 falls and consequently so does the flow rate. Thus it takes progressively longer for each unit mass of gas to flow from the tank. In principle, this problem could be treated in a manner similar to that in Section 10.1; however, the complexity of equation 6.108 makes this impracticable.

A suitable method of calculation is to divide the period of flow into a number of short intervals so that conditions change only slightly during each interval. One method of doing this is to specify the length of each time interval and calculate the mass of gas flowing from the tank in that interval. From this the mass of gas remaining in the tank can be calculated and hence the pressure at the end of the interval determined. Conditions for the next interval can then be calculated from this pressure.

In Example 10.2, the range of pressure in the supply tank is specified and a more convenient method of calculation is to split the pressure range into a number of increments. After calculating the mass flow rate, the mass remaining in the tank is determined from the pressure at the end of the increment and hence the required time may be determined. A small refinement is to base the flow rate calculation on the mean pressure in the increment rather than on that at the beginning.

Example 10.2

Nitrogen contained in a $10\,\mathrm{m}^3$ tank at a pressure of 200 000 Pa and an initial temperature of 300 K flows to a second tank through a convergent nozzle with a 15 mm throat diameter. The pressure in the second tank and at the throat of the nozzle is constant at 140 000 Pa. Calculate the time required for the pressure in the first tank to fall to 180 000 Pa. For nitrogen $\gamma = C_p/C_v = 1.39$.

Calculations

The initial conditions are the same as in Example 6.3, where the following values were calculated:

$$\text{throat area } S_t = 1.767 \times 10^{-4}\,\mathrm{m}^2$$

$$\text{for } P_0 = 200\,000\,\mathrm{Pa},\ V_0 = 1/\rho_0 = 0.4451\,\mathrm{m}^3/\mathrm{kg}$$

$$\text{mass of gas in tank } W = (10\,\mathrm{m}^3)/(0.4451\,\mathrm{m}^3/\mathrm{kg}) = 22.467\,\mathrm{kg}$$

For the first increment P_0 falls from 200000 Pa to 198000 Pa. The mean pressure in the first vessel during the first increment is $\bar{P}_0 = 199000$ Pa.

For isentropic expanson, the mean specific volume is given by

$$\bar{V}_0 = 0.4451\left(\frac{200}{199}\right)^{\frac{1}{1.39}} = 0.44670 \text{ m}^3/\text{kg} .$$

consequently,

$$(\gamma\bar{P}_0/\bar{V}_0)^{1/2} = 786.9 \text{ kg/(m}^2\text{s)}$$

and

$$\frac{P_t}{P_0} = \frac{140}{199} = 0.7035$$

From equation 6.108

$$\psi = 0.5390$$

From equation 6.107

$$M = 0.07495 \text{ kg/s}$$

At the end of increment, $P_0 = 198000$ Pa

Expansion of the gas in the tank must be nearly adiabatic because heat transfer from the walls will be very slow. Furthermore, the gradual expansion will be almost frictionless and consequently isentropic expansion may be assumed. Thus

$$W = 22.467\left(\frac{198}{200}\right)^{\frac{1}{1.39}} = 22.305 \text{ kg}$$

The reduction of mass in the tank is given by

$$\Delta W = 22.467 - 22.305 = 0.162 \text{ kg}$$

and the required time by

$$\Delta t = \frac{\Delta W}{M} = \frac{0.162}{0.07495} = 2.16 \text{ s}$$

The calculation can now be repeated for the second increment in which the pressure falls from 198000 Pa to 196000 Pa, conditions in the first tank being evaluated at the mean pressure of 197000 Pa. Repeating the calculation until the pressure has fallen to 180000 Pa gives the results shown in Table 10.1.

Table 10.1

$\bar{P}_0$ kPa	M kg/s	W kg	ΔW kg	Δt s	$\Sigma\Delta t$ s
initially		22.467			
199	0.07495	22.305	0.162	2.16	2.16
197	0.07381	22.143	0.162	2.20	4.36
195	0.07265	21.980	0.163	2.24	6.60
193	0.07146	21.817	0.163	2.28	8.88
191	0.07023	21.653	0.164	2.34	11.22
189	0.06898	21.489	0.164	2.38	13.59
187	0.06770	21.324	0.165	2.44	16.03
185	0.06638	21.159	0.165	2.49	18.52
183	0.06502	20.993	0.166	2.55	21.07
181	0.06362	20.827	0.166	2.61	23.68

10.3 Time for a solid spherical particle to reach 99 per cent of its terminal velocity when falling from rest in the Stokes regime

Consider a spherical particle of diameter d_p and density ρ_p falling from rest in a stationary fluid of density ρ and dynamic viscosity μ. The particle will accelerate until it reaches its terminal velocity u_t. At any time t, let u be the particle's velocity. Recalling that the drag force acting on a sphere in the Stokes regime is of magnitude $3\pi d_p \mu u$, application of Newton's second law of motion can be written as

$$\left(\frac{\pi d_p^3 \rho_p}{6}\right)\frac{du}{dt} = \frac{\pi d_p^3(\rho_p - \rho)g}{6} - 3\pi d_p \mu u \tag{10.11}$$

Noting that the terminal velocity is given by

$$u_t = \frac{d_p^2(\rho_p - \rho)g}{18\mu} \tag{9.8}$$

equation 10.11 can be written as

$$\frac{du}{dt} = \left(1 - \frac{\rho}{\rho_p}\right) g \left(1 - \frac{u}{u_t}\right) \tag{10.12}$$

Integrating equation 10.12 gives

$$-u_t \ln\left(1 - \frac{u}{u_t}\right) = \left(1 - \frac{\rho}{\rho_p}\right) gt + C \tag{10.13}$$

where C is a constant. The initial condition is $u = 0$ at $t = 0$, therefore $C = 0$. Consequently, equation 10.13 becomes

$$\ln\left(1 - \frac{u}{u_t}\right) = -\left(1 - \frac{\rho}{\rho_p}\right) \frac{gt}{u_t} \tag{10.14}$$

which can be written as

$$\frac{u}{u_t} = 1 - \exp\left[-\left(1 - \frac{\rho}{\rho_p}\right) \frac{gt}{u_t}\right] \tag{10.15}$$

Neglecting the trivial case $u_t = 0$ when $\rho_p = \rho$, equation 9.8 can be used to substitute for u_t so that equation 10.15 becomes

$$\frac{u}{u_t} = 1 - \exp\left(-\frac{18\mu t}{d_p^2 \rho_p}\right) \tag{10.16}$$

10.4 Suddenly accelerated plate in a Newtonian fluid

A very large horizontal plate is held in a Newtonian fluid which is at rest. At time $t = 0$ the plate is suddenly set in motion in its own plane with a constant velocity u_0. Determine the motion of the fluid as a function of time and distance from the plate assuming that the flow remains laminar.

The positive x-direction will be taken as the direction of motion of the plate and y as the distance from the surface of the plate. As the plate is very large, the motion will be independent of x except close to the edges. The pressure is independent of x because the plate moves in its own plane producing only a shearing action. The pressure varies in the y-direction due only to the hydrostatic head: this does not affect the motion. There is only one non-zero velocity component v_x and this is a function of y and t.

Consider a fixed element of space with unit area in the x–z plane and having its surfaces at distances y and $y + \delta y$ from the plate. Using the negative sign convention for stress components (which coincides with the directions in which the stress physically acts in this case), the shear forces acting on the lower and upper surfaces of the element are $(1)\tau_{yx}|_y$ acting in the positive x-direction and $(1)\tau_{yx}|_{y+\delta y}$ acting in the negative x-direction respectively. As the flow is the same at all values of x, the momentum flow rates into and out of the element are equal and the rate of change of momentum is just the rate of change of momentum of the fluid instantaneously in the element. Thus

$$\frac{\partial}{\partial t}[\rho(1)\delta y v_x] = (1)\tau_{yx}|_y - (1)\tau_{yx}|_{y+\delta y} \tag{10.17}$$

In the limit $\delta y \to 0$, this leads to

$$\rho\frac{\partial v_x}{\partial t} = -\frac{\partial \tau_{yx}}{\partial y} \tag{10.18}$$

for incompressible flow.

For an incompressible Newtonian fluid, the shear stress is given by

$$\tau_{yx} = -\mu\frac{\partial v_x}{\partial y} \tag{1.44b}$$

and hence equation 10.18 can be written as

$$\frac{\partial v_x}{\partial t} = \nu\frac{\partial^2 v_x}{\partial y^2} \tag{10.19}$$

where $\nu = \mu/\rho$ is the kinematic viscosity of the fluid.

Equation 10.19 will be recognized as an example of the diffusion equation. One-dimensional molecular diffusion of component A is described by the equation

$$\frac{\partial C_A}{\partial t} = \mathcal{D}\frac{\partial^2 C_A}{\partial y^2} \tag{10.20}$$

where $\mathcal{D}$ is the molecular diffusivity. Comparing equations 10.19 and 10.20 shows that the kinematic viscosity ν can be interpreted as the diffusivity of momentum of the fluid.

Equation 10.19 can be solved in various ways including using the Laplace transformation, as employed in the first edition of this book. For the benefit of readers unfamiliar with Laplace transforms, an alternative method will be used here. For a few simple problems, it is possible to find a combination of the independent variables which can be treated as a single variable, thus transforming the partial differential equation into an ordinary differential equation. This is applicable when the problem has an open range, in this case $y > 0$.

The method of combination of variables requires that a suitable combination of y and t can be found. Dimensionally, equation 10.19 can be written as

$$[v_x][y^2/\nu t] = [v_x] \tag{10.21}$$

This suggests that the group $y/\sqrt{(\nu t)}$ can be used as a combined variable. For convenience, a factor of 2 may be introduced and the combined variable η defined as

$$\eta = \frac{y}{\sqrt{4\nu t}} \tag{10.22}$$

It is necessary to replace the partial derivatives in equation 10.19 by ordinary derivatives with respect to η.

$$\frac{\partial v_x}{\partial t} = \frac{\mathrm{d}v_x}{\mathrm{d}\eta}\frac{\partial \eta}{\partial t} = \frac{\mathrm{d}v_x}{\mathrm{d}\eta}\left(-\frac{y}{2\sqrt{4\nu}}t^{-3/2}\right) \tag{10.23}$$

$$= \frac{\mathrm{d}v_x}{\mathrm{d}\eta}\left(-\frac{\eta}{2t}\right)$$

Similarly

$$\frac{\partial v_x}{\partial y} = \frac{\mathrm{d}v_x}{\mathrm{d}\eta}\frac{\partial \eta}{\partial y} = \frac{\mathrm{d}v_x}{\mathrm{d}\eta}\left(\frac{1}{\sqrt{4\nu t}}\right) \tag{10.24}$$

and consequently the second derivative is given by

$$\frac{\partial^2 v_x}{\partial y^2} = \frac{\partial}{\partial y}\left(\frac{\partial v_x}{\partial y}\right) = \frac{\partial}{\partial y}\left(\frac{\mathrm{d}v_x}{\mathrm{d}\eta}\frac{1}{\sqrt{4\nu t}}\right)$$

$$= \frac{\mathrm{d}^2 v_x}{\mathrm{d}\eta^2}\frac{\partial \eta}{\partial y}\left(\frac{1}{\sqrt{4\nu t}}\right)$$

Thus

$$\frac{\partial^2 v_x}{\partial y^2} = \frac{\mathrm{d}^2 v_x}{\mathrm{d}\eta^2}\left(\frac{1}{4\nu t}\right) \tag{10.25}$$

Substituting for the partial derivatives in equation 10.19 allows it to be written as

$$\frac{\mathrm{d}^2 v_x}{\mathrm{d}\eta^2} + 2\eta\frac{\mathrm{d}v_x}{\mathrm{d}\eta} = 0 \tag{10.26}$$

This simple ordinary differential equation can be solved by making the substitution

$$\phi = \frac{dv_x}{d\eta} \tag{10.27}$$

so that equation 10.26 becomes

$$\frac{d\phi}{d\eta} + 2\eta\phi = 0 \tag{10.28}$$

Integrating equation 10.28 gives

$$\ln\phi = -\eta^2 + C_1 \tag{10.29}$$

Therefore

$$\frac{dv_x}{d\eta} = \phi = C_2 e^{-\eta^2} \tag{10.30}$$

Integrating again gives the velocity v_x as

$$v_x = C_2 \int e^{-\eta^2} d\eta + C_3 \tag{10.31}$$

The integral in equation 10.31 cannot be evaluated analytically but it can be written in terms of the error function erf(η) defined as

$$\mathrm{erf}(\eta) = \frac{2}{\sqrt{\pi}} \int_0^{\eta} e^{-s^2} ds \tag{10.32}$$

Note that s is a dummy variable: the value of the integral depends only on the value of the upper limit. Tables of the error function are available and values can be calculated from power series [Dwight (1961), Kreyszig (1988)]. The error function has the properties erf(0) = 0 and erf(∞) = 1.

Equation 10.31 can be written in terms of the error function as

$$v_x = A\,\mathrm{erf}(\eta) + B \tag{10.33}$$

The boundary conditions are

(i)	$v_x = 0$,	$y > 0$,	$t = 0$
(ii)	$v_x = u_0$,	$y = 0$,	$t > 0$
(iii)	$v_x \to 0$,	$y \to \infty$,	$t > 0$

These boundary conditions can be expressed in terms of the combined variable η as

$$v_x = u_0, \quad \eta = 0$$
$$v_x \to 0, \quad \eta \to \infty$$

The last statement represents both conditions (i) and (iii).

Substituting the boundary conditions, using the properties of the error function given above, shows that $-A = B = u_0$. The velocity field is therefore

$$v_x = u_0\left[1 - \operatorname{erf}\left(\frac{y}{\sqrt{4\nu t}}\right)\right] \tag{10.34}$$

It is clear from equation 10.34 that the time taken for the fluid at a given distance y from the surface of the plate to reach a specified fraction of the plate's velocity is proportional to y^2 and inversely proportional to the fluid's kinematic viscosity ν. This is illustrated in Figure 10.2.

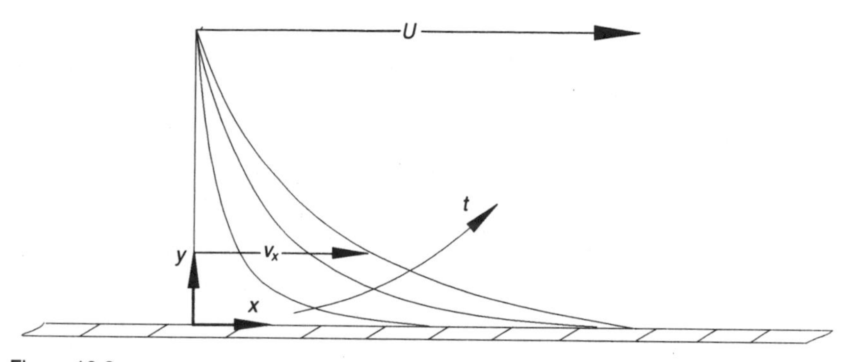

Figure 10.2
Fluid velocity as a function of time and distance from the plate

The following numerical example indicates how long may be required for the motion of the plate to diffuse through the liquid. Consider the case of a fluid of density 1000 kg/m^3 and dynamic viscosity 0.1 Pa s. What is the velocity of the liquid, as a fraction of the plate's velocity, at a location 0.1 m away from the plate 25 s and 2500 s after the plate is set in motion?

$$\nu = \mu/\rho = 1 \times 10^{-4}\ \text{m}^2/\text{s}$$

For $t = 25$ s, $\eta = y/\sqrt{(4\nu t)} = 1.0$ and from error function tables [Kreyszig (1988)],

$$\operatorname{erf}(\eta) = 0.8427 \text{ so that } v_x/u_0 = 0.1573$$

For $t = 2500$ s, $\eta = y/\sqrt{(4\nu t)} = 0.10$ and $\operatorname{erf}(\eta) = 0.1125$, so that $v_x/u_0 = 0.8875$.

Thus, for these conditions, the fluid attains a velocity of only 15.7 per cent of the plate's velocity after 25 s and requires 2500 s to attain 88.75 per cent of it.

10.5 Pressure surge in pipelines

Consider the consequences of closing the flow control valve V in the pipeline shown in Figure 10.3. The momentum of the liquid in a length L

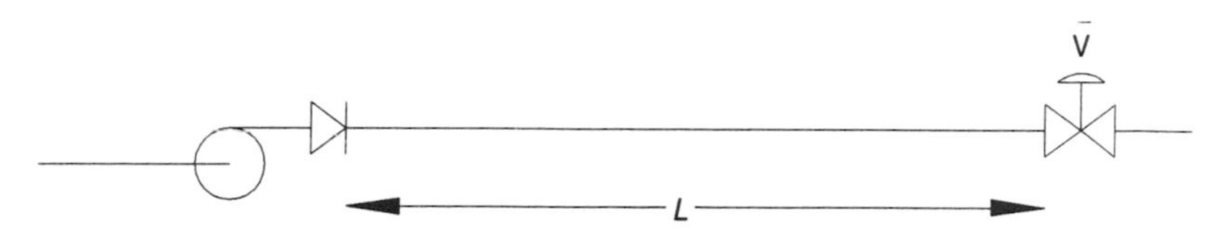

Figure 10.3
Pipeline with check valve and flow control valve

of pipe is equal to $\pi r_i^2 L\rho u$ if the flow is assumed to be turbulent with volumetric average velocity u and to fill the pipe. For water flowing at 2 m/s in a pipe of diameter 0.3 m and length 1000 m, the momentum is equal to 1.4×10^5 N s. If valve V is gradually closed, the momentum of the liquid will be destroyed and this requires that the valve exert a force on the liquid. This appears as a rise in pressure: an equal and opposite force is exerted by the liquid on the valve. In this example, if the valve is closed over a period of 10 s, the average rate of change of momentum will be 1.4×10^4 N, requiring a force of this magnitude or a pressure rise of 1.98×10^5 Pa, ie, 2 bar.

If the valve is closed more quickly, the pressure rise will be correspondingly greater. It might be thought that if the valve were closed instantly the pressure rise would be infinite. This is not the case. When a valve is closed suddenly, a pressure wave propagates upstream at approximately the speed of sound in the fluid and only the fluid through which the pressure wave has passed is decelerated; thus the pressure rise is finite because the speed of sound is finite.

The speed of sound c in a fluid is related to the compressibility by

$$c^2 = \left(\frac{\partial P}{\partial \rho}\right)_s \tag{10.35}$$

Equation 10.35 is the form of equation 6.65 for the particular case of isentropic conditions, denoted by the subscript s. Being less compressible than gases, liquids transmit sound at higher speeds. The speed of sound in water is about 1400 m/s, compared with the value of 340 m/s for dry air under ambient conditions.

It is convenient to introduce the bulk modulus K of the fluid, defined by

$$K = -V\left(\frac{\partial P}{\partial V}\right)_s \tag{10.36}$$

Replacing the specific volume V by $1/\rho$, equation 10.36 can be written as

$$K = \rho\left(\frac{\partial P}{\partial \rho}\right)_s \tag{10.37}$$

Thus, the speed of sound is related to the bulk modulus by

$$c = \sqrt{K/\rho} \tag{10.38}$$

When a pressure wave propagates through fluid in a pipe, the pipe expands slightly as a result of its elasticity. This has the effect of reducing the speed a with which the pressure wave is propagated through the fluid in the pipe. A convenient way of expressing the propagation speed a of the pressure wave is as given in equation 10.39 [Streeter and Wylie (1983), Watters (1979)].

$$a = \frac{(K/\rho)^{1/2}}{\left[1 + \frac{K}{E}\frac{d_i}{t_w}(C)\right]^{1/2}} \tag{10.39}$$

The value of C in equation 10.39 depends on the way in which the pipe is restrained but for practical purposes a value of unity is adequate. In this equation, E is Young's modulus of elasticity of the pipe, d_i the internal diameter of the pipe and t_w its wall thickness. The value of E for steel is about 2×10^5 MPa and K for water is about 2×10^3 MPa; thus K/E is about 10^{-2}. It will be seen that the elasticity of the pipe has a negligible effect with thick-wall pipes but with thin-wall ones (say $d_i/t_w > 40$) the propagation speed a will typically be reduced to about 70 per cent of the speed of sound c in the liquid.

The pressure rise resulting from the sudden closing or partial closing of a valve can be determined from the change of fluid momentum across the pressure wave. It will be assumed that the fluid is a liquid so that changes in density are negligible. If the fluid's velocity upstream of the pressure wave is u_1 and that downstream is u_2, the change of momentum per unit volume is $\rho(u_1 - u_2)$. The pressure wave passes through the fluid at a speed a so that the rate of change of momentum per unit area is equal to $\rho(u_1 - u_2)a$. Consequently the pressure rise is given by

$$\Delta P = \rho(u_1 - u_2)a \quad (10.40)$$

When the valve is suddenly closed completely, $u_2 = 0$ and equation 10.40 becomes

$$\Delta P = \rho u_1 a \quad (10.41)$$

Equation 10.41 shows that although the pressure rise is finite it may be very large. For water ($\rho = 1000\,\text{kg/m}^3$) flowing at 2 m/s, taking a as 1400 m/s, the pressure rise will be equal to 2.8×10^6 Pa, ie 28 bar. In a thin-wall pipe, a might be as low as 1000 m/s but the pressure rise would be reduced only to 20 bar. Pressure surges of this magnitude cause severe damage.

Pressure surge can be controlled by the use of surge vessels connected to the pipeline with a gas space above the liquid. If a pressure surge passes along the pipeline, liquid flows into the vessel through a check valve and the gas is compressed. Another valve then lets out the liquid at a controlled rate. This type of arrangement is used in large hydraulic lines.

The main method of avoiding surges, particularly in chemical plants, is by adjusting valves sufficiently slowly.

Consider the pressure wave that results from partially closing valve V shown in Figure 10.3. The compression wave travels upstream at speed a and therefore reaches the check valve at a time L/a after valve V was adjusted. The high pressure at the check valve, and therefore at the outlet of the centrifugal pump, causes the discharge from the pump to fall. At this point, the liquid in the pipe is compressed slightly and the pipe distended. The liquid at the upstream end of the pipe begins to expand and the pipe to contract so an expansion wave now starts to propagate downstream at speed a. Consequently, the expansion wave and the reduced flow rate reach valve V at a time $2L/a$ after that valve was adjusted.

The time period $2L/a$ is known as the characteristic time of the pipeline. If a change of valve setting is made in a time less than the characteristic time, the adjustment is effectively instantaneous and the pressure rise is given by equation 10.40. In order for the valve adjustment to be effectively gradual, it must be made over a period of time that is long compared with the characteristic time $2L/a$.

Example 10.3
In order to protect a relief valve from corrosive materials, it is sometimes preceded by a bursting disc in the vent pipe. Similarly, two bursting discs may be used in series. It is known that, with these arrangements, if the first disc fails the resulting pressure surge will cause the second device to relieve even though the pressure may be significantly below its rated pressure. On the basis of experiments done with air ($\gamma = 1.40$ for dry air), it has been suggested that to prevent the second of a pair of bursting discs from relieving, the upstream one should be rated at about 75 per cent of the pressure of the second disc. Can this finding be substantiated by simple analysis?

Derivations
Under the conditions encountered, the pressure difference across the bursting disc is large enough to ensure that choking occurs. As a result, on failure of the disc, gas flows at the local sonic speed towards the second device. The transient flow will occupy the whole cross section, a *vena contracta* forming later. When the sonic gas flow meets the second device there will be a pressure rise just as if the flow had been steady and an obstruction were placed across the flow.

Under these conditions, the gas speed is the sonic speed c and the pressure wave may be assumed to propagate at the sonic speed so that equation 10.41 can be written as

$$\Delta P = \rho c^2 \tag{10.42}$$

For an ideal gas, the sonic speed c is given by equation 6.70:

$$c = \sqrt{\gamma P_* V_*} \tag{6.70}$$

where P_* and V_* are the pressure and specific volume at the sonic conditions.

Consequently, the pressure rise ΔP is given by

$$\Delta P = \rho_* \gamma P_* V_* = \gamma P_* \tag{10.43}$$

The total pressure P^+ exerted by the gas on the disc or valve face is the sum of the static pressure P_* and the impact pressure rise ΔP:

$$P^+ = P_* 2 + \Delta P = (1 + \gamma) P_* \tag{10.44}$$

For an upstream pressure P_0, the critical pressure producing the sonic speed is given by equation 6.110

$$\frac{P_*}{P_0} = \left(\frac{2}{\gamma+1}\right)^{\gamma/(\gamma-1)} \tag{6.110}$$

The total pressure P^+ is therefore related to the upstream pressure P_0 by

$$P^+ = P_0(1+\gamma)\left(\frac{2}{\gamma+1}\right)^{\gamma/(\gamma-1)} \tag{10.45}$$

Equation 10.45 shows that P^+/P_0 is a very weak function of γ.

If the second device is to be on the point of just relieving when the upstream disc ruptures, the required ratings of the two devices must be in the ratio P_0/P^+. From equation 10.45 this ratio has the following values at various values of γ.

γ	1.10	1.20	1.40
P_0/P^+	0.81	0.80	0.79

This analysis is in excellent agreement with the experimental results of Beveridge and Jones (1984). In their measurements using air ($\gamma = 1.40$), they measured the bursting pressures for a downstream disc following rupture of a lower-rated upstream disc. Their measured values of the pressure ratio, equivalent to P_0/P^+, at which bursting of the second disc started was 0.79(1), (trial no. 3).

References

Beveridge, H.J.R. and Jones, C.G., Shock effects on a bursting disk in a relief manifold, in *The Protection of Exothermic Reactors and Pressurised Storage Vessels*, I Chem E Symposium Series No. 85, pp. 207–14 (1984).

Dwight, H.B., *Tables of Integrals and Other Mathematical Data*. Fourth edition, New York, Macmillan Company, 1961.

Kreyszig, E., *Advanced Engineering Mathematics*, Sixth edition, New York, John Wiley and Sons Inc., 1988.

Streeter, V.L. and Wylie, E.B., *Fluid Mechanics*, First SI edition, Singapore, McGraw-Hill Book Company Inc, p. 499 (1983).

Watters, G.Z., *Modern Analysis and Control of Unsteady Flow in Pipelines*, Ann Arbor, Ann Arbor Science Publishers, p. 34 (1979).

Appendix I
The Navier–Stokes equations

Rather than setting up a force–momentum balance for a particular flow problem as was done in Chapter 1, general equations, known as the Navier–Stokes equations, may be formulated. Before discussing the Navier–Stokes equations, it is necessary to consider some related matters.

Differentiation following the flow

Let ϕ represent some property of the flow, for example the velocity, temperature or density of the fluid. In general ϕ is a function of the time t and the spatial coordinates x, y, z. Then the total derivative of ϕ with respect to t is given by

$$\frac{\mathrm{d}\phi}{\mathrm{d}t} = \frac{\partial \phi}{\partial t} + \frac{\mathrm{d}x}{\mathrm{d}t}\frac{\partial \phi}{\partial x} + \frac{\mathrm{d}y}{\mathrm{d}t}\frac{\partial \phi}{\partial y} + \frac{\mathrm{d}z}{\mathrm{d}t}\frac{\partial \phi}{\partial z} \tag{A.1}$$

As x, y, z and t are independent variables, $\mathrm{d}x/\mathrm{d}t$, $\mathrm{d}y/\mathrm{d}t$, $\mathrm{d}z/\mathrm{d}t$ have to be defined.

Denoting these derivatives by w_x, w_y, w_z, respectively, equation A.1 can be written as

$$\frac{\mathrm{d}\phi}{\mathrm{d}t} = \frac{\partial \phi}{\partial t} + w_x\frac{\partial \phi}{\partial x} + w_y\frac{\partial \phi}{\partial y} + w_z\frac{\partial \phi}{\partial z} \tag{A.2}$$

If w_x, w_y, w_z are the velocity components of an observer, equation A.2 gives the total rate of change of ϕ measured by that observer. At a given location, the total rate of change $\mathrm{d}\phi/\mathrm{d}t$ is the sum of the rate of change at that location $\partial\phi/\partial t$ and the rate of change due to the observer's motion through a spatial gradient.

In the special case when the velocity components are those of the fluid, the total rate of change of ϕ is denoted by $\mathrm{D}\phi/\mathrm{D}t$, which is known as the substantive derivative:

$$\frac{\mathrm{D}\phi}{\mathrm{D}t} = \frac{\partial \phi}{\partial t} + v_x \frac{\partial \phi}{\partial x} + v_y \frac{\partial \phi}{\partial y} + v_z \frac{\partial \phi}{\partial z} \qquad \text{(A.3)}$$

In equation A.3, v_x, v_y, v_z are the velocity components of the fluid. Thus, $\mathrm{D}\phi/\mathrm{D}t$ gives the rate of change of ϕ for a *material element* as it flows along. This is known as differentiation following the flow.

In cylindrical coordinates (r,θ,z) the substantive derivative is given by

$$\frac{\mathrm{D}\phi}{\mathrm{D}t} = \frac{\partial \phi}{\partial t} + v_r \frac{\partial \phi}{\partial r} + \frac{v_\theta}{r} \frac{\partial \phi}{\partial \theta} + v_z \frac{\partial \phi}{\partial z} \qquad \text{(A.4)}$$

Continuity equation

A material balance on a fixed rectangular region of space having sides δx, δy, δz can be written as

rate of accumulation + rate of efflux − rate of influx = 0

$$\frac{\partial}{\partial t}(\rho \delta x \delta y \delta z) + (\rho v_x \delta y \delta z)\big|_{x+\delta x} + (\rho v_y \delta z \delta x)\big|_{y+\delta y} + (\rho v_z \delta x \delta y)\big|_{z+\delta z}$$

$$- (\rho v_x \delta y \delta z)\big|_x - (\rho v_y \delta z \delta x)\big|_y - (\rho v_z \delta x \delta y)\big|_z = 0 \qquad \text{(A.5)}$$

Thus

$$\frac{\partial \rho}{\partial t} + \frac{(\rho v_x)\big|_x^{x+\delta x}}{\delta x} + \frac{(\rho v_y)\big|_y^{y+\delta y}}{\delta y} + \frac{(\rho v_z)\big|_z^{z+\delta z}}{\delta z} = 0 \qquad \text{(A.6)}$$

In the limit δx, δy, $\delta z \to 0$, this becomes

$$\frac{\partial \rho}{\partial t} + \frac{\partial}{\partial x}(\rho v_x) + \frac{\partial}{\partial y}(\rho v_y) + \frac{\partial}{\partial z}(\rho v_z) = 0 \qquad \text{(A.7)}$$

Equation A.7 represents conservation of mass for a general flow in rectangular Cartesian coordinates.

For incompressible flow, the density ρ is constant and equation A.7 reduces to

$$\frac{\partial v_x}{\partial x} + \frac{\partial v_y}{\partial y} + \frac{\partial v_z}{\partial z} = 0 \qquad \text{(A.8)}$$

Thus, for incompressible flow the net rate of expansion is zero. Note that the flow need not be steady for equation A.8 to hold: the time derivative of ρ disappears because ρ is constant but the velocity components in equation A.8 may change with time.

Expanding equation A.7,

$$\frac{\partial \rho}{\partial t} + v_x \frac{\partial \rho}{\partial x} + \rho \frac{\partial v_x}{\partial x} + v_y \frac{\partial \rho}{\partial y} + \rho \frac{\partial v_y}{\partial y} + v_z \frac{\partial \rho}{\partial z} + \rho \frac{\partial v_z}{\partial z} = 0 \qquad \text{(A.9)}$$

which can be written as

$$\frac{1}{\rho} \frac{\mathrm{D}\rho}{\mathrm{D}t} + \frac{\partial v_x}{\partial x} + \frac{\partial v_y}{\partial y} + \frac{\partial v_z}{\partial z} = 0 \qquad \text{(A.10)}$$

Equation A.10 shows that the fractional rate of change of the density of a fluid element is equal to minus its net rate of expansion.

In cylindrical coordinates (r,θ,z) the continuity equation is

$$\frac{\partial \rho}{\partial t} + \frac{1}{r} \frac{\partial}{\partial r} (r\rho v_r) + \frac{1}{r} \frac{\partial}{\partial \theta} (\rho v_\theta) + \frac{\partial}{\partial z} (\rho v_z) = 0 \qquad \text{(A.11)}$$

Equation of motion

Consider the fluid's x-component of motion in a rectangular Cartesian coordinate system. By following the flow, the rate of change of a fluid element's momentum is given by the substantive derivative of the momentum. By Newton's second law of motion, this can be equated to the net force acting on the element. For an element of fluid having volume $\delta x \delta y \delta z$, the equation of motion can be written for the x-component as follows:

$$\frac{\mathrm{D}}{\mathrm{D}t} (\rho v_x \, \delta x \delta y \delta z) = B_x + S_x \qquad \text{(A.12)}$$

where B_x and S_x are respectively the body force and the surface force (ie that due to the relative motion of the fluid) acting on the fluid element in the positive x-direction.

It will be assumed that the only body force is that due to gravity:

$$B_x = \rho g_x \delta x \delta y \delta z \qquad \text{(A.13)}$$

where g_x is the component of the gravitational acceleration acting in the positive x-direction.

Using the negative sign convention for stress components, the surface force S_x can be written as

$$S_x = (\tau_{xx}|_x - \tau_{xx}|_{x+\delta x})\delta y \delta z + (\tau_{yx}|_y - \tau_{yx}|_{y+\delta y})\delta z \delta x + (\tau_{zx}|_z - \tau_{zx}|_{z+\delta z})\delta x \delta y \qquad \text{(A.14)}$$

Substituting equations A.13 and A.14 into equation A.12 gives the result

$$\frac{\mathrm{D}}{\mathrm{D}t}(\rho v_x) = \rho g_x - \frac{\tau_{xx}\big|_x^{x+\delta x}}{\delta x} - \frac{\tau_{yx}\big|_y^{y+\delta y}}{\delta y} - \frac{\tau_{zx}\big|_z^{z+\delta z}}{\delta z} \tag{A.15}$$

Now the momentum of a fluid element depends on its velocity and not on the spatial distribution of its mass, so ρ can be taken outside the derivative. In the limit δx, δy, $\delta z \rightarrow 0$, equation A.15 becomes

$$\rho \frac{\mathrm{D}v_x}{\mathrm{D}t} = \rho g_x - \frac{\partial \tau_{xx}}{\partial x} - \frac{\partial \tau_{yx}}{\partial y} - \frac{\partial \tau_{zx}}{\partial z} \tag{A.16}$$

Equation A.16 is valid for any fluid.

It is convenient to decompose the stress component τ_{ij} as follows:

$$\tau_{ij} = P\delta_{ij} + \sigma_{ij}, \qquad i, j = x, y, z \tag{A.17}$$

where δ_{ij} is known as the Kronecker delta and has the properties

$$\delta_{ij} = \begin{cases} 1 & \text{if} \quad i=j \\ 0 & \text{if} \quad i \neq j \end{cases} \tag{A.18}$$

Each σ_{ij} is known as a deviatoric stress component: it is the amount by which the stress component deviates from the static pressure. From equation A.18, the stress components are related to the deviatoric stress components as follows:

$$\begin{aligned} \tau_{xx} &= P + \sigma_{xx} \\ \tau_{yx} &= \sigma_{yx} \\ \tau_{zx} &= \sigma_{zx} \end{aligned} \tag{A.19}$$

It is the deviatoric stress components that are related to the rate of deformation, ie to the flow.

Navier–Stokes equations

For an incompressible Newtonian fluid, Newton's law of viscosity can be written as

$$\begin{aligned} \sigma_{xx} &= -2\mu \frac{\partial v_x}{\partial x} \\ \sigma_{yx} &= -\mu\left(\frac{\partial v_x}{\partial y} + \frac{\partial v_y}{\partial x}\right) \\ \sigma_{zx} &= -\mu\left(\frac{\partial v_x}{\partial z} + \frac{\partial v_z}{\partial x}\right) \end{aligned} \tag{A.20}$$

Equation A.20 is a generalization of equation 1.44b.

Substituting for the stress components in equation A.16 using equations A.19 and A.20 gives

$$\rho\frac{\mathrm{D}v_x}{\mathrm{D}t} = \rho g_x - \frac{\partial P}{\partial x} + \mu\left(\frac{\partial^2 v_x}{\partial x^2} + \frac{\partial^2 v_x}{\partial y^2} + \frac{\partial^2 v_x}{\partial z^2}\right) + \mu\frac{\partial}{\partial x}\left(\frac{\partial v_x}{\partial x} + \frac{\partial v_y}{\partial y} + \frac{\partial v_z}{\partial z}\right) \quad \text{(A.21)}$$

By continuity, the last term in equation A.21 vanishes for incompressible flow and equation A.21 can be written as

$$\rho\left(\frac{\partial v_x}{\partial t} + v_x\frac{\partial v_x}{\partial x} + v_y\frac{\partial v_x}{\partial y} + v_z\frac{\partial v_x}{\partial z}\right)$$
$$= \rho g_x - \frac{\partial P}{\partial x} + \mu\left(\frac{\partial^2 v_x}{\partial x^2} + \frac{\partial^2 v_x}{\partial y^2} + \frac{\partial^2 v_x}{\partial z^2}\right) \quad \text{(A.22)}$$

Equation A.22 is the Navier–Stokes equation for the x-component of motion in rectangular Cartesian coordinates. The corresponding equations for the y and z components are obvious.

The terms on the left hand side of equation A.22 represent inertial stresses, the first due to acceleration and the others to advection. The first and second terms on the right hand side are the component of the gravitational force and the pressure gradient. The remaining terms represent the viscous stress components acting in the x-direction.

In cylindrical coordinates (r,θ,z) the Navier–Stokes equations are

$$\rho\left(\frac{\partial v_r}{\partial t} + v_r\frac{\partial v_r}{\partial r} + \frac{v_\theta}{r}\frac{\partial v_r}{\partial \theta} - \frac{v_\theta^2}{r} + v_z\frac{\partial v_r}{\partial z}\right) = \rho g_r - \frac{\partial P}{\partial r}$$
$$+ \mu\left[\frac{\partial}{\partial r}\left(\frac{1}{r}\frac{\partial}{\partial r}(r v_r)\right) + \frac{1}{r^2}\frac{\partial^2 v_r}{\partial \theta^2} - \frac{2}{r^2}\frac{\partial v_\theta}{\partial \theta} + \frac{\partial^2 v_r}{\partial z^2}\right] \quad \text{(A.23)}$$

$$\rho\left(\frac{\partial v_\theta}{\partial t} + v_r\frac{\partial v_\theta}{\partial r} + \frac{v_\theta}{r}\frac{\partial v_\theta}{\partial \theta} + \frac{v_r v_\theta}{r} + v_z\frac{\partial v_\theta}{\partial z}\right) = \rho g_\theta - \frac{1}{r}\frac{\partial P}{\partial \theta}$$
$$+ \mu\left[\frac{\partial}{\partial r}\left(\frac{1}{r}\frac{\partial}{\partial r}(r v_\theta)\right) + \frac{1}{r^2}\frac{\partial^2 v_\theta}{\partial \theta^2} + \frac{2}{r^2}\frac{\partial v_r}{\partial \theta} + \frac{\partial^2 v_\theta}{\partial z^2}\right] \quad \text{(A.24)}$$

$$\rho\left(\frac{\partial v_z}{\partial t} + v_r\frac{\partial v_z}{\partial r} + \frac{v_\theta}{r}\frac{\partial v_z}{\partial \theta} + v_z\frac{\partial v_z}{\partial z}\right) = \rho g_z - \frac{\partial P}{\partial z}$$
$$+ \mu\left[\frac{1}{r}\frac{\partial}{\partial r}\left(r\frac{\partial v_z}{\partial r}\right) + \frac{1}{r^2}\frac{\partial^2 v_z}{\partial \theta^2} + \frac{\partial^2 v_z}{\partial z^2}\right] \quad \text{(A.25)}$$

There is no general solution of the Navier–Stokes equations, which is due in part to the non-linear inertial terms. Analytical solutions are possible in cases when several of the terms vanish or are negligible. The skill in obtaining analytical solutions of the Navier–Stokes equations lies in recognizing simplifications that can be made for the particular flow being analysed. Use of the continuity equation is usually essential.

Some of the simplifications that may be possible are illustrated by the case of steady, fully-developed, laminar, incompressible flow of a Newtonian fluid in a horizontal pipe. The flow is assumed to be axisymmetric with no swirl component of velocity so that derivatives wrt θ vanish and $v_\theta = 0$. For fully-developed flow, derivatives wrt z are zero. With these simplifications and noting that the flow is incompressible, the continuity equation (equation A.11) reduces to

$$\frac{\partial}{\partial r}(rv_r) = 0$$

As $v_r = 0$ at the wall it follows that $v_r = 0$ everywhere.

The z-component Navier–Stokes equation is equation A.25. Each of the inertial terms is zero, the reasons being respectively that the flow is steady, $v_r = 0$, the flow is axisymmetric and the flow is fully developed. The second and third viscous terms vanish because the flow is axisymmetric and fully-developed. The flow being horizontal, $g_z = 0$.

Thus, equation A.25 reduces to

$$\mu\left[\frac{1}{r}\frac{\mathrm{d}}{\mathrm{d}r}\left(r\frac{\mathrm{d}v_z}{\mathrm{d}r}\right)\right] = \frac{\partial P}{\partial z}$$

The derivative wrt r has been changed to an ordinary derivative because it has been established that v_z is independent of θ, z and t. Integrating this equation twice gives

$$\mu\frac{\mathrm{d}v_z}{\mathrm{d}r} = \frac{r}{2}\frac{\partial P}{\partial z} + \frac{A}{r}$$

and

$$\mu v_z = \frac{r^2}{4}\frac{\partial P}{\partial z} + A\ln r + B$$

These last two equations correspond to equations 1.55 and 1.56. Note that $\partial P/\partial z = -\Delta P/L$.

Potential flow

Consider the two-dimensional flow shown in Figure A.1. If the velocity gradient $\partial v_x/\partial y$ is positive it tends to cause the element to rotate in the clockwise direction. Similarly, if $\partial v_y/\partial x$ is positive it tends to cause rotation in the anti-clockwise direction. Thus, the quantity $\partial v_y/\partial x - \partial v_x/\partial y$ gives the net rate of rotation in the anti-clockwise direction as viewed. It is the clockwise direction about a line parallel to the z-coordinate as viewed in the positive z-direction. This quantity is the z-component of the fluid's vorticity ω:

$$\omega_z = \frac{\partial v_y}{\partial x} - \frac{\partial v_x}{\partial y} \tag{A.26a}$$

The x and y components are

$$\omega_x = \frac{\partial v_z}{\partial y} - \frac{\partial v_y}{\partial z} \tag{A.26b}$$

and

$$\omega_y = \frac{\partial v_x}{\partial z} - \frac{\partial v_z}{\partial x} \tag{A.26c}$$

When all three components of the vorticity are zero the flow is said to be irrotational. In irrotational flow the effects of viscosity disappear as will be

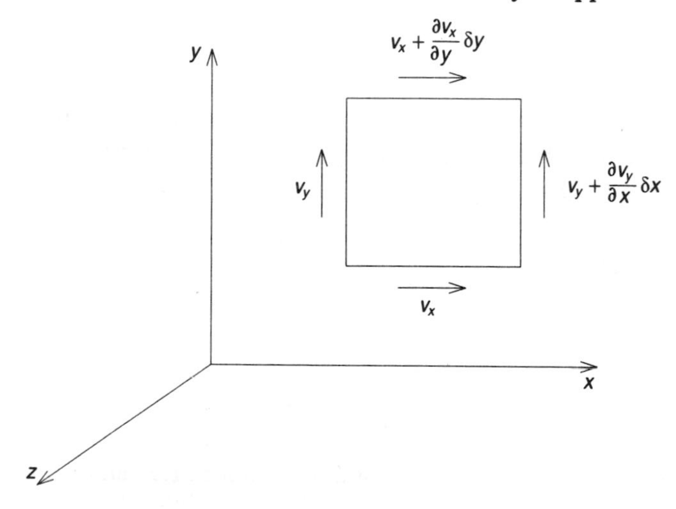

Figure A.1
Velocity components for a two-dimensional flow

seen. Laminar flow in a pipe is rotational everywhere except on the centre-line. The only non-zero velocity gradient for fully-developed flow is $\partial v_z/\partial r$, which has a maximum magnitude at the wall and falls to zero on the centre-line. A small neutrally-buoyant sphere placed in the fluid would be seen to rotate owing to the higher velocity of the fluid nearer the centre-line than that nearer the wall. Turbulent flow outside boundary layers often exhibits a negligible velocity gradient and can be treated as being irrotational. For irrotational flow the Navier–Stokes equations can be simplified considerably as follows.

Dividing equation A.22 throughout by ρ gives

$$\frac{\partial v_x}{\partial t} + v_x\frac{\partial v_x}{\partial x} + v_y\frac{\partial v_x}{\partial y} + v_z\frac{\partial v_x}{\partial z}$$
$$= g_x - \frac{1}{\rho}\frac{\partial P}{\partial x} + \nu\left(\frac{\partial^2 v_x}{\partial x^2} + \frac{\partial^2 v_x}{\partial y^2} + \frac{\partial^2 v_x}{\partial z^2}\right) \qquad \text{(A.27)}$$

If the flow is irrotational, then from equations A.26

$$\frac{\partial v_x}{\partial y} = \frac{\partial v_y}{\partial x} \qquad \text{and} \qquad \frac{\partial v_x}{\partial z} = \frac{\partial v_z}{\partial x}$$

so that the inertial terms in equation A.27 can be expressed as

$$\left(\frac{\partial v_x}{\partial t} + v_x\frac{\partial v_x}{\partial x} + v_y\frac{\partial v_y}{\partial x} + v_z\frac{\partial v_z}{\partial x}\right) = \frac{\partial v_x}{\partial t} + \frac{1}{2}\frac{\partial}{\partial x}\left(v_x^2 + v_y^2 + v_z^2\right) \qquad \text{(A.28)}$$

The same substitutions can be made in the viscous terms. For example, $\partial^2 v_x/\partial y^2$ can be written as follows

$$\frac{\partial^2 v_x}{\partial y^2} = \frac{\partial}{\partial y}\left(\frac{\partial v_x}{\partial y}\right) = \frac{\partial}{\partial y}\left(\frac{\partial v_y}{\partial x}\right) = \frac{\partial}{\partial x}\left(\frac{\partial v_y}{\partial y}\right) \qquad \text{(A.29)}$$

The change in the order of differentiation to give the last term in equation A.29 is permissible because the velocity field will satisfy the sufficient conditions, namely that the two mixed partial derivatives are continuous.

Equation A.27 can now be written as

$$\frac{\partial v_x}{\partial t} + \frac{\partial}{\partial x}\left(\frac{v^2}{2}\right) = g_x - \frac{1}{\rho}\frac{\partial P}{\partial x} + \nu\frac{\partial}{\partial x}\left(\frac{\partial v_x}{\partial x} + \frac{\partial v_y}{\partial y} + \frac{\partial v_z}{\partial z}\right) \qquad \text{(A.30)}$$

where $v^2 = v_x{}^2 + v_y{}^2 + v_z{}^2$ is the square of the fluid's speed.

For incompressible flow, the last expression in equation A.30 vanishes by virtue of equation A.8; also ρ is constant. Putting

$$g_x = -g\frac{\partial h}{\partial x} \tag{A.31}$$

where h is the height above an arbitrary datum, equation A.30 can be written as

$$\frac{\partial v_x}{\partial t} + \frac{\partial}{\partial x}\left(\frac{v^2}{2} + \frac{P}{\rho} + gh\right) = 0 \tag{A.32}$$

For steady flow, $\partial v_x/\partial t = 0$ so that the quantity $v^2/2+P/\rho+gh$ must be independent of x.

Similar equations can be written for the y and z components of the velocity so it can be concluded that

$$\frac{v^2}{2} + \frac{P}{\rho} + gh = \text{constant} \tag{A.33}$$

This is a statement of Bernoulli's theorem: the quantity $v^2/2+P/\rho+gh$ is constant throughout the fluid for steady, irrotational flow. Equation A.33 is the same as equation 1.11. It will be recalled that, for rotational flow with friction, the engineering form of Bernoulli's equation applies only along a streamline and allowance must be made for frictional losses.

Velocity potential

If Φ is a sufficiently differentiable function, then

$$\frac{\partial}{\partial x}\left(\frac{\partial \Phi}{\partial y}\right) - \frac{\partial}{\partial y}\left(\frac{\partial \Phi}{\partial x}\right) = 0 \tag{A.34}$$

For irrotational flow

$$\frac{\partial v_y}{\partial x} - \frac{\partial v_x}{\partial y} = 0 \tag{A.35}$$

Comparing equations A.34 and A.35, the following relationships can be written:

$$v_x = -\frac{\partial \Phi}{\partial x} \tag{A.36a}$$

and

$$v_y = -\frac{\partial \Phi}{\partial y} \tag{A.36b}$$

Similarly

$$v_z = -\frac{\partial \Phi}{\partial z} \tag{A.36c}$$

The function Φ is known as the velocity potential. In equation A.36 the minus sign is arbitrary but is usually incorporated so that flow is from a high value of the velocity potential to a low value.

Substituting for the velocity components in equation A.8 using equation A.36 gives

$$\frac{\partial^2 \Phi}{\partial x^2} + \frac{\partial^2 \Phi}{\partial y^2} + \frac{\partial^2 \Phi}{\partial z^2} = 0 \tag{A.37}$$

which shows that the velocity potential satisfies Laplace's equation in the case of incompressible flow.

For potential flow, ie incompressible, irrotational flow, the velocity field can be found by solving Laplace's equation for the velocity potential then differentiating the potential to find the velocity components. Use of Bernoulli's equation then allows the pressure distribution to be determined. It should be noted that the no-slip boundary condition cannot be imposed for potential flow.

Appendix II
Further problems

(The numbers refer to the relevant chapter.)

1–1 An incompressible fluid flows upwards in steady state in a cylindrical pipe at an angle θ with the horizontal. Assume that the head loss due to friction is negligible.

(a) Derive an expression for the pressure gradient in the pipe.

(b) Derive an expression for the length of pipe over which the pressure is reduced by half.

(c) Calculate the length of pipe L over which the pressure is reduced by half if the gravitational acceleration $g = 9.81\ \mathrm{m/s^2}$, $\theta = 30°$, the liquid density $\rho = 1200\ \mathrm{kg/m^3}$, and the initial pressure $P_1 = 200\,000$ Pa.

1–2 Water flows at a speed of 2 m/s through the pipe–work shown in the following diagrams. The pipe diameter is 0.10 m throughout.

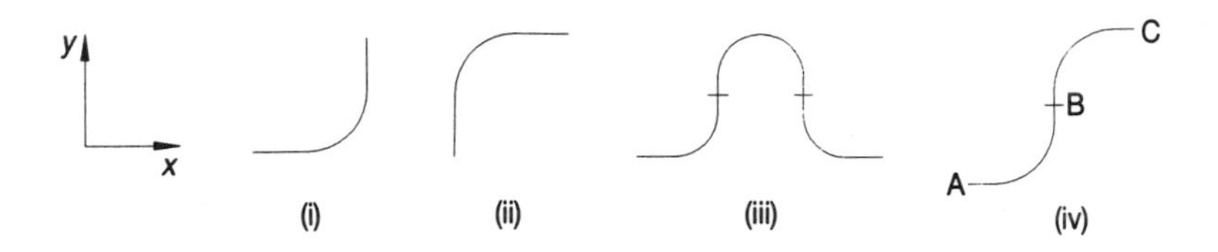

(a) Calculate the components of the reaction on the pipe–work due to the change of the fluid's momentum in each case.

(b) What is the tension in the bolts through the flanges in diagram (iii)?

(c) Determine the above forces for the case of steam at 20 atm (absolute) flowing at a speed of 20 m/s.

(d) Show that, for the double bend shown in diagram (iv), the overall force due to the change of the fluid's momentum is zero but the pipe–work is subject to a moment about B and calculate its value.

(e) Assuming the pressure drop to be negligible, calculate the moment on the double bend caused by the steam's pressure.

Data: density of water = 1000 kg/m^3, density of 20 atm steam = 10.0 kg/m^3

1–3 A corrosive liquid is to be transferred from one tank to a higher tank without using a pump but by pressurizing the space above the liquid in the lower tank. The frictional head loss in the pipe–work is equal to 1000 velocity heads [ie $h_f = 1000(u^2/2g)$] and the density of the liquid is 800 kg/m^3. What pressure difference will be required to cause the liquid to flow at a speed of 0.3 m/s when the liquid surface in the supply tank is 7 m below that in the discharge tank?

2–1 (a) Derive an expression for the mean velocity u of a liquid in steady state turbulent flow in a smooth cylindrical tube in terms of the pressure gradient $\Delta P_f/L$, the liquid density ρ, the inside diameter of the tube d_i and the liquid dynamic viscosity μ.

(b) Calculate u if

$$\Delta P_f/L = 528 \text{ Pa/m}$$

$$\rho = 1200 \text{ kg/m}^3$$

$$\mu = 0.01 \text{ Pa s}$$

and

$$d_i = 0.05 \text{ m}$$

2–2 (a) Derive an expression for the frictional pressure gradient $\Delta P_f/L$ for a liquid in steady state turbulent flow in a rough cylindrical pipe in terms of the liquid density ρ, the mean velocity u, the inside pipe diameter d_i and the roughness e.

(b) Use this to calculate $\Delta P_f/L$ for

$$\rho = 1200 \text{ kg/m}^3$$

$$d_i = 0.0526 \text{ m}$$

$$u = 1.160 \text{ m/s}$$

$$\mu = 0.01 \text{ Pa s}$$

and

$$e = 0.000045 \text{ m}$$

2–3 Calculate the frictional pressure gradient $\Delta P_f/L$ for a liquid in steady state turbulent flow in a coil of inside tube diameter $d_i = 0.02$ m and coil diameter $D_C = 2$ m if the liquid density

$\rho = 1200$ kg/m^3, the liquid dynamic viscosity $\mu = 0.001$ Pa s and the mean velocity $u = 2$ m/s.

2–4 A liquid flows in a steady state in a cylindrical pipe of inside diameter $d_i = 0.05$ m at a flow rate $Q = 2 \times 10^{-3}$ m^3/s. Calculate the head loss and the pressure drop for a sudden expansion to a pipe of inside diameter 0.1 m, if the liquid density $\rho = 1000$ kg/m^3.

2–5 A Newtonian liquid flows in steady state in a cylindrical pipe.

(a) Calculate point velocities v_x at the centre and at a radius of one quarter of the pipe diameter if the inside diameter of the pipe $d_i = 0.05$ m, the liquid dynamic viscosity $\mu = 0.15$ Pa s, the liquid density $\rho = 1100$ kg/m^3 and the pressure gradient $\Delta P_f/L = 3000$ Pa/m.

(b) Calculate the volumetric flow rate Q through the pipe.

2–6 Plot laminar and turbulent velocity profiles for steady state flow in a cylindrical pipe for a maximum velocity $v_{max} = 5$ m/s using the radial positions $2r/d_i = 0, 0.2, 0.4, 0.6$ and 0.8.

2–7 The magnitude of the time–averaged wall shear stress in turbulent flow is given by

$$\overline{\tau_{yx}} = \rho(\nu + \varepsilon)\frac{d\bar{v}_x}{dy}$$

where y is the distance from the wall. Rewrite this equation in terms of the dimensionless velocity v^+ and the dimensionless distance from the wall y^+, and show that

$$1 = \left(1 + \frac{\varepsilon}{\nu}\right)\frac{dv^+}{dy^+}$$

Using the time–averaged velocity profiles in the three parts of a turbulent boundary layer given by equations 2.58, 2.70 and 2.71, derive expressions for ε/ν for these three regions. Calculate ε/ν for $y^+ = 5, 15, 25, 50, 200$ and 500.

3–1 Calculate the frictional pressure gradient $\Delta P_f/L$ for a time independent non-Newtonian fluid in steady state flow in a cylindrical tube if

the liquid density	$\rho = 1000$ kg/m^3
the inside diameter of the tube	$d_i = 0.08$ m
the mean velocity	$u = 1$ m/s
the point pipe consistency coefficient	$K' = 2$ Pa s$^{0.5}$
and the flow behaviour index	$n' = 0.5$

3–2 For Newtonian fluids in which the dynamic viscosity μ is a function only of temperature, ie $\mu = f(T)$, the expression $\phi = \mu_w/\mu$ raised to some power is used to correct isothermal equations for non-isothermal conditions. Suggest an analogous correction for non-Newtonian power law fluids flowing in pipes in which the apparent dynamic viscosity μ_a is a function of shear stress τ, shear rate $\dot{\gamma}$ and temperature T, ie $\mu_a = f(\tau, \dot{\gamma}, T)$.

3–3 The laminar flow velocity profile in a pipe for a power law liquid in steady state flow is given by the equation

$$v_x = u\left(\frac{3n+1}{n+1}\right)\left[1 - \left(\frac{2r}{d_i}\right)^{(n+1)/n}\right]$$

where n is the power law index and u is the mean velocity. Use this to derive the expression

$$\left(\frac{-\mathrm{d}v_x}{\mathrm{d}r}\right) = \left(\frac{8u}{d_i}\right)\left(\frac{3n+1}{4n}\right)$$

for the velocity gradient at the pipe wall.

3–4 The following measurements were made on the inner cylinder of a coaxial cylinders viscometer having inner and outer diameters of 24 mm and 26 mm, and an effective cylinder length of 35 mm. Using these data, determine the values of the shear stress, shear rate and apparent viscosity of the sample.

Torque (10^{-5} N m)	*Speed* (r.p.m.)
20.0	36.8
40.0	90.6
60.0	153.4
80.0	223.0

3–5 Show that the volumetric average velocity u of a homogeneous, time–independent liquid in laminar flow through a straight pipe can be written as

$$u = \frac{r_i}{\tau_w^3}\int_0^{\tau_w} \tau^2(-\dot{\gamma})\,\mathrm{d}\tau$$

If the material whose viscometric properties were determined in question 3–4 were pumped through a 25 mm diameter pipe so that the wall shear stress had the value corresponding to the last measurement in that question, what would be the volumetric average velocity and what value of pressure gradient would be required?

3–6 A plastic film is to be produced by extruding it through a narrow slit of depth $2h$. The width of the slit is much greater than $2h$ and the slit is long enough for end effects to be neglected. The flow is laminar. Starting from first principles, derive the following equation, which is a form of the Rabinowitsch–Mooney equation for this geometry:

$$\dot{\gamma}_w = \dot{\gamma}_{wN}\left[\frac{2}{3} + \frac{1}{3}\frac{\partial \ln Q}{\partial \ln \tau_w}\right]$$

where Q is the volumetric flow rate and $\dot{\gamma}_w$, $\dot{\gamma}_{wN}$ are respectively the true wall shear rate and the wall shear rate for a Newtonian fluid with the same volumetric flow rate.

4–1 Figure 1.5 diagrammatically represents the heads in a liquid flowing through a pipe. Redraw this diagram with a pump placed between points 1 and 2.

4–2 Calculate the available net positive section head NPSH in a pumping system if the liquid density $\rho = 1200$ kg/m^3, the liquid dynamic viscosity $\mu = 0.4$ Pa s, the mean velocity $u = 1$ m/s, the static head on the suction side $z_s = 3$ m, the inside pipe diameter $d_i = 0.0526$ m, the gravitational acceleration $g = 9.81$ m/s^2, and the equivalent length on the suction side $\Sigma L_{es} = 5.0$ m.

The liquid is at its normal boiling point. Neglect entrance and exit losses.

4–3 A centrifugal pump is used to pump a liquid in steady turbulent flow through a smooth pipe from one tank to another. Develop an expression for the system total head Δh in terms of the static heads on the discharge and suction sides z_d and z_s respectively, the gas pressures above the tanks on the discharge and suction sides P_d and P_s respectively, the liquid density ρ, the liquid dynamic viscosity μ, the gravitational acceleration g, the total equivalent lengths on the discharge and suction sides ΣL_{ed} and ΣL_{es} respectively, and the volumetric flow rate Q.

4–4 A system total head against mean velocity curve for a particular power law liquid in a particular pipe system can be represented by the equation

$$\Delta h = (0.03)(100^n)(u^n) + 4.0 \quad \text{for } u \leqslant 1.5 \text{ m/s}$$

where

Δh is the total head in m

u is the mean velocity in m/s

and

n is the power law index.

A centrifugal pump operates in this particular system with a total head against mean velocity curve represented by the equation

$$\Delta h = 8.0 - 0.2u - 1.0u^2 \quad \text{for } u \leqslant 1.5 \text{ m/s}$$

(this is a simplification since Δh is also affected by n).

(a) Determine the operating points for the pump for
  (i) a Newtonian liquid
  (ii) a shear thinning liquid with $n = 0.9$
  (iii) a shear thinning liquid with $n = 0.8$.
(b) Comment on the effect of slight shear thinning on centrifugal pump operation.

4–5 A volute centrifugal pump has the following performance data at the best efficiency point:

volumetric flow rate	$Q = 0.015$ m^3/s
total head	$\Delta h = 65$ m
required net positive suction head	NPSH $= 16$ m
liquid power	$P_E = 14000$ W
impeller speed	$N = 58.4$ rev/s
impeller diameter	$D = 0.22$ m

Evaluate the performance of an homologous pump which operates at an impeller speed of 29.2 rev/s but which develops the same total head Δh and requires the same NPSH.

4–6 Two centrifugal pumps are connected in series in a given pumping system. Plot total head Δh against capacity Q pump and system curves and determine the operating points for

(a) only pump 1 running
(b) only pump 2 running
(c) both pumps running

on the basis of the following data:

operating data for pump 1

Δh_1 m,	50.0	49.5	48.5	48.0	46.5	44.0	42.0	39.5	36.0	32.5	28.5
Q m³/h,	0	25	50	75	100	125	150	175	200	225	250

operating data for pump 2

Δh_2 m,	40.0	39.5	39.0	38.0	37.0	36.0	34.0	32.0	30.5	28.0	25.5
Q m³/h,	0	25	50	75	100	125	150	175	200	225	250

data for system

Δh_s m,	35.0	37.0	40.0	43.5	46.5	50.5	54.5	59.5	66.0	72.5	80.0
Q m³/h,	0	25	50	75	100	125	150	175	200	225	250

4–7 Two centrifugal pumps are connected in parallel in a given pumping system. Plot total head Δh against capacity Q pump and system curves for both pumps running on the basis of the following data:

operating data for pump 1

Δh m,	40.0	35.0	30.0	25.0
Q_1 m³/h,	169	209	239	265

operating data for pump 2

Δh m,	40.0	35.0	30.0	25.0
Q_2 m³/h,	0	136	203	267

data for system

Δh m,	20.0	25.0	30.0	35.0
Q_s m³/h,	0	244	372	470

4–8 (a) Name some types of pumps which are seriously affected by misalignment.

(b) What is the shape of the total head against capacity characteristic curve of a gear pump?

(c) If very hot fluid is pumped with a gear pump, what difficulty might occur?

(d) Gear pumps can be small liquid cavity high speed pumps or large liquid cavity low speed pumps. Which type would you use to pump

(i) a shear thinning liquid?
(ii) a shear thickening liquid?
(iii) a slurry?

5–1 Calculate the theoretical power in watts for a 0.25 m diameter, six-blade flat blade turbine agitator rotating at $N = 4$ rev/s in a tank system with a power curve given in Figure 5.10. The liquid in the tank is shear thinning with an apparent dynamic viscosity dependent on the impeller speed N and given by the equation $\mu_a = 25(N)^{n-1}$ Pa s where the power law index $n = \frac{1}{2}$ and the liquid density $\rho = 1000$ kg/m^3.

5–2 For laminar flow of a Newtonian liquid in a stirred tank, the power P_a is given by the equation

$$P_A = \mu C N^2 D_A^3$$

where

μ is the liquid dynamic viscosity
N is the agitator speed
D_A is the agitator diameter

and

C is a constant for the system.

A shear thinning liquid has an apparent dynamic viscosity given by the equation

$$\mu_a = K(N)^{n-1}$$

where the consistency coefficient $K = \mu$ at a power law index $n = 1$. Show that for the same power the shear thinning liquid can be agitated at a higher agitator speed N_1 given by the equation

$$N_1 = N^{2/(n+1)}$$

5–3 Solute-free liquid at a volumetric flow rate Q is used to purge off quality solute from a stirred tank of volume V. Show that if three equal size tanks are used in series, the removal of solute is n times more effective after a time t where n and t are related by the equation

$$t = \frac{V[(2n-1)^{1/2} - 1]}{Q}$$

6–1 An ideal gas in which the pressure P is related to the volume V by the equation $PV = 75\ m^2/s^2$ flows in steady isothermal flow along a horizontal pipe of inside diameter $d_i = 0.02$ m. The pressure drops from 20000 Pa to 10000 Pa in a 5 m length. Calculate the mass flux assuming that the Fanning function factor $f = 9.0 \times 10^{-3}$.

6–2 Ethylene flows through a pipeline 10 km long to a receiving station A. At a point 3 km from A, a spur leads off the main pipeline and runs 5 km to a receiving station B. The internal diameter of the main pipeline is 0.20 m and that of the spur is 0.15 m. The flow rates into A and B are regulated by valves at these locations. If the pressure immediately upstream of valve A is 3.88 bar (absolute) and that at B is 3.69 bar when the flow rate into B is 0.63 kg/s, calculate the pressure at the beginning of the main pipeline, assuming that flow in the pipeline is isothermal at a temperature of 20 °C.

Data: specific volume of ethylene at 20 °C, 1 bar = 0.870 m^3/kg, Fanning friction factor = 0.0045.

6–3 Calculate the air velocity in m/s required to cause a temperature drop of 1 K on a conventional thermometer given that for the air at atmospheric pressure and 373 K, the thermal capacity per unit mass at constant pressure $C_p = 1006$ J/(kg K).

6–4 An ideal gas flows in steady state adiabatic flow along a horizontal pipe of inside diameter $d_i = 0.02$ m. The pressure and density at a point are $P = 20000$ Pa and $\rho = 200\ kg/m^3$ respectively. The density drops from 200 kg/m^3 to 100 kg/m^3 in a 5 m length. Calculate the mass flux assuming that the Fanning friction factor $f = 9.0 \times 10^{-3}$ and the ratio of heat capacities at constant pressure and constant volume $\gamma = 1.40$.

6–5 Show that the work required to compress an ideal gas adiabatically from a pressure P_1 to a pressure P_2 in a compressor with two equal stages is $[(P_2/P_1)^{(\gamma-1)/4\gamma} + 1]/2$ greater than in a compressor with four equal stages where γ is the ratio of heat capacities at constant pressure and constant volume.

6–6 Air flows from a large reservoir where the temperature and pressure are 25°C and 10 atm, through a convergent–divergent nozzle and discharges to the atmosphere. The area of the nozzle's exit is twice that of its throat. Show that under these conditions a shock wave must occur. ($\gamma = 1.4$.)

6–7 Air at a pressure of 5 bar in a closed tank is to be vented by allowing it to discharge through a convergent nozzle straight to the atmosphere. Show that the mass flow rate M is given by

$$M = S\left[\frac{\gamma P_0}{V_0}\left(\frac{2}{\gamma + 1}\right)^{\frac{\gamma+1}{\gamma-1}}\right]^{1/2}$$

where S is the exit area of the nozzle and P_0, V_0 are the pressure and specific volume of the gas in the tank.

If the pressure is to be reduced from 5 bar to 2 bar, will the mass flow rate be constant during the venting operation?

6–8 Nitrogen is to be vented to the atmosphere from a closed tank at a pressure of 2 atm gauge and a temperature of 20°C through a convergent nozzle with an exit diameter of 15 mm.

(a) Explain why a shock wave will occur at the nozzle exit.

(b) To what value must the pressure in the tank fall before the shock wave disappears?

(c) Without using equation 6.107, calculate the mass flow rate initially and at the point when the shock wave disappears.

Data: for nitrogen, $\gamma = 1.39$, relative molecular mass = 28.02.

7–1 For a particular bubble column, $u_t = 0.5$ m/s and n (in the Richardson–Zaki equation) may be approximated as 2. Determine the value of the void fraction α for each of the following conditions:

(a) $Q_G/S = 0.06$ m/s, $Q_L/S = 0.06$ m/s (co-current)

(b) $Q_G/S = 0.06$ m/s, $Q_L/S = -0.06$ m/s (counter-current)

(c) when flooding occurs for $Q_L/S = -0.06$ m/s.

7–2 Air and water flow at 0.08 kg/s and 0.32 kg/s respectively in a horizontal tube of inside diameter 25 mm. The mean pressure is 9.90 bar (absolute) and the pressure drop across a length of 5 m is

0.215 bar. What is the value of the friction factor? Assume isothermal conditions.

Data: at 9.90 bar, $V_G = 8.28 \times 10^{-2}$ m³/kg, $V_L = 1.02 \times 10^{-3}$ m³/kg.

7–3 A mixture of gas and liquid flows through a pipe of internal diameter $d_i = 0.02$ m at a steady total flow rate of 0.2 kg/s. The pipe roughness $e = 0.000045$ m. The dynamic viscosities of the gas and liquid are $\mu_G = 1.0 \times 10^{-5}$ and $\mu_L = 3.0 \times 10^{-3}$ Pa s respectively. The densities of the gas and liquid are $\rho_G = 60$ kg/m³ and $\rho_L = 1000$ kg/m³ respectively. The weight fraction of gas is 0.149. Calculate the pressure gradient in the pipe using the Lockhart–Martinelli correlation

7–4 Saturated water flows into a horizontal, uniformly heated, smooth tube at a rate of 0.2 kg/s. The tube is 5 m long and has an inside diameter of 50 mm. The inlet pressure is 3 bar and the exit quality 40 per cent. Use the Martinelli–Nelson correlation to estimate the pressure drop across the tube.

Data: liquid density = 930 kg/m³, liquid viscosity = 2.0×10^{-4} Pa s

8–1 A Pitot tube is used to measure point velocities in water. The reading on a mercury manometer attached to the Pitot tube is 1.6 cm. Calculate the water velocity given that the specific gravity S.G. = 13.6 for mercury.

8–2 Calculate the volumetric flow rate of water in m³/s through a pipe with an inside diameter of 0.2 m fitted with an orifice plate containing a concentric hole of diameter 0.1 m given the following data:

1. a difference in level of 0.5 m on a mercury manometer connected across the orifice plate
2. a mercury specific gravity S.G. = 13.6
3. a discharge coefficient $C_d = 0.60$.

8–3 Water flows upwards at a speed of 2 m/s in a vertical pipe. A Venturi meter having a throat diameter equal to half the pipe diameter is fitted in the pipe and has pressure taps connected to a mercury manometer. The distance between the pressure taps is 50 mm. If the discharge coefficient of the Venturi is 0.98, what will be the manometer reading and what is the pressure difference between the two taps? The specific gravity of mercury is 13.6.

8–4 Calculate the volumetric flow rate in m^3/s through a V-notch weir when the height of liquid above the weir is 0.15 m given that the notch angle $\theta = 20°$ and the discharge coefficient $C_d = 0.62$.

8–5 Show that a flowmeter with a square root scale has an error $50/Q$ times that for a linear scale where the maximum volumetric flow rate $Q_{max} = 100$ per cent.

9–1 A gas of density $\rho = 1.25$ kg/m^3 and dynamic viscosity $\mu = 1.5 \times 10^{-5}$ Pa s flows steadily through a bed of spherical particles 0.005 m in diameter. The bed has a height of 5.00 m and a voidage of $\frac{1}{3}$. The pressure drop is 150 Pa. Calculate the superficial velocity.

10–1 Calculate the time in seconds and in hours for a liquid to fall in a tank from a height $z_1 = 9$ m to a height $z_2 = 4$ m above a discharge hole of diameter $d_i = 0.02$ m given the following data:

tank diameter	$D_T = 2$ m
dimensionless correction factor	$\alpha = 1$
gravitational acceleration	$g = 9.81$ m/s^2

the pressure over the liquid in the tank is equal to the pressure at the outlet.

10–2 Show that the time to reach 50 per cent of the terminal velocity for a spherical particle falling from rest in laminar flow in a fluid is

$$t = 0.071 u_t$$

where u_t is the terminal velocity.

Answers to problems

1–1 (a) $(P_1 - P_2)/L = \rho g \sin\theta$

(b) $L = P_1/(2\rho g \sin\theta)$

(c) 17 m

1–2 (a) (i) $R_x = 31.4$ N, $R_y = -31.4$ N

(ii) $R_x = -31.4$ N, $R_y = 31.4$ N

(iii) $R_x = R_y = 0$

(b) Total tension = 62.8 N (31.4 N on each flange)

(c) Same values because ρu^2 has the same value

(d) 37.7 N m (anticlockwise)

(e) 18 150 N m (anticlockwise)

1–3 $P_1 - P_2 = 0.91$ bar

2–1 (a) $$u = \left(\frac{\mu}{\rho d_i}\right)\left[\left(\frac{\Delta P_f}{L}\right)\left(\frac{\rho d_i^{\,3}}{0.1584\,\mu^2}\right)\right]^{4/7}$$

(b) 1.10 m/s

2–2 (a) $$\frac{\Delta P_f}{L} = \left(\frac{2\rho u^2}{d_i}\right)\left[4.06 \log\left(\frac{d_i}{e}\right) + 2.16\right]^{-2}$$

(b) 286.5 Pa/m

2–3 1584 Pa/m

2–4 293 Pa

2–5 Confirm laminar flow

(a) $v_{max} = 3.13$ m/s

$v_x = 2.35$ m/s at $d_i/4$

(b) 3×10^{-3} m^3/s

2–7 0, 2, 4, 19, 79, 199. (Discontinuity at $y^+ = 30$ physically impossible)

3–1 1000 Pa/m

3–2 $$\phi = \frac{(\mu_{ap})_w}{(\mu_{ap})_m} = \left(\frac{K_w}{K_b}\right)\left[\frac{2n}{(n+1)}\right]$$

(m—at mean stress, b—at bulk temperature)

3–4 At highest speed, $\tau = 25.3$ Pa, $\dot{\gamma} = 280\ \text{s}^{-1}$, $\mu_a = 0.0901$ Pa s

3–5 $u = 0.82$ m/s, $\Delta P/L = 4042$ Pa/m

4–2 1.038 m

4–3 $\Delta h = k_1 + k_2 Q^{1.75}$

where the constant

$k_1 = (z_d - z_s) + (P_d - P_s)/(\rho g)$

and the constant

$$k_2 = \frac{(\Sigma L_{es} + \Sigma L_{ed})(0.239)}{(g d_i^{4.75})}\left(\frac{\mu}{\rho}\right)^{0.25}$$

4–4 (a) (i) $\Delta h = 6.88$ m, $u = 0.96$ m/s
(ii) $\Delta h = 6.28$ m, $u = 1.21$ m/s
(iii) $\Delta h = 5.60$ m, $u = 1.45$ m/s

4–5 $D_2 = 0.44$ m
$Q_2 = 0.060\ \text{m}^3/\text{s}$
$P_{E2} = 56000$ W

5–1 250 W

6–1 621 kg/(m^2 s)

6–2 6.71 bar (pressure at junction = 4.5 bar)

6–3 47.8 m/s

6–4 623 kg/(m^2 s)

6–6 Show that $S\psi$ has different values at the throat and at the exit.

6–7 Although choked flow, flow rate falls because supply pressure falls.

6–8 (b) 1.89 atm
(c) 0.124 kg/s and 0.0836 kg/s

7–1 (a) 0.12
(b) 0.18
(c) 0.35

7–2 $f = 0.00461$ ($\Delta P_a = 239$ Pa, $\Delta P_f = 21261$ Pa)

7–3 2775 Pa/m

7–4 $\Delta P = 2.25$ kPa ($f_{LO} = 0.0061$)

8–1 1.99 m/s

8–2 5.42×10^{-2} m^3/s

8–3 $\Delta z_m = 0.234$ m, $\Delta P = 31730$ Pa

8–4 2.25×10^{-3} m^3/s

9–1 2.315×10^{-2} m/s

10–1 4510 s, 1.253 h

Conversion factors

area	1 ft^2	= 0.092903 m^2
density	1 lb/ft^3	= 16.018 kg/m^3
	1 lb/UK gal	= 99.779 kg/m^3
	1 lb/US gal	= 119.83 kg/m^3
dynamic viscosity	1 cP	= 0.001 Pa s (or N s/m^2)
	1 lb/(h ft)	= 4.1338×10^{-4} Pa s
	1 lb/(s ft)	= 1.4882 Pa s
energy	1 Btu	= 1055.06 J
	1 ft pdl	= 0.042139 J
flow rate, mass per unit time	1 lb/h	= 1.2600×10^{-4} kg/s
flow rate, volume per unit time	1 ft^3/s	= 0.028317 m^3/s
	1 ft^3/min	= 4.7195×10^{-4} m^3/s
	1 UK gal/min	= 7.5766×10^{-5} m^3/s
	1 US gal/min	= 6.3089×10^{-5} m^3/s
heat capacity per unit mass	1 Btu/(lb F)	= 4186.8 J/(kg K)
kinematic viscosity	1 ft^2/s	= 0.092903 m^2/s
length	1 ft	= 0.3048 m
linear velocity	1 ft/s	= 0.3048 m/s
mass	1 lb	= 0.45359 kg
mass flux	1 lb/(h ft^2)	= 1.3562×10^{-3} kg/(s m^2)
pressure	1 atm	= 101325 Pa (or N/m^2)
	1 pdl/ft^2	= 1.4882 Pa
	1 psi	= 6894.8 Pa
pressure gradient	1 (pdl/ft^2)/ft	= 4.8824 Pa/m
power	1 ft pdl/s	= 0.04214 W
	1 hp (British)	= 745.7 W
	1 ton refrigeration	= 3516.9 W
specific volume	1 ft^3/lb	= 0.062428 m^3/kg
surface tension	1 dyne/cm	= 0.001 N/m
temperature difference	1 F	= 0.5556 K
volume	1 ft^3	= 0.028317 m^3
	1 UK gal	= 0.0045460 m^3
	1 US gal	= 0.0037853 m^3

Friction factor flow charts

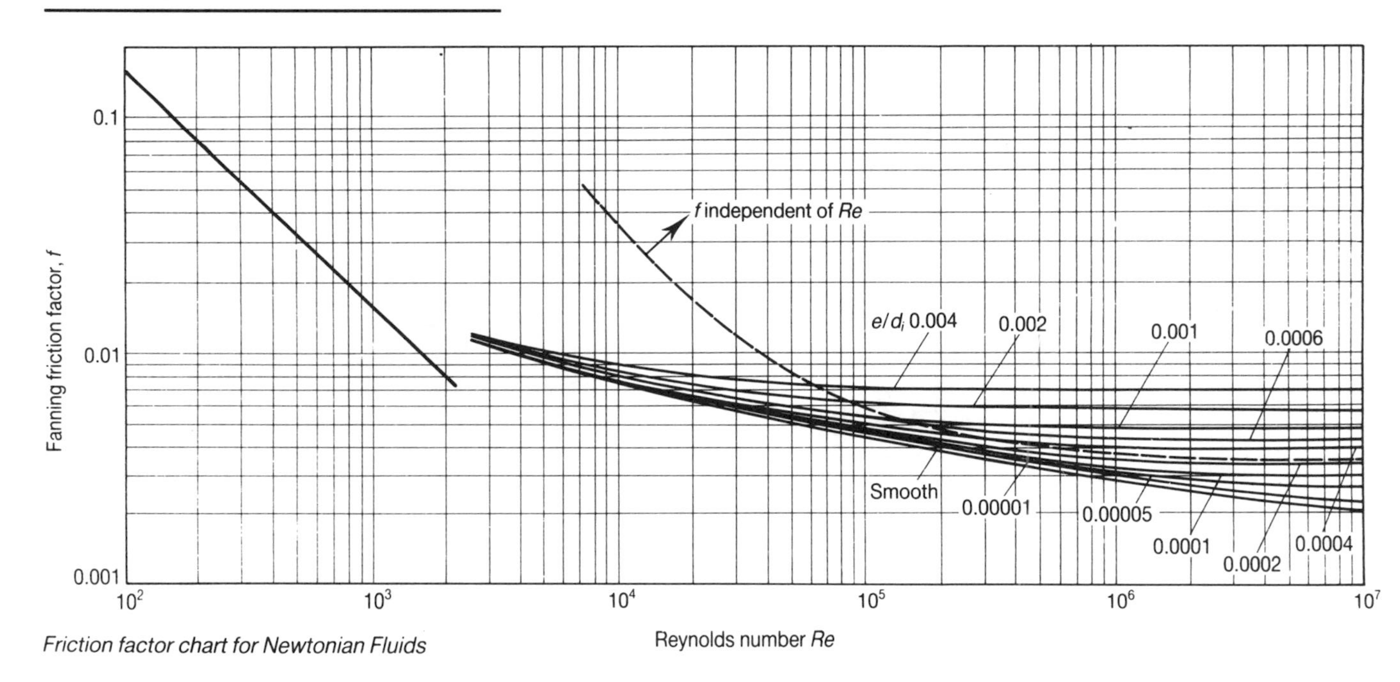

Friction factor chart for Newtonian Fluids

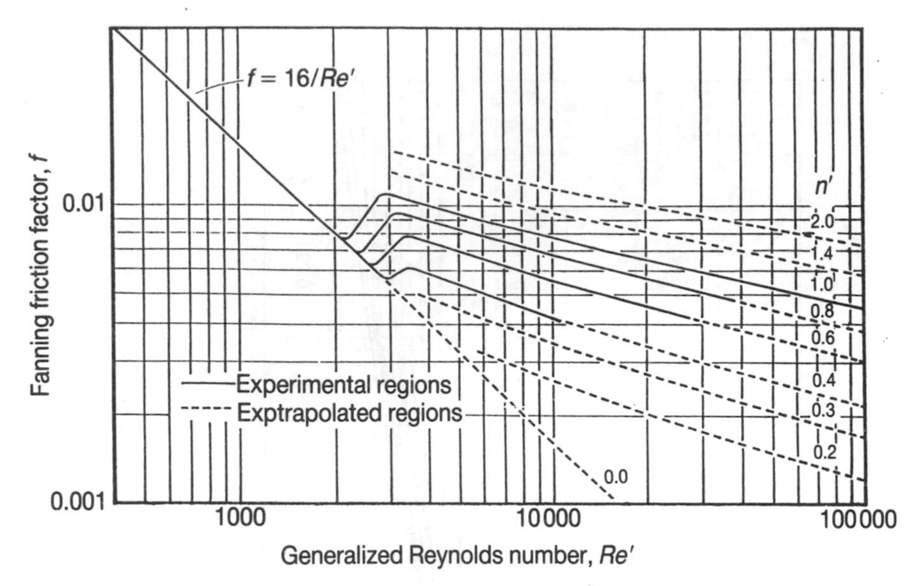

Friction factor chart for purely viscous non-Newtonian fluids

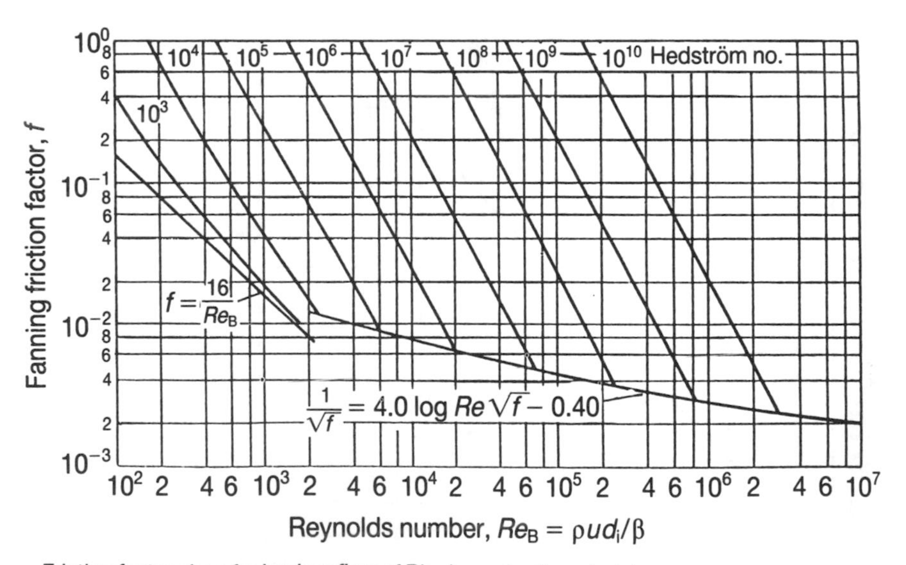

Friction factor chart for laminar flow of Bingham plastic materials

Index